Monographs on Endocrinology

Volume 31

Edited by

F. Gross (†), Heidelberg · M. M. Grumbach, San Francisco
A. Labhart, Zürich · M. B. Lipsett (†), Bethesda
T. Mann, Cambridge · L. T. Samuels (†), Salt Lake City
J. Zander, München

Fred A. Kincl

Hormone Toxicity in the Newborn

With 26 Illustrations

Springer-Verlag
Berlin Heidelberg New York London
Paris Tokyo Hong Kong Barcelona

Fred A. Kincl, Ph.D., DSc., FRSC (Deceased)
Professor Emeritus
The College of Staten Island of the
City University of New York
Staten Island, New York, USA

ISBN 3-540-51153-9 Springer-Verlag Berlin Heidelberg New York
ISBN 0-387-51153-9 Springer-Verlag New York Berlin Heidelberg

Library of Congress Cataloging in Publication Data.
Kincl, Fred A.
Hormone toxicity in the newborn / Fred A. Kincl.
(Monographs on endocrinology ; v. 31)
Includes bibliographical references.
ISBN 0-387-51153-9 (U.S.)
1. Fetus--Effect of drugs on. 2. Infants (Newborn)--Effect of
drugs on. 3. Hormones--Toxicology. 4. Sex (Biology)
5. Endocrinology, Experimental. I. Title. II. Series.
[DNLM: 1. Fetus--drug effects. 2. Hormones--adverse effects.
3. Infant, Newborn. 4. Steroids--adverse effects. W1 M057 v. 31 /
WS 420 K51h]
RG627.6.D79K56 1989
618.3'268--dc20
DNLM/DLC

© Springer-Verlag Berlin Heidelberg 1990
Printed in Germany

The use of registered names, trademarks, etc. in this publications does not imply, even in the absence of a specific statement, that such names are exempt from the relevant protective laws and regulations and therefore free for general use.

Product liability: The publisher can give no guarantee for information about drug dosage and application thereof contained in this book. In every individual case the respective user must check its accuracy by consulting other pharmaceutical literature.

Typesetting: Appl, Wemding; Offsetprinting: Saladruck, Berlin;
Bookbinding: B. Helm, Berlin.
2127/3020-543210 – Printed on acid-free Paper

Preface

The account of "neonatal sterilization" is the story of the advocates of direct effect of steroids on the gonads and those who believed in the indirect influence, mediated through the hypothalamus and/or the pituitary gland. As often happens in biology, both convictions represent the same image seen from different perspectives.

Prof DC Johnson (Kansas City, KS) reminisced the beginning of the story in a letter to me. I am paraphrasing parts of the letter with his permission.

"As a starting point we could pick the life-long research of Emil Steinach..." Steinach recognized the influence of testes on the development of accessory sex organs in 1894, described virilization of females and feminization of males in 1913, and identified the controlling influence of the hypophysis on the gonads in 1928.

He reviewed his work in a book *Sex and Life, Forty Years of Biological and Medical Experience* (E Steinach and L Loebel; Faber and Faber, London, 1940). He got on the wrong road in later years and that is the reason everybody seems to have forgotten him. He presented his hypothesis that estrogen has a direct effect upon the testes, i. e. hormone antagonism, at the 1st International Congress on Sex Research in 1926.

Carl Moore went to Chicago to study embryology under Frank Lillie. Lillie kept Moore after graduation for the faculty and convinced him that he could make a name for himself if he could answer the riddle of the freemartin. Lillie thought that the female was "sterilized" by the hormones of the male; I think he was influenced by Steinach. However, Moore could not confirm the direct antagonism Steinach was talking about and found that androgens were as damaging to the male as were estrogens, but that chorionic gonadotropin could reverse the action of both. He presented his research at the 2nd International Congress on Sex Research in 1930 (Price was not a co-author). You will notice there is no mention of "feedback" in the 1932 paper by Moore and Price, but one of the conclusions reached was that the concept of sex hormone antagonism was wrong. Unfortunately the preoccupation with the indirect action of steroids prevailed for 40 years and only recently have we gained respect for the "old" direct action.

The next step in the trail leads to LeRoy Goodman (Ant Rec 59:233, 1934) who, as a medical student, did a project involving transplantation of ovaries into the anterior chamber of the eye of female and male rats. He noted that they did not ovulate nor luteinize if in males. Witschi was interested in finding differences between male and female gonadotropins and saw the importance of Goodman's experiments. Remember that at this time there was a controversy over the number of gonadotropins. Witschi firmly believed (until 1961) that there were three gonadotropins; LH, ICSH, and FSH. Males secreted only ICSH and FSH but females secreted all three. The presence of testis at birth set the pattern for the loss of LH. Witschi gave Pfeiffer the project, but apparently he was aided by RT Hill who was at the time working on parabiosis. Upon graduation Hill went to Wisconsin and started RK Meyer on a long trail of parabiotic research while Pfeiffer went to Yale. He met up with Hamilton, Young and Wilson who were studying the effects of repeated injections of steroids on rats. They had noted that injections started soon after birth caused permanent damage and mentioned the similarity to the effects produced by testicular transplants as shown by Pfeiffer. However, they mentioned that personal communication with Pfeiffer indicated that androgen did not have the same effect as testes; Pfeiffer had used testosterone and not the propionate. Pfeiffer did not follow up on the research which bothered Witschi a great deal but he felt that he should not "steal" a project from one of this students.

Little attention is given to the tremendous potency of the testes for "androgenization". Even in cases of ovo-testis when the androgen level is too low to stimulate the Wolffian ducts, and the production of Müllerian inhibiting substance is too low to prevent production of uteri, the animal is still "androgenized". Gynandromorphic mice, which have very poor testicular development, are always anovulatory. Apparently the presence of a Y chromosome is the key to "masculinization" and "androgenization" is something else entirely.

Finally, let me get back to the original idea which started the research to begin with; aging. You did not take this up in your review but it is a pet of mine. I think that exposure of the neonatal female rat, and maybe other species too, causes "damage" to the CNS which results in a premature aging. Swanson and Van der Werff ten Bosch proposed that the delayed anovulatory syndrome was aging and there have been several others who have picked up on the idea. The androgenized female rat has several of the features of the aged female. Furthermore, fetal anoxia or X-irradiation on the fetus produces premature aging with loss of estrous cycles (constant estrus). The big question is what do these agents (steroids, DDT, caffeine, etc.) do to the CNS? I think that their main effect is upon rhythms. Barraclough, and particularly Gorski, have concentrated upon "tonic" gonadotropin release in males and "cyclic" function in females. However, this may be backwards. Males "cycle" on a daily basis-lots of data support this at least for ACTH, corticosterone, LH, prolactin, but not FSH. Females seem to cycle only at 4–5 day periods, but they can be made to cycle daily after ovariectomy. Jerry Yochim has shown that the rat behaves as if it has two clocks, one set internally with a 20 hour day and the other entrained to the 24 hour light-dark cycle. Interactions between the two produce a cycle of about 5 days. He can manipulate the exogenous light cycle to various lengths and change the ratio of internal to external time so that he can have 7 or 10 day cycles. When both clocks are set for a 20 hour day the animal loses cyclicity; the estrus or diestrus phase will be about 150+ days long. Yochim believes that aging of the CNS gradually lengthens the internal clock so that eventually the two are phased so that acyclicity is established. Note that "old" female rats are locked into either pseudopregnant diestrus, or estrus.

Although we can produce a female rat with the androgenization syndrome by such diverse treatment as constant light or rather mild diabetes mellitus, we do not consider them "masculinized". Both of these treatments cause loss of daily cyclic pituitary function and again focus attention upon loss of this as a causative factor . . .

I became interested in the sterilizing effects of steroid hormones in the late 1950's when I met, during a congress in Miami, a young Hungarian scientist, Béla Flerkó. He lectured in part on the effect of androgens in the neonate. The company I worked for was developing oral contraceptives, and I found the subject fascinating. To achieve a permanent sterilization by a single injection looked the up-and-coming method to have. I had thought that we could employ this particular activity in veterinary medicine to castrate male piglets "chemically". We tested synthetic estrogens in rats and injected several newborn piglets. The results were encouraging, but it was more convenient, and easier, to castrate the piglet surgically. Next, we turned to rodents. The indigenous rat population in the macadamia nut plantations in Hawaii were causing great crop damage. We reasoned that if we could sterilize a large proportion the damage might become acceptable.

We were using the most effective synthetic estrogen, mestranol, with encouraging results. However, the management decided against the trials. They reasoned that it would not look good to use the same estrogen as a constituent of an oral contraceptive for humans and also as a "rat poison".

I have surveyed in the monograph *Hormone Toxicity in the Newborn* the teratological effects of hormones in the neonate and in the fetus. The first chapter is a historical introduction. I describe in the second chapter the control of reproduction as we understand it today; the purpose is to recount the great complexity which controls the process. The third chapter surveys the effects of steroid hormones on the embryo and the fourth chapter their effects in the neonate. Nonsteroidal hormones are discussed in Chapter 5. Changes in behavior and possible effects on psychosexual development in humans are covered in Chapter 6. I have summarized the evidence in the last chapter.

The appendix includes abbreviations used in the text, structures and biological activity of the principal hormones mentioned, and a brief outline of biological tests used to assay steroid hormones.

I wish to thank the many friends and colleagues who gave me considerable help, read part of the manuscript and offered valuable suggestions: S Cekan (Stockholm, Sweden); IJ Clarke (Melbourne, Australia); G Dörner (Berlin, DDR); RA Gorski (Los Angeles, CA); JL Hill (Bethesda, MD); T Iguchi (Yokohama, Japan); DC Johnson (Kansas City, KS); S Kawashima (Hiroshima, Japan); J Kinsel (Bethesda, MD); CR Kramer (Staten Island, NY); RD Lisk (Princeton, NJ); L Macho (Bratislava, Czechoslovakia); RA Maurer (Iowa City, IA); FL Meyer-Bahlburg (New York, NY); T Mori (Tokyo, Japan); S Nilsson (Falun, Sweden); WH Rooks II (Palo Alto; CA); DM Sheehan (Jefferson, AZ); A Raynaud (Vabre, France); JM Reinisch (Bloomington, IN); HW Rudel (Elizabeth, NJ); P Södersten (Göteborg, Sweden); J Schreiber (Praha, Czechoslovakia); J Vreeburg (Rotterdam, The Netherlands); J Weisz (Hershey, PA). S Kawashima (Hiroshima, Japan) further helped by sending reprints of Japanese publications. T Iguchi, T Mori and A Raynaud sent material for illustrations. M Gregory (Staten Island, NY) reproduced aged slides into presentable microphotographs and PA Kincl (New York, NY) helped with illustrations. My wife, Lada Kincl, proofread all the references and helped to correlate these with text. B Bevan, P Cosumano and F Pollutri typed parts of text, references and tables.

Staten Island, NY   Fred A. Kincl
October 1989

Springer-Verlag regrets to announce that the author died shortly before publication of his book.

Contents

1 Introduction

Two classes of regulatory substances, enzymes and hormones, control life processes in multicellular organisms. Chemical reactions, taking place in dilute aqueous solutions within the cell, that would normally proceed too slowly are catalyzed by enzymes. Enzyme activity is often induced by the other class of regulatory agents, the hormones. Hormones, produced by specific ductless organs (endocrine glands), diffuse into the blood and are distributed throughout the body. These regulatory agents may act on one specific tissue (target organ) or stimulate a variety of tissues. Hormones have a variety of chemical structures, from water soluble simple peptides (three amino acid residues) to complex glycoproteins (molecular weight in the range of 30,000 daltons) to lipids of low molecular weight (steroid hormones and prostanoids).

Several modalities influence hormone action on target organs:

(i)   the rate of synthesis and secretion of the stored hormone(s) from the endocrine gland;
(ii)   the biological half-life (the time it takes a given quantity of hormone circulating in blood to decrease by one-half, $t\frac{1}{2}$);
(iii) the presence of binding (or transport) proteins in plasma;
(iv)   the presence of hormone receptor proteins on target cell membranes, or within cells;
(v)   removal of the hormone from the circulation (metabolism) which may take place within the target tissue, liver, or kidneys, and
(vi)   by elimination (usually in urine). Often one hormone may stimulate and another inhibit a given function.

Hormones appear in blood in rising and falling concentrations superimposed upon a background of basal levels. Rapid metabolism (biological $t_{1/2}$ is usually a few minutes) prevents excessive accumulation. When a physiological need arises, the endocrine gland responds by producing larger amounts of the needed hormone. Increased production is restricted to resting levels by the inhibitory action of the hormone, *the feedback mechanism,* on the endocrine gland. Often the feedback may influence several tissues, including the brain. For example, in males increased production of testosterone partially inhibits the production of gonadotropin releasing hormone from the hypothalamus and of gonadotropic hormones from the pituitary gland. Decreased titer of these hormones leads to a decrease of gonadal function. Accumulation of biosynthetic intermediates may further reduce testosterone output.

The control by feedback, and the resulting homeostatic balance, is essential for the proper maintenance of physiological functions. When homeostasis is destroyed and high concentrations of gonadal hormones persist (for example by supplying exogenous androgens), the testes will atrophy. Much evidence shows that in the adolescent and the adult, such repression is transitory; removal of the suppressive influences results in the resumption of pituitary and gonadal functions.

The fetus and the neonate are different in this respect. Following a brief exposure during fetal development or neonatal period, gonadal hormones, corticosteroid hormones, thyroid hormones and a variety of other chemicals have been found to provoke permanent changes in endocrine systems and target organs.

Moore and Price (1932) recognized that excessive amounts of hormones may be toxic. They injected repeatedly newborn rats with testosterone propionate and noted infertility and atrophy of the gonads in the adult. Pfeiffer (1936) found that ovarian grafts, transplanted into the anterior chamber of the eye of adult male rats did not luteinize. In contrast, similar procedures in neonatally castrated males resulted in the formation of corpora lutea. He ascribed the difference to the "masculinizing effect" of testicular hormones on the pituitary gland during early development. Bradbury (1941) observed that injections of testosterone propionate induced a condition similar to that observed by Pfeiffer (1936). Mme Dantchakova, a Russian emigree associated with the University of Kaunas (Lithuania) and later in Bratislava (Czechoslovakia), studied the effects of gonadal hormones on the development of embryonic guinea pigs and noted a permanent injury in both sexes (Dantchakoff, 1938 a, b). In Paris, A. Raynaud explored the effects of androgens and estrogens on the developing fetus in rats (Raynaud, 1942).

Leathem and Barraclough (Barraclough and Leathem, 1954; Leathem, 1956; Barraclough, 1961) established that a single injection of an androgen (testosterone propionate) was sufficient to induce sterility in female rodents and coined the phrase "androgenized rat". The authors stressed that treatment should be given before about the tenth day of life to be effective. This established the concept of a "critical period". The work of Harris (1948), Segal and Johnson (1959), Gorski (1963), and Barraclough (1966) focused attention on the role of the brain (hypothalamic centers) rather than the pituitary gland.

The first reports (from Japan) on the effect of gonadal hormones on development were published from the Institute of Zoology of Tokyo University (Takasugi, 1952). During the 1960's, Japanese workers using mainly *in utero* technique made important contributions describing effects of estrogens on peripheral organs of female rodents.

Post-natal development of the phenotype has been the subject of many articles and reviews, and two international symposia. The material of a symposium held in 1967 in Czechoslovakia was published as a monograph (Kazda and Denenberg, 1970). Only a short report was devoted to a symposium held in Japan in 1972 (Bern *et al.*, 1973).

I found in this monograph deleterious effects of gonadal hormones in the fetus and the neonate. The term "neonatal" is ambiguous; it refers to an ill defined period after birth. In rats, the neonatal period may last about the first ten days of life, or 20-25% of the prepubertal period. In lambs, the period may be only a few days, a much shorter portion of life before puberty. Those species whose offspring are born well developed (e.g. guinea pigs) are held by some to lack a neonatal sensitivity within this narrow interpretation. I use the term "neonatal" to describe that period of life after birth during which the developing individual is abnormally sensitive to hormonal influences. The period may be deemed an extension of intrauterine life in that various physiological mechanisms are still maturing and can be modified. The parameters may include enzymes directing synthesis and metabolism of physiologically important substances in the nervous system, the liver, kidneys and other organs.

We have known for years that various hormonal and environmental influences modulate the development of the brain. Testosterone may influence earlier maturation of the left hemisphere and the resulting trend toward language abilities, right handedness, and spatial coordination in boys. We also know gonadal hormones influence neuron growth and synapse formation, and in rodent brain sex-linked biochemical differences. Thus, not surprisingly the original thought was that an androgenized female rat resulted from alterations of hypothalamic functions that are destined to control reproductive functions in later life. Later research has shown that the changes are due not only to differences in the brain but also to lesions in the gonads, and alterations in the fate of endogenous hormones, biological half-life and metabolism.

References

Barraclough CA (1961) Production of anovulatory, sterile rats by single injection of testosterone propionate. Endocrinology 68: 62–67

Barraclough CA (1966) Modification in the CNS regulation of reproduction after exposure of prebubertal rats to steroid hormones. Rec Prog Horm Res 22: 503–529

Barraclough CA, Leathem JH (1954) Infertility induced in mice by a single injection of testosterone propionate. Proc Soc Exp Biol Med 85: 673–674

Bern HA, Gorski RA, Kawashima S (1973) Long-term effects of perinatal hormone administration. Science 181: 1889–1990

Bradbury T (1941) Permanent after-effects following masculinization of the infantile rat. Endocrinology 28: 101–106

Dantchakoff V (1938a) Sur les effets de l'hormone male dans une jeune cobaye femmelle traite depuis un stade embryonnaire (inversions sexuelle). C R Soc Biol 127: 1255–1258

Dantchakoff V (1938b) Sur les effets de l'hormone male dans un jeune cobaye male traite depuis un stade embryonnaire (production d'hypermales). C R Soc Biol 127: 1259–1262

Gorski RA (1963) Effects of low dosages of androgen on the differentiation of hypothalamic regulatory control of ovulation in the rat. Endocrinology 73: 210–216

Harris GW (1948) Neural control of the pituitary gland. Physiol Rev 28: 139–179

Kazda S, Denenberg VH (eds) (1970) The post-natal development of phenotype. Academia, Prague

Leathem JH (1956) The influence of steroids on prepubertal animals. Proc III Int Congr Anim Reprod, Cambridge

Moore CR, Price D (1932) Gonadal hormone formation and the reciprocal influence between gonads and hypophysis with its bearing on the problem of sex hormone antagonism. Am J Anat 50: 13–71

Pfeiffer CA (1936) Sexual differences of the hypophysis and their determination by the gonads. Am J Anat 58: 195–225

Raynaud A (1942) Modification expérimentale de la differentiation sexuelle des embryons de souris par action des hormones androgenes and estrogenes. In: Actualites scientifique et industrielles. No. 925 et 926, Hermann and Co, Paris, pp 1464

Segal SJ, Johnson DC (1959) Inductive influence of steroid hormones on the neural system: ovulation controlling mechanism. Arch Anat Micr Morph Exptl 48: 261-274

Takasugi N (1952) Einflüsse von Androgen und Estrogen auf die Ovarien der neugeborenen und reifen, weiblichen Ratten. Annot Zool Japon 25: 120-131

2 Control of Reproductive Function in the Adult

In lower vertebrates, fishes and frogs, reproduction is energy expensive. Most females produce large numbers of eggs (fertilized outside of the body) to insure that at least a few survive. Mammals have developed a more efficient process: only a few eggs are produced which are fertilized and develop internally.

Reproductive functions must follow within a prescribed temporal relationship to assure maintenance of the species. The structure of the reproductive organs facilitates sperm delivery to the female reproductive tract; germ cells must mature within the gonads (testes and ovary); prior to mating distinct behavioral patterns must signal that both partners are ready to breed; the descent of fertilized ova, implantation and the growth of the fetus take place while specific hormones maintain a well defined milieu; birth occurs when the embryo signals that its development has been completed; and nursing, and nursing behavior, are hormone directed. Adequate function of other glands with internal secretion (thyroid, adrenal, growth hormone) and good nutrition are of paramount importance.

The sexes differ in a major respect: females are born with a given set of germs cells (oocyte). When they reach senescence, they lose their reproductive ability. Preservation of the individual dictates the loss. Pregnancy imposes a severe physiological stress which an aged female may not tolerate. In contrast, males have the ability to produce continuously new crops of gametes (sperm) well into old age.

There are more than 4000 species of mammals living in the world (see May, 1988), our knowledge of reproductive physiology is based on the study of a mere few, and even between those differences exist. Some ruminants cycle regularly while others breed only once a year; most rodents have regular cycles but in the rabbit, as in some carnivores, ovulation takes place in response to copulation; in the nine-banded armadillo a single fertilized egg divides into four identical quadruplets which contain identical set of genes; some non-human primates are seasonal breeders while others breed throughout the year. Despite the differences in breeding strategy there is a basic endocrine commonality among the basic principles which govern reproduction. An independent process, generated within the brain, directs the function of a "master" endocrine gland (the pituitary) which in turn stimulates the gonads to produce germ cells and hormones which influence the role of the brain and the pituitary. It is the aim of this chapter to explore the principles which control reproduction in the adult male and female mammals.

2.1 The Reproductive Cycle – An Overview

Mature females undergo during a reproductive cycle changes easily followed by monitoring cellular changes in the vaginal or uterine (endometrial) epithelium. At the end of each cycle the uterine lining in primates is sloughed off and bleeding (menstrual flow) results. In all other species, whether seasonal or cyclic such changes are referred as *estrus*. The length of the cycle varies from species to species: it is 4–5 days in small rodents (rats, mice, hamsters) and about a month in ruminants (cow) and primates (monkeys, humans). Many species (sheep, dog, deer) breed only a few times a year. Eckstein and Zuckerman (1962) described in detail estrus cycles of many mammalian species.

The brain is the primary center involved in the reproductive process in both sexes. External and internal signals activate noradrenergic neurons which trigger the release of a neurohormone from the hypothalamus; dopaminergic nerve terminals are inhibitory during this phase. The neurohormone (gonadotropin stimulating hormone, GnRH) stimulates the release, and the synthesis, of pituitary tropic hormones (follicle stimulating hormone, FSH, and luteinizing hormone, LH). A sudden increase of pituitary hormones, especially of LH (the preovulatory peak) triggers ovulation in females. The peak is absent in males. Gonadotropins in their turn arouse the production of gonadal (steroid) hormones produced by specialized cells. In males the cells (Leydig cells) are located between the seminiferous tubules. The main "male" hormone produced is testosterone; in some tissues testosterone becomes active only if converted into the reduced form (5α-dihydrotestosterone). In females the growing follicle produces estrogenic hormone (estradiol). After ovulation the follicle converts into a new endocrine organ, the corpus luteum (CL) which produces progesterone. In some species (primates) the CL has also the ability to secrete estrogens.

An elaborate system of restraints and balances tunes the process: the brain hormone stimulates not only the pituitary gland but acts also directly on the gonads, the steroid hormones of the gonads provide both stimulatory and inhibitory influences on the brain, the pituitary and the gonad itself.

2.1.1 Marsupials

The reproductive physiology of marsupials is more primitive than that of other mammals. Fertilized ova travel rapidly through the oviduct; the dividing egg reaches the uterus within 24 h after fertilization. Gestation is short, and most of fetal growth and development takes place outside the uterus, in the pouch (a few marsupials even lack this structure). In macropodia (kangaroos) gestation lasts the length of the cycle (30 days). It is even shorter (about 60% of the cycle) in opossums. After birth the poorly developed offspring becomes attached to a teat in the pouch where it continues to grow and develop (the lactation period). During gestation and lactation new pregnancy is not possible. Two different mechanisms establish the block.

(i)   in opossums ovulation does not take place until lactation ceases;
(ii) kangaroos ovulate during gestation, but when the fertilized egg reaches the uterus it does not implant. The embryonic growth is postponed after lactation has ceased. Ova descent and blastocyst development need adequate progesterone production (a functional CL). The function of the corpus luteum is independent of pituitary function. The development of Graafian follicles and ovulation requires pituitary function. There is no seasonal variation either in FSH or LH concentrations in the blood. Preovulatory progesterone peak does occur and may be needed for ovulation. Estradiol contributes inhibitory feedback.

The cycle of a tammar wallaby is illustrative (Tyndale-Biscoe *et al.*, 1986). In the southern hemisphere the majority of females give birth about 6 weeks after the winter solstice; in the northern hemisphere the cycle is reversed. The female comes into estrus soon after giving birth, and if fertilization has taken place and the blastocyst has reached the uterus, it will remain dormant for the next 8 months. During this time the single young suckles in the pouch. When the young has matured, or if it is removed prior to emergence, the blastocyst and the CL resume development. Reproductive quiescence before summer solstice is maintained by the suckling impulse (lactational quiescence) and after the summer solstice by seasonal factors (seasonal quiescence). The environmental cue is transmitted through the pineal gland, and melatonin is the hormone involved. The suprachiasmatic nucleus located in the hypothalamus regulates the pineal rhythm (Section 2.3.3).

The male tammar is unresponsive to photoperiod changes and can breed the year around. However, it does breed cyclically, and the testes undergo periods of quiescence and active spermatogenesis. The male cycle corresponds to the changes in females. An olfactory cue (a production of a pheromone), influenced either by progesterone, or estradiol, triggers the onset (Section 2.3.3).

2.1.2 Eutherians

Eutherians either breed cyclically (primates, rodents, and some ungulates), seasonally (some rodents, ungulates, and carnivores) or ovulate in response to mating (some rodents and carnivores). In eutherians fetal sojourn in the uterus is long, and the offspring are born more fully developed.

2.1.2.1 Cyclic Breeders

In cyclic breeders seasonal changes may modulate but do not control breeding ability. In rodents (rats) light transmitted to the brain gives the cue to begin a new cycle; an intact connection between the preoptic area and median hypothalamus is a must. In primates estrogens of ovarian origin stimulate directly the pituitary; ovulation can take place even when the connection between the two neural areas is interrupted (Figure 2.1).

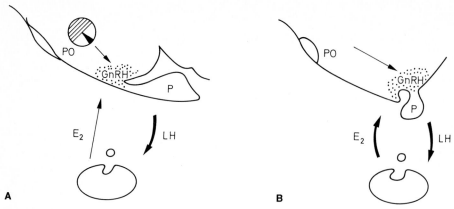

Figure 2.1. Ovulation Control in Primates and Rodents. **A:** In rodents (rats) the cue to begin a new cycle originates in the preoptic area (PO). The event is regulated by dark-light cycles (stippled-open areas in the circle). The impulse (dark band) which triggers the release of gonadotropin releasing hormone (GnRH) lasts only about 2 h. GnRH induces the release of luteinizing hormone (LH) from the pituitary gland (P), a necessary event to induce ovulation. Estrogens (E_2) synthesized in the growing follicle (not shown) only modulate the function of the hypothalamus. **B:** In primates estradiol (E_2) originating in the growing follicle provides the main stimulus to the median eminence (dotted area) to release GnRH and subsequently LH. Other hypothalamic nuclei (not shown) modulate GnRH synthesis and/or release.

Rats

Females maintained on regular light-dark schedules (14 h light, 10 h darkness) are receptive to males during late proestrus or early estrus. The appearance of nucleated cells in the wash obtained from the vagina heralds pro-estrus; the event lasts only about 12 h. The "critical period" for LH release lasts about 2 h in the afternoon of proestrus. During this period brain impulses which set into motion the ovulation procedure can be blocked by central nervous system depressants (Section 2.3.1). Females will breed during this time. About 24 h later the vaginal epithelium undergoes cornification in response to the growing biosynthesis of estrogens by the graafian follicle; the cells become large and flat. This is the sign of estrus. The appearance of small cells (leukocytes) in vaginal lavage heralds the occurrence of ovulation, formation of corpora lutea and progesterone synthesis. This period, which last about 48 h, is termed diestrus. If fertilization has taken place corpora lutea continue to secrete progesterone, and implantation will take place. If fertilization has not occurred the cycle is repeated. The cycle may last 4 or 5 days, depending on strain (Figure 2.2).

During the cycle concentrations of brain (GnRH), pituitary (FSH and LH) and ovarian (estradiol and progesterone) hormones fluctuate. The levels are high during proestrus and estrus and lower during the diestrus (Table 2.1) released in high and low pulses (Section 2.2).

The organ that controls the reproductive rhythms ("Zeitgeber") in rodents is the brain. Ovulation is blocked by:

(i)   removal of the pituitary (hypophysectomy) or administration of central nervous system depressants; barbiturates, ether, morphine, atropine will block ovulation but only if given prior to the critical period;

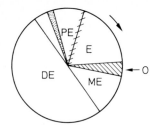

Figure 2.2. Chronology of a Reproductive Cycle in Rodents. Schematic representation of a four day cycle. During proestrus (PE) epithelial (nucleated) cells appear in vaginal lavage. The chain of events which leads to ovulation is set during the "critical" period (narrow dotted area) triggered by light-dark periods; females are receptive to males later (stippled area). During estrus (E) estrogen concentration in blood is at a peak; nucleated cells in the vagina are transformed into cornified (large, flat) cells; the uterus is hyperemic, enlarged; ovulation (O) takes place toward the end of this period. During metestrus (ME) cornified cells are still present and leukocytes begin to appear; the uterus becomes smaller; corpora lutea begin to form, and ova are present in the oviduct. Diestrus (DE) is characterized by the presence of ova in the oviduct and in the uterus, formation of corpora lutea and increasing progesterone concentration in blood; the uterus is small, pale; cornified cells have disappeared, and only leukocytes and some nucleated cells will be present in vaginal lavage.

Table 2.1 Concentrations of GnRH in the Hypothalamus and of Various Hormones in Serum of Female Rats During the "Critical" Period and During Diestrus.

	GnRH[a]	FSH[b]	LH[b]	Estradiol[c]	Progesterone[b]
Proetrus	116 ± 12	208 ± 42	2154 ± 92	40 ± 7.3	19 ± 3.1
Diestrus	71 ± 9.8	136 ± 6.5	54 ± 11	8.1 ± 0.6	14 ± 1.5

After Barraclough *et al.* (1984). a, pg/μg protein in median eminence; b, ng/ml of serum; c, pg/ml of serum.

(ii) electrical stimulation of the medial preoptic area will elicit ovulation in blocked rats;

(iii) shifts of light-dark periods will postpone, or advance, the critical period;

(iv) exposure of female rats to constant light abolishes the cycle leading to a persistent cornification of vaginal epithelium (increased estrogens).

Steroid hormones play a modulatory role:

(i) gonadectomized animals show high titer of both FSH and LH which is decreased by exogenous steroid hormones;

(ii) injection of estradiol (on day 3) to 5 day cycle rats shortens the cycle while injection of progesterone to 4-day rats on day 2 lengthens the cycle to 5 days (Figure 2.3).

Ruminants

Ruminants breed either the whole year or seasonally. Cows ovulate about once a month; others, such as sheep, only during a short period once a year. Ovulation control of regularly cycling ruminants resembles that of primates, but the control of the gametogenic process differs. Follicular growth is functionally dominant;

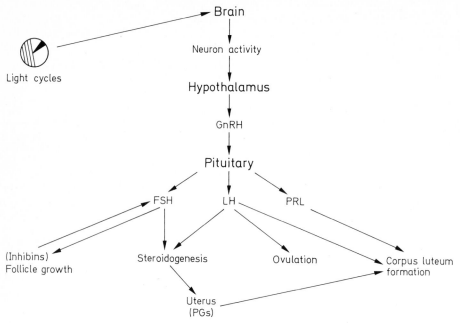

Figure 2.3. Hormonal Influences Which Contribute to the Control of the Reproductive Cycle in Female Rats. The release of pituitary hormones is governed by the frequency of hypothalamic GnRH. Follicle stimulating hormone (FSH) release is modulated by polypeptide hormones (inhibins) originating in the growing follicle; steroidogenesis is stimulated by FSH and by luteinizing hormone (LH) and corpus luteum (CL) formation by LH and prolactin (PRL). Prostaglandins (PGs) originating in the uterus modulate the function of CL.

removal of the growing follicle stimulates the growth of a new one within 2–3 days in sheep and cows, and in 6 days in pigs. During the luteal phase the secretion of progesterone keeps the follicular growth quiescent. Another difference resides in the steroidogenic capability of the corpus luteum. In ruminants the CL lacks the ability to secrete estrogens.

Primates

The primates cycle lasts on average 28 days. The ovary regulates the rhythmicity, and the brain plays only a permissive (yet obligatory) role (see Knobil, 1980; Goodman and Hodgen, 1983; Weick *et al.*, 1983). Primates are mainly monovular, i.e. only one follicle ripens and ovulates. Intra- and extraovarian components regulate folliculogenesis. It is the rising production of inhibin and estradiol that blocks the maturation of other follicles. Extirpation of a follicle (the removal of the estrogen source) will postpone the growth of the next crop by one cycle. Rising estrogen concentration in blood also modulates the activity of the pituitary gland. When the concentration reaches more than 150 pg/ml, and persists for 1½ days, a threshold is reached, and both FSH and LH suddenly surge. The preovulatory peak does not require changes in the pattern of GnRH discharge. The crucial role of estradiol in inducing the surge has been demonstrated in animals in which the basal median hypothalamus has been destroyed; a bolus of exogenous estradiol will result in LH surge. LH surge will not occur under similar conditions in ovariectomized animals. In contrast, cautery of the already formed CL (decreased progesterone production) will advance the cycle.

The luteal phase lasts about the same as the follicular phase (14 days). Estradiol (Section 2.4.3) and the hormonal milieu of the antecedent follicular phase control the life span of the nonfertile CL. A CL which developed from a deficient follicular phase does not respond well to LH stimulation since it is deficient in LH receptors (Section 2.6). If the cycle was fertile and pregnancy ensues the CL maintains early pregnancy before the placenta assumes the role (see Pasqualini and Kincl, 1985).

During the luteal phase progesterone prevents follicular growth. When the CL atrophies, progesterone influence fades, and the cycle repeats. Progesterone also provides a negative feedback on the hypothalamus-pituitary axis (Section 2.5.2). Estradiol-induced LH surge is blocked by simultaneous administration of progesterone.

Males

The neuroendocrine mechanism which regulates the reproductive process in males is very similar to that seen in females. Males show a female-like pattern of gonadotropin secretion and the LH surge in response to estradiol stimulation (Section 2.2.1). The cyclicity of gonadotropin release in male rhesus monkeys is sufficiently similar to that of females. Norman and Spies (1986) transplanted ovarian tissue into an abdominal pouch and observed that the tissue released estradiol and progesterone in a cyclic manner.

2.1.2.2 Seasonal Breeders

During the period of anestrus the concentrations of pituitary and gonadal hormones in peripheral plasma are low and the gonads atrophic. It is only during the rut that the hypothalamic-pituitary axis is active, females become receptive and the males aggressive, and they mate. Changes in the length of the day (Figure 2.4) adjust the onset of the reproductive period.

The "coming on heat" or "coming in season" (estrus) is preceded by a period of proestrus. During this period reproductive organs (uterus) begin to grow, and follicular development begins. In males testicular activity resumes. Breeding and successful fertilization take place only during the estrus when the female is ready to receive the male. The period may last several days, and various cycles may be repeated. Ovulation usually occurs towards the end of estrus. Often, both periods are not easily distinguishable from each other and are referred to as "heat".

Hormonal influences which control the reproductive process are similar to those operating in cyclic breeders. During the inhibitory photoperiod stage (anestrus) the brain inhibits reproduction. During this period exogenous gonadal hormones have no effect on inducing the pituitary to function. In estrus the brain function is reversed, CNS drugs are no longer active in suppressing ovulation, and estradiol induces preovulatory LH surge. Possibly, prostaglandin $F_{2\alpha}$ synthesized

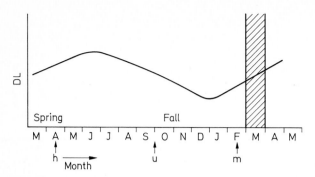

Figure 2.4. Reproductive Patterns of Seasonal Breeders. Seasonal breeders use the length of the day (DL) to synchronize mating so that birth takes place during spring when the conditions for offspring survival are most clement (stippled area). The onset of estrus is shown by arrows. Species with long gestation period (h, horses 11 months) breed during April (A). Those with shorter gestation (6 months) such as ungulates (u, deer and sheep) may breed in September (S). Rodents (mice, m) gestate less than a month and breed in February (F).

in the hypothalamus, is obligatory to modulate estradiol effect leading to the surge (McCracken, 1974).

2.1.2.3 Induced Ovulators

Several species of rodents and carnivores (rabbit, vole, short-tailed shrew, ground squirrel, hare, cat, ferret, weasel and mink) ovulate only in response to coital stimulation. In the wild induced ovulators have a defined breeding season. Rabbits will breed mainly beginning from January till June. Seasonal restriction is food, not photoperiod, dependent. If suitable nutritional conditions exists (domestication) the doe is in estrus and ready to accept the buck throughout the year. Periods of anestrus may be present but are usually short and infrequent.

During estrus follicular cycles of about 7 to 10 days' duration follow each other. The stimulus of coitus induces the follicles to mature, and ovulation will take place within 10 to 12 hours (Everett, 1961). The ovulation reflex is initiated by afferent impulses of multiple origin, and the integrity of the lumbar cord is prerequisite; paralysis of the lower part of the body will prevent ovulation (Brooks, 1937; 1938). Copulation, or artificial stimulation of the vagina, results in changes in the electrical activity of the brain in cats (Porter *et al.,* 1957) and rabbits (Sawyer and Kawakami, 1959). Gonadotropins and steroid hormones modify the threshold activity (Kawakami and Sawyer, 1959).

Season variations influence the incidence of spontaneous ovulation (Sawyer, 1959).

Ovulation can be also induced by electrical stimulation of the anterior hypothalamus, by intravenous injections of small quantities of aqueous solution of copper acetate, or directly by stimulating the ovaries with LH (or hCG). Progesterone, acting on the brain blocks ovulation induced by mating (or stimulation of the vagina) but not if chorionic gonadotropin is used. This indicates the insensitivity of the ovary to the feedback (Section 2.5.1).

2.2 Patterns of Hormone Release

Hormones are released in periodic bursts; the duration, frequency and amplitude of the bursts is hormone specific and changes with reproductive status. The episodic events may be short (20 to 90 minutes, circhoral occurrence), diurnal (twice a day), circadian (about a day long), lunar, or annual. The biological clock is located in the hypothalamus (Section 2.3.3.1).

The episodic hormone release is the result of fluctuating electrical activity of nerve cells. The activity of a group of neurons (multiunit activity) is synchronized. The nature of the trigger is not known. Spike activity increases prior to hormonal events (Figure 2.5). In ovariectomized rats, electrical frequency in the arcuate region increases from a basal 1000 spikes/min to 4500 to 5000 spikes/min prior to LH release. Electrical activity also increases in the mediobasal hypothalamus of ovariectomized rhesus monkeys from a basal ≤ 1000 spikes/min to about 3000

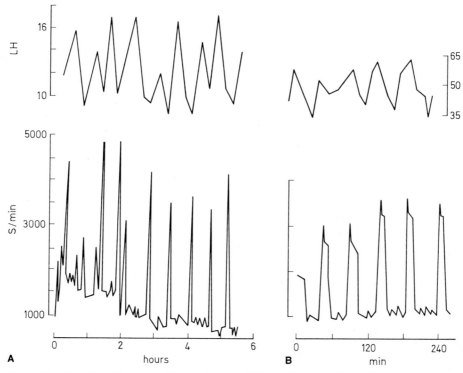

Figure 2.5. Neural Activity in the Hypothalamus Is Followed by the Release of Luteinizing Hormone. **A:** Electrical multiunit activity (MUA), spikes per min (S/min), of cells in the arcuate nucleus of ovariectomized rat fluctuates from 1000 to about 5000. Increases in the activity (1–5 min duration) are followed by increases in the concentration of luteinizing hormone (LH, ng/ml) in peripheral blood. **B:** In ovariectomized rhesus monkey MUA in the mediobasal hypothalamus lasts about 11 min and recurs every 47 min; the increases are followed by rises in circulating LH. Redrawn with permission from Lincoln *et al.* (1985).

spikes/min (see Lincoln *et al.*, 1985). As a result LH concentration in peripheral plasma rose from about 35 ng/ml to about 65 ng/ml (Wilson *et al.*,1984). Differences in pulsatile release between sexes may be contingent upon progesterone modulation. Lin and Ramirez (1988) reported on the experiments in rabbits. In females GnRH pulses developed once every hour with a mean release of 0.91 ± 0.13 pg/10 min ($\pm$ SE). Pulses of progesterone (10 ng/ml) increased the amplitude to 1.66 ± 0.20 pg/10 min. Pulses in males were about the same (1 every 60 min), but progesterone did not affect the amplitude of spontaneous pulses.

The ability of GnRH releasing neurons to release the neuropeptide is inherent. Isolated cell groups retain the competence to deliver the hormone in about 20 min intervals (Melrose *et al.*, 1987).

2.2.1 Males

In the older literature, gonadotropin release in the males has been usually referred to as "tonic" (non-fluctuating) while females were held to exhibit a "cyclic" release. The availability of more sensitive analytical methods which allow hormone detection in a small volume of blood (and more frequent sampling) have established cyclic variations in males. Episodic fluctuations of reproductive hormones develop in male rabbits and rats (Ellis and Desjardins, 1982; Sisk and Desjardins, 1986), hamsters (Chappel *et al.*, 1984b), rams (Lincoln and Short, 1980), bulls and reindeer bulls (see Turek *et al.*, 1984), and primates (Karsch *et al.*, 1973; Stearns *et al.*, 1973; Steiner *et al.*, 1976; 1980; Plant, 1981). Naftolin *et al.* (1973) and Kicovic *et al.* (1980) reported episodic and diurnal fluctuation in the concentration of gonadal hormones in men.

In rats and primates LH is released in single or a train of two to three pulses lasting 5–10 min followed by a gradual decline lasting 50–70 min. The episodic release of LH is followed within 30–60 min by a testosterone spike. The rise and fall of plasma testosterone may span 3–6 h and is followed by a nadir in LH concentration. The absence of LH episodes would be repeated 2 to 4 times a day. The authors found high individual variations in amplitude and timing within a particular animal on different days and stress the need to use large groups to obtain meaningful data (Ellis and Desjardins, 1982). In men LH pulses with a frequency of about 120 min and an amplitude of almost 10 mIU/ml (Crowley *et al.*, 1985).

In mice, bulls, rams and monkeys LH and testosterone pulses occur at the same time. In men, rats and ferrets several LH pulses are needed to produce a single rise of testosterone in blood. In castrated animals the absence of testosterone triggers an increase in the frequency of LH pulses (Ellis and Desjardins, 1984).

Testosterone in peripheral blood of men pulses with a frequency of 2–4 h; it shows a diurnal variation, and annual fluctuation (Figure 2.6).

Hormone activity of seasonal breeders is slow during non-breeding periods. During rut the activity increases. Lincoln and Short (1980) detected a ten fold increase in the frequency of LH (from about once every 24 h to about once every 2 h) in the peripheral plasma of Soay rams. The amplitude of the peaks increased almost threefold (Table 2.2). Fluctuating concentrations of testosterone in blood, in patterns similar to that of LH, duplicated the changes in the concentration of

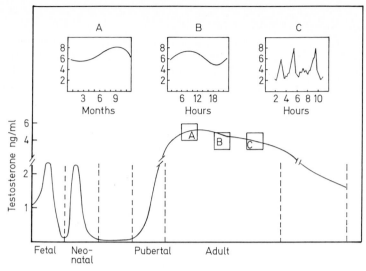

Figure 2.6. Episodic, Diurnal and Annual Fluctuations of Testosterone in Peripheral Plasma of the Human Male at Different Times in the Life Cycle. The peak of testosterone in the peripheral blood of the fetus occurs between 12 and 18 weeks of gestation (lower left corner; gestational age not shown). In the neonate the peak occurs at about 2 months of age. During puberty (between 12 and 17 years) testosterone in peripheral plasma begins to increase and reaches a peak during the second or third decade of life. Inset **A** shows the annual rhythm. The peak and nadir occur in the fall and spring, respectively. Inset **B** shows diurnal variations. The peak in peripheral plasma occurs in the morning, the nadir in the evening. Inset **C** shows the irregular episodic fluctuations. Redrawn from Ewing and Zirkin (1983) with permission.

Table 2.2 Frequency and Amplitude of Luteinizing Hormone Release from the Pituitary Gland of Rams During Rut and Non-breeding Season.

	Rut	Non-breeding
Mean LH frequency per 24 h	12.5 ± 0.4*	1.1 ± 0.3
Mean amplitude of LH peaks, ng/ml	4.2 ± 0.7	1.5 ± 0.1

* Mean ± SEM; from Lincoln and Short (1980).

LH. During the breeding period the concentration shifted from ≤ 10 ng/ml to 20 to 30 ng/ml (Lincoln and Short, 1980).

2.2.2 Females

In females LH release is periodic, in bursts of 20 to 40 min. Episodic secretion is functionally more efficient in eliciting the release of estradiol and progesterone from ovaries than continuous release.

2.2.2.1 Rodents

In juvenile females (27 to 29 days old) LH pulses occur every 30 min. In peripubertal animals (30 to 38 days old) the frequency remains the same, but the amplitude increases about two-fold as the time of the first estrus approaches (Urbanski and Ojeda, 1985a). In the adult LH pulses change with the fluctuation of gonadal hormones during the estrus cycle (Gay and Sheth, 1972; Gall, 1980; 1981; Higuchi and Kawakami, 1982; Fox and Smith, 1985; Condon *et al.*, 1986). Levine and Ramirez (1982) tried to detect changes in GnRH frequency during a cycle in rats but were unable to do so despite very frequent sampling (six per hour). Higher concentrations present during proestrus made it possible to detect intervals of 48 min and a significant proestrus surge. Kamel *et al.* (1987a) used an *in vitro* pituitary cell preparation and were able to show that increases in pulse amplitude elevated the amount of LH liberated. The increases also led to a rise in the synthesis of LH. However, rapid frequencies (3 or more per h) were followed by a decrease in the amount of freed LH per pulse.

Stress influences pulse frequency (Blake, 1975).

2.2.2.2 Primates

In the female rhesus monkey both GnRH and LH are released episodically (Marut *et al.*, 1981; Norman *et al.*, 1984). LH pulses about once every hour and reaches a peak before ovulation. During the preovulatory peak the pulse interval increases (35 min) during the ascending phase and slows to one pulse in 45 min during the descending aspect. Pulse amplitude rises from 45 ng/ml of LH (prior the peak) to about 125 ng/ml during the ascending phase, 250 ng/ml at peak and about 70 ng/ml during the descending aspect. Episodic infusion, 1 µg/min of GnRH for 6 min every hour, mimics the circhoral pattern (Teresawa *et al.*, 1987).

Human
LH release during the human menstrual cycle is pulsatile (Yen *et al.*, 1972; Reame *et al.*, 1984). The data of Backström *et al.* (1982) illustrate the fluctuations of LH, FSH, PRL, estradiol and progesterone during a menstrual cycle. During the early follicular phase LH pulses once every 105 min; during the late follicular phase the frequency increases to about once every 80 min and slows to about once every 200 min during the luteal phase; FSH and PRL frequency and amplitude follow that of LH albeit to a lesser degree. Estradiol pulse frequency declines from 80 min during mid-follicular phase to 200 min during the luteal phase. Crowley *et al.* (1985) reported LH pulses with a frequency of about 90 min during the follicular phase and more than 200 min during the luteal phase; the amplitudes during the luteal phase are higher. Progesterone concentration in blood paralleled closely episodic release of LH rising several times from a nadir (about 10 ng/ml) to more than 30 ng/ml within a 24 h period. The concentrations of FSH in either sex remained stable (Table 2.3).

In men testosterone concentration fluctuates daily (Rosenfield and Hilke, 1974).

Table 2.3 Pulse Frequency and Amplitude of Gonadotropin Releasing Hormone in Women During a Menstrual Cycle.

	Late Follicular Phase	Luteal Phase	
		Early	Late
Pulse Frequency, min	72	101	303
Amplitude mU/ml	10.4 ± 3.9*	14.7 ± 1.7	11.5 ± 1.4

* Mean ± SEM; data from Crowley *et al.* (1985).

2.2.3 Seasonal Breeders

Species breeding seasonally include badgers (Maurel and Boissin, 1982; Audy *et al.*, 1985), bats (Gustafson and Damassa, 1986), ferret (Sisk and Desjardins, 1986), foxes (Mondain-Monval *et al.*, 1979; Maurel and Boissin, 1982), hares (Davis and Meyer, 1973), kangaroos (Tyndale-Biscoe *et al.*, 1986), roe deer (Sempere and Lacroix, 1982) and sheep. Sheep reproduce in the northern hemisphere during the fall when daylight begins to decrease. During the sexually nonactive period, episodic release of hormones is infrequent and of low amplitude. During the rut brain activity increases. In the estrous ewe the frequency of GnRH appearance in blood varies with cycle status. During the follicular phase (preceding ovulation) GnRH peaks every 70 min. The rate decreases to one peak every 220 min during the luteal phase (Baird *et al.*, 1976; Hauger *et al.*, 1977). Karsch *et al.* (1984) describe LH peaking every 3–4 h during the follicular phase. The frequency increases to one pulse every 30 min prior to the onset of the ovulatory surge.

In estrous sheep estrogens enhance pulse frequency (Karsch *et al.*, 1985).

2.2.4 Annual Cycles

In many species cyclic reproduction is tied to food availability. Species which reproduce seasonally in the temperate zone will breed without restriction in the equatorial zone where food is abundant the year round. The bonnet monkey from southern India is not a seasonal breeder whereas its close relative, the rhesus monkey from northern India, shows seasonal breeding patterns. In this animal gonadal hormones (testosterone, 5α-dihydrotestosterone and dehydroepiandrosterone) fluctuate annually (Taranov and Goncharov, 1985). The barbary sheep is a round-year breeder whereas sheep living in moderate zones reproduce seasonally.

In male laboratory rats the weights of testes and accessory sex organs and the concentrations of gonadotropins and testosterone fluctuate annually (Mock and Frankel, 1978a,b; 1980).

Even in humans living in a harsh environment (Bush people of the Kalahari desert) fertility may peak during most clement conditions; most children are born during the rainy period (Frisch, 1980). Takagi (1986) observed an increase in testosterone plasma concentration during the period of darkness (June) in Antarctic expedition members but no seasonal fluctuation in Japan. Vancauter *et al.* (1981) described seasonal variations in adrenal functions (ACTH and cortisol concentrations) in humans.

In addition to light cue other environmental factors, such as temperature, food availability and individual sensitivity may determine reproductive response (Desjardins and Lopez, 1983).

2.3 Neuroendocrine Control of Reproduction

Exteroceptive (auditory, visual, olfactory, tactile), interoceptive (changes in body chemistry such as variations in the concentration of hormones, glucose, water, salts, acidity), or emotional states produce cues, processed and integrated by the central nervous system. The CNS responds by generating electrical impulses. The impulses, conducted through only partially chartered pathways, reach the hypothalamus, an area located at the base of the brain. In the hypothalamus the neural signals are transduced into endocrine signals by cells with the ability to produce a specific neurohormone, the gonadotropin releasing hormone (GnRH). In response to GnRH stimulus the hypophysis produces specific gonadotropic hormones (FSH and LH). These hormones stimulate the growth and function of the reproductive tract and the production of gonadal (steroid) hormones. The internal signals (hormones) impinge in turn upon the same areas of the brain or subordinate zones (the pituitary, the gonads) stimulating, modulating, or inhibiting the process.

The recognition of the intimate relation between the central nervous system and reproductive functions was gradual. In 1932 German researchers, Hohlweg and Junkmann, postulated that the brain controls sex functions and coined the term *das Sexualcentrum* (they did not identify the area). Two American endocrinologists, Everett and Sawyer, followed the idea in the late 1940s. The British endocrinologist Harris (1948; 1964) evolved the concept that hormonal control of the anterior pituitary is neural, transmitted through the hypophysial portal system. In Hungary, Szentagothai's school (Szenthagothai *et al.,* 1968) introduced the concept of the *hypophysiotropic* area, a portion of the hypothalamus which is capable of supporting pituitary function by locally produced hormone(s). The neurohormone (GnRH) has been isolated and its structure reported in 1971 by two American teams lead by Schally and Guillemin, respectively.

The hormone produced by hypothalamic neurons is a decapeptide (see Appendix). Carried through blood capillaries (portal vessels) within the pituitary stalk, the hormone evokes the release of two glycoprotein hormones (gonadotropins) from the anterior lobe of the hypophysis: the follicle stimulating hormone (FSH) and the luteinizing hormone (LH). A third glycoprotein of similar structure involved in the reproductive process, prolactin (PRL) is produced in response to the removal of an inhibitory influence (Section 2.4.2).

A variety of neurotransmitters and neuropeptides influences the functions of GnRH producing neurons. Stimulatory agents are norepinephrine (NE), epinephrine, acetylcholine, vasointestinal peptides and substance P. Brain opioids, serotonin, γ-aminobutyric acid, neurotensin and gastrin are inhibitory. Dopamine (DA) may play a dual role. Some agents may exert their effects by action on pri-

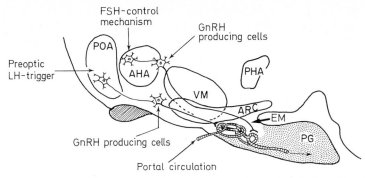

Figure 2.7. Hypothalamic Nuclei Important in the Control of Pituitary Function. Sagittal projection of the rat brain. AHA, anterior hypothalamus nucleus; ARC, arcuate nucleus; PHA, posterior hypothalamus area; POA, preoptic area; VM, ventromedial nucleus. The arrows indicate blood flow in the hypophyseal portal system, the special capillary loops that penetrate the median eminence (EM).

mary stimulants: opioids may suppress the action of DA and NE; NE may stimulate the neurons by suppressing the release of brain opioids. Estradiol alone, or in synergism with progesterone, may either affect directly the responsiveness of the neurons to neurotransmitters and neuropeptides or indirectly modulate the responsiveness to other agents (Barraclough *et al.*, 1984). Estrogens also alter the density of α_1-receptors in nuclei which control LH release (Drouva *et al.*, 1982; Weiland and Wise, 1987). Prostaglandins E may mediate the response (Heaulme and Dray, 1984).

To delineate the controlling centers involved the use of many experimental procedures: severance of neural pathways between various areas of the hypothalamus; measurements of GnRH concentrations in hypothalamic nuclei; fluorescent histochemical methods utilizing specific antisera; testing with α-blockers; and administration of estradiol to variously prepared animals. The basal medial hypothalamus and the arcuate nucleus are the two areas most intimately associated with reproductive control (Figure 2.7).

Two different controlling mechanisms are present (see Figure 2.1). In *primates*, the presence of adequate estradiol in blood (fully functioning ovaries) is essential for successful ovulation. Interruption of pituitary function does not interfere with basal circhoral rhythm of LH and FSH discharge. Lesions in the arcuate region inhibit the cycle. Administration of estradiol to lesioned animals induces LH discharge indicating a direct influence of the ovary on the pituitary gland (Knobil, 1980). *Rodents (rats)* need an intact connection between the preoptic area, the anterior hypothalamus and the MBH for successful ovulation.

2.3.1 Neural Control of Gonadotropin Release

The base for our knowledge of neural pathways involved in the regulation of reproductive function is the work in rats and rabbits. There is no reason, however, to suspect that similar mechanisms may not be operative in other mammalian species, including primates.

Rats maintained under regular light/dark cycles exhibit a critical period on the day of proestrus, from about 14:00 to about 16:00 h during which the brain initiates a chain of events which result in ovulation 12–18 h later (see Figure 2.3). Drugs which depress the CNS, such as barbiturates (Everett and Sawyer, 1950; Barraclough and Sawyer, 1957; Meyerson and Sawyer, 1968), morphine (Barraclough and Sawyer, 1955; Pang *et al.*, 1977), adrenergic blockers (Everett *et al.*, 1944; Sawyer *et al.*, 1950) and α-adrenergic receptor antagonists (chlorpromazine), and reserpine (which depletes catecholamines from presynaptic terminals) will delay ovulation by 24 h. The drugs must be administered during the critical period, otherwise ovulation will take place. In diestrus rats LH surge can be induced by exogenous estradiol; the surge can be blocked by pentobarbital.

Many CNS acting drugs increase the metabolism of testosterone (to 5α-reduced products) and of monoamineoxidase (Kaneyuki *et al.*, 1979). This may be one of the influences by which CNS drugs affect reproduction.

The cue which triggers the brain activity is photic. Changes in the timing of the light regimen will shift the critical period (Everett and Sawyer, 1950). The preoptic area is intimately associated with the transmittal of the signal since electrical stimulation of the area will restore the LH surge (Halasz *et al.*, 1965; Halasz and Gorski, 1967).

Brain catecholamines are the main transmitting agents involved. The theory rests on the following observations:

(i)   intraventricular injection of epinephrine (EP) or norepinephrine (NEP) induce ovulation in estrous rabbits (Sawyer, 1952);

(ii) in rats LH release is stimulated by intracerebroventricular administration of either NE or NEP but not by dopamine (Rubinstein and Sawyer, 1970);

(iii) administration of agents that block synthesis of epinephrine and norepinephrine, or α-adrenergic receptors, will also abolish ovulation (Kalra *et al.*, 1972) and disrupt episodic LH release (Gallo, 1980);

(iv) administration of α-adrenergic agonist, clonidine, will result in increased hormone concentration (Estes *et al.*, 1982);

(v)   both GnRH and catecholamines are present in the same cell population in the median eminence (McNeill and Sladek, 1978).

Experiments in castrated animals confirm the critical role of EP and NEP in ovulation control. After ovariectomy secretion of LH, FSH and PRL increases but the increase diminishes following the depletion of NEP stores (Anton-Tay *et al.*, 1970; Drouva and Gallo, 1976; Gnodde and Schuiling, 1976; Weick, 1978; Honna and Wutke, 1980). Negro-Vilar *et al.* (1982) analyzed three hypothalamic areas (median eminence ME, arcuate-ventromedial and suprachiasmatic-medial preoptic region, SMP) for norepinephrine (NE), dopamine (DA) and GnRH following LH surge in peripheral plasma of ovariectomized rats. Prior to LH surge NE levels increased in SMP while DA and GnRH decreased in the median eminence. Barraclough (1984) reported an inverse relationship between the turnover rate of NE and GnRH in medial preoptic, suprachiasmatic, and arcuate nuclei, and in the median eminence. During the time that GnRH accumulates in the median eminence (09:00 to 11:00 h on proestrus) NE turnover rate is low but increases during the afternoon hours when GnRH becomes depleted.

Aged anovulatory rats, which often exhibit a constant estrus and in which the ovulatory surge LH is absent, provide additional evidence of the role of NE as ovulation stimulant. Increases in LH concentrations can be achieved in such animals by NE treatment, albeit not fully (Estes and Simpkins, 1982).

Serotonin, acetylcholine and prostaglandins may play a role (MacLeod and Login, 1977; Weiner and Ganong, 1978; Krulich, 1979; Tuomisto and Mannistö, 1985). When slices of median eminence, obtained from proestrus rats, are perfused *in vitro*, addition of serotonin will produce a rapid release of GnRH into the medium, the result of direct stimulation of GnRH terminals (Vitale *et al.*, 1986).

The dominant role of the CNS in rats is not wholly independent of ovarian function. In the median eminence estradiol modulates the activity of NEP- and EP-dependent neurons and promotes GnRH release (Honna and Wutke, 1980; Wise *et al.*, 1981; DePaolo *et al.*, 1982). Further suppport for the view follows from the observation that ovariectomy abolishes in rats the cyclic LH surge. Instead, LH is released daily in response to the modulatory influence of estrogens transmitted through the brain (Legan and Karsch, 1975). In diestrous rats exogenous estrogens induce a LH wave, and the surge can be blocked by barbiturates.

The contribution of the second ovarian hormone progesterone is also of importance. Ovariectomized rats treated with estradiol and two days later with progesterone lose epinephrine stores in MBH indicating both ovarian hormones stimulate the rise of GnRH during the afternoon of proestrus (Adler *et al.*, 1983).

2.3.1.1 Regulation of Prolactin Function

Prolactin is an essential reproductive hormone in rodents (rats); it maintains the luteal function. In primates prolactin is mainly needed during lactation (see Hart, 1975). A PRL-inhibiting factor blocks the release of prolactin from the anterior hypophysis; removal of the inhibitory influence releases the hormone. The neurons which produce the inhibitor(s) are located in the lateral preoptic area. The regulation of PRL secretion further involves synergistic-antagonistic interaction of ovarian steroids (Labrie *et al.*,1980) with hypothalamic releasing (Tashjian *et al.*, 1971; Ruberg *et al.*, 1978) and inhibiting factors (MacLeod and Lehmeyer, 1974; Shaar and Clemens, 1974; Martin *et al.*, 1978). The role of the CNS in regulating PRL release has been postulated from clinical observations that in women treated with chlorpromazine or reserpine, ovulation is inhibited. Rudel and Kincl (1966) suggested that increased prolactin secretion was the cause of the problem.

Dopamine (DA) is one of the inhibitory factors (Weiner and Ganong, 1978). The concentration of DA in portal blood fluctuates during the estrus cycle reaching a nadir during proestrus (Ben-Jonathan *et al.*, 1977). Barraclough and Sawyer (1959) speculated that pseudopregnancy induced in cycling rats by chlorpromazine, or reserpine, is the result of blocked DA receptors. Blockade of dopamine receptors releases prolactin (Ojeda *et al.*, 1977a). The assertion that the number of dopaminergic receptors changes during the estrus cycle provides indirect evidence (Heiman and Ben-Jonathan, 1982).

TRH, neurotensin and substance P cause PRL elevation in peripheral blood (Rivier *et al.*, 1977). Nikolics *et al.* (1985) suggested that a portion of the human placental precursor of GnRH, a 56 amino acid peptide (GnRH associated peptide), may be the hypothalamic inhibitory factor. Other factors influencing PRL release could be a vasoactive intestinal peptide (Kato *et al.*, 1978; Vijayan *et al.*, 1979) and a histidine-isoleucine polypeptide (Werner *et al.*, 1983; Kaji *et al.*, 1984).

The posterior pituitary contains a polypeptide of less than 5 kD which stimulates PRL secretion from perfused anterior pituitary cells in the presence of physiological concentrations of dopamine (Hyde and Ben-Jonathan, 1988).

Serotonergic terminals in the anterior hypothalamus, mediated through the paraventricular nucleus (Minamitani *et al.*, 1987), control prolactin release during lactation (Parisi *et al.*, 1987). Greef *et al.* (1987) believe thyrotropin releasing hormone (a tripeptide), not serotonin, is the releasing agent.

α-Adrenergic neurons govern episodic PRL release in male rats (Terry and Martin, 1981) and male monkeys (Butler *et al.*, 1975; Diefenbach *et al.*, 1976). The observation that in males infusion of dopamine (0.1 μg/kg/min) significantly depressed PRL concentrations in peripheral blood lends credence to the hypothesis (Neill *et al.*, 1981).

2.3.1.2 Endorphins and Enkephalins

Reproductive function in narcotic addicts is low. In men testosterone secretion declines, and spermatogenesis is abnormal (Cicero *et al.*, 1975; Mendelson *et al.*, 1975). Women may become amenorrhoeic and infertile. Morphine increases the sensitivity of castrated male rats to the inhibitory feedback of testosterone (Gabriel *et al.*, 1985), estradiol, and DHT (Gabriel *et al.*, 1986) due to a reduced pituitary responsiveness to GnRH pulses (Kalra *et al.*, 1988). In females a direct relationship was demonstrated between the degree of endogenous opioid tone and the magnitude of preovulatory LH peak. A continuous infusion (2 mg/h) of endogenous opioid antagonist (naloxone) on the morning of proestrus provoked a decrease in opioid concentration and a premature LH surge (Allen and Kalra, 1986).

In males the gonads play a permissive role in the inhibitory action of endogenous opiates on gonadotropin secretion (Cicero *et al.*, 1979). After long term castration the sensitivity of opioid receptors is lost.

Corticotropin releasing factor (CRF) augments the release of endogenous opioids (Almeida *et al.*, 1988).

Opioid peptide-containing neurons located in hypothalamic nuclei may play a role in the modulation of episodic LH release (Kalra and Simpkins, 1981). The assumption rests on the observation that in monkeys, β-endorphin concentrations in hypophyseal portal blood rise from about 700 pg/ml during the mid- to late follicular phase to about 1800 pg/ml during luteal phase; endorphins are undetected (≤ 130 pg/ml) at menstruation (Wehrenberg *et al.*, 1982). Enkephalins are also present in the pituitary gland (Panula and Lindberg, 1987). Opioid peptides may also adjust the sensitivity of the hypothalamus to the negative feedback of gonadal hormones (Gabriel *et al.*, 1986). Leydig cells synthesize β-endorphin which may modulate testosterone secretion (Gerendai *et al.*, 1984).

2.3.1.3 Steroid Hormones

The brain is a biochemically active endocrine gland. Enzymes, located throughout the white matter of the brain, are able to cleave the cholesterol side chain and synthesize pregnenolone, dehydroepiandrosterone and estrogens independently of peripheral sources. Testosterone and estradiol stimulate, or inhibit, the release of GnRH, modulate electrical activity of hypothalamic nuclei, affect cerebral blood flow, brain oxygen and glucose consumption, phosphorus, protein and fat metabolism and regulate brain excitability (see Woodbury and Vernadakis, 1966). Certain 21-hydroxy pregnane derivatives are hypnotic and anesthetic agents (see P'an and Laubach, 1964).

Section 2.2 covers the effects of gonadal hormones on pulsatile release and Section 2.5.2 the influences on the negative and positive feedbacks.

2.3.1.4 Other Peptide Hormones

Corticotropin releasing factor (CRF), the main neuropeptide which regulates the pituitary-adrenal axis, also influences reproductive functions. In rats the secretion of CRF is mediated by neural pathways activated by stress (Plotsky and Vale, 1984). Increased secretion of CRF inhibits GnRH release into the portal circulation, an action independent of basal GnRH release (Petraglia *et al.*, 1987). This results in lower LH (and also growth hormone) concentrations (Ono *et al.*, 1984; 1985), decreased reproductive function (Rivier and Vale, 1984; 1985) and sexual receptivity (Sirinathsinghji *et al.*, 1983).

Growth hormone releasing factor may influence behavior (Tannenbaum, 1984).

Section 2.4.4 covers the function of peptides originating in the gonads (inhibin, activins and follistatins).

2.3.2 Neural Pathways Regulating Reproduction

Intact midbrain catecholaminergic innervation of the preoptic-hypothalamic area is important. Bilateral interruption of midbrain ascending pathways to the anterolateral and midlateral hypothalamus will block the proestrus rise of LH, FSH and PRL (Kawakami and Ando, 1981). Lower brain stem catecholaminergic systems are also involved (Martinovic and McCann, 1977; Clifton and Sawyer, 1979; Kawakami *et al.*, 1979). Soper and Weick (1980) employed cuts and lesions in different hypothalamic areas and concluded that two pathways can stimulate independently episodic LH release. One involves the rostral portion of the arcuate nucleus and surrounding areas, and the other extrahypothalamic structures which enter the hypothalamus anteriorly. The arcuate nucleus plays a pivotal role in both sexes. Placing lesions in this area abolishes the release of GnRH into the median eminence, while lesions in the preoptic area or in the suprachiasmatic nucleus produce only a partial decrease (Krey and Silverman, 1981). There is no sex difference, and lesioned male rats exhibit the same response (Krey and Silverman, 1978).

In hamsters the neuroendocrine signal that triggers the FSH pulse originates within the rostral hypothalamus and travels to the arcuate-median eminence region during the afternoon of proestrus. The event may be extinguished by the transection of neural connections between the rostral and basal hypothalamus or by electrical ablation of the arcuate-median eminence area (Chappel et al., 1977; 1979). Administration of phenobarbital abolishes the surge (Siegel et al., 1976).

Neural pathways which regulate behavior responses are discussed in Chapter 6.

2.3.3 Photic Stimulation

The observation that some species breed only once a year has been made since ancient times yet it was only in 1930 that a noted English physiologist, F.H.S.Marshall, connected changes in day length and breeding patterns in a report titled "Light as a factor in sexual periodicity". Estrus behavior occurs in the dark and shifts with varying day length in the guinea pig (Dempsey et al., 1936; Dempsey, 1937) and the mouse (Snell et al., 1940). Ovulation can be blocked in rats maintained under regular light/dark cycles by brain depressants if given during the "critical" period on the day of proestrus (Section 2.3.1). In the white footed mouse male (Peromyscus leucopus) reduction of the light period induces atrophy of the testes and seminal vesicles, decreases in the circulating LH and LH pituitary stores, and plasma testosterone. The activity of hypothalamic amines and neuropeptides is altered; the metabolism of serotonin and dopamine and the concentration of β-endorphin in mediobasal hypothalamus increase while the concentration of GnRH is low (Glass et al., 1988).

In addition to a light cue other environmental factors, such as temperature, food availability and individual sensitivity may determine reproductive response (Section 2.2.4).

Visible changes accompany the initiation of the reproductive season: females become sexually receptive and attractive to males; ovulation takes place and proceeds for several cycles. Males become aggressive; some elaborate musk-like pheromone, noticeable even to humans. Testes begin to grow and testosterone concentration in blood increases while prolactin decreases. Muscle mass (neck) may increase despite decreased, or zero, food intake (Figure 2.8).

Breeding can take place in the spring when the amount of light begins to increase ("long day breeders") or during the autumn when the days shorten ("short day breeders"). In the northern hemisphere, beaver, rabbit, racoon, cat, woodchuck and other rodent species which have a short gestation period procreate in late winter or early spring. Animals with long gestation period such as sheep, deer, horse breed in the summer or fall, and the young are born the next spring (see Amoroso and Marshall, 1960 and Figure 2.4). Transporting seasonal breeders to the Southern hemisphere the changing light pattern will adjust the breeding season accordingly: sheep will breed in the spring and lambs will be dropped in the fall. During translocation males of some species (sheep, deer) respond more rapidly to photoperiod change than females. Even so, endocrine changes are slow. It takes several weeks for rams to adjust to an artificial light schedule (Lincoln and Short, 1980).

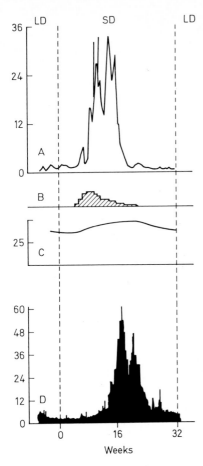

Figure 2.8. Change from Long to Short Days Triggers a Breeding Response in Rams. A ram was first exposed for 16 weeks to long days (16 h light, 8 h darkness) and then to 16 weeks of short days (8 h light, 16 h darkness). Responses measured were: A, concentration of testosterone in plasma, ng/ml; B, sexual flush, scale 0–5; C, neck girth, cm; D, aggressive index, counts/day x 100. Drawn from Lincoln and Short (1980) by permission.

Kangaroos transferred to the northern hemisphere readjust their annual cycles and give birth in July and August, rather than after the winter solstice (Berge, 1966). In female kangaroos the breeding season is synchronized by photic cues while the males are probably unresponsive to photoperiodic changes. The males respond to appearance of estrus in females, and a pheromone is suspected to provide a signal (Tyndale-Biscoe *et al.*, 1986).

During the quiescent period the hypothalamic-pituitary-gonadal axis is partially responsive. The axis will respond to castration by increases in gonadotropin release (Lincoln and Short, 1980).

2.3.3.1 Location of the Biological Clock

The hypothalamus regulates vital physiological functions: eating, sleeping, drinking, locomotor activity, heart rate, endocrine functions and the processing of sensory cues. The suprachiasmatic nucleus is the area which processes light signals into endocrine messages (Section 2.3.1). In hamsters lesions in this area render the animals asynchronous to day length. The animals will not breed seasonally. In humans melatonin receptors were found in the same area (Reppert et al., 1988).

The biological clock begins to function during the perinatal period in rats (Fuchs and Moore, 1980; Hiroshige et al., 1982; Takahashi et al., 1982), hamsters (Davis and Gorski (1985), and monkeys (Reppert and Schwartz, 1984). The sensitivity to photoperiod influence is communicated from the mother to the fetus in the Djungarian hamster (Horton, 1984; Stetson et al., 1986; Weaver and Reppert, 1986) and monkeys (Viswanathan and Chandrashekaran, 1984). The nature of the signal is not clear. Removal of maternal pituitary, adrenals, thyroid-parathyroid, ovaries or the pineal gland does not abolish maternal coordination of fetal circadian clock (Reppert and Schwartz, 1986). Possibly, other hormonal signals, maternal behavior, feeding patterns or temperature fluctuations may be sensed by the suprachiasmatic nuclei of the developing embryo.

Exposing lambs to longer light periods influences the timing of puberty. Lambs, exposed early to short days, display altered responsiveness to estradiol inhibition of LH secretion. Puberty is either delayed or advanced, and as adults they show abnormal reproductive cycles (Foster, 1983). Rhesus monkeys housed outdoors become seasonal breeders. During puberty seasonal environment overrides the developing neuroendocrine system. If first ovulation does not occur during the fall or winter months following menarche, the females become anovulatory during the spring and summer until the next fall (Wilson et al., 1986).

2.3.3.2 Contribution of the Pineal Gland

Epiphysis cerebri (pineal gland), a small organ within the brain modulates endocrine functions in seasonal ovulators. It furnishes a specific bioamine, melatonin synthesized from serotonin (5-hydroxytryptamine), an abundant brain indoleamine. Two enzymes are needed: N-acetyl transferase (acetylation of nitrogen in the indole ring) and hydroxyindole-O-methyltransferase (methylation of the hydroxyl group in the ring). The metabolic activity is low during the day and high during the night. Melatonin production is high during the night. The concentration is low during the day in cerebrospinal fluid and blood (Reppert et al., 1979). Light-dark changes provide the cue; light of sufficient intensity and duration during dark periods inhibits melatonin biosynthesis (Lewy et al., 1987). Melatonin circadian cycles are not abolished in continuous darkness.

Brain opiates modulate the activity of the pineal gland; concentrations of β-endorphin increase during elevated melatonin levels. Melatonin may modulate pituitary sensitivity to estradiol; in castrated animals the response of the pituitary gland is increased by exogenous melatonin (see Hoffman, 1973).

The pineal gland influences marginally, if at all, reproductive functions in cyclic breeders. In albino rats removal of the gland within 24 h of birth has no influence on reproduction of either sex, maternal behavior and lactation (Kincl and Benagiano, 1967). Minor changes include earlier vaginal opening, increases in body weight in both sexes, advanced bone calcification and an increase in corticosterone concentration in peripheral blood (Henzl et al., 1970). Pinealectomy of adult females results in increased ovarian weights (Simonnet et al., 1951) and increased incidence of cornified vaginal smears (Chu *et al.*, 1964). Blake (1976) found no effect of pinealectomy on the timing or magnitude of LH, FSH, and PRL release at proestrus and the length of the estrus cycle. Daily administration of melatonin does not inhibit sexual maturation in the male rat (Lang *et al.*, 1984).

Pinealectomy or melatonin administration (the dose and timing are important) influence reproductive function in seasonal breeders. In Syrian (golden) hamsters light periods shorter than 12 hours will induce anestrus in females and testes atrophy in males (Reiter and Fraschini, 1969). Testes regression induced by dark can be counteracted by pinealectomy or sympathetic denervation of the gland (Reiter, 1968) or exogenous melatonin (Hoffman, 1974). The gonads attain normal status only when plasma melatonin levels are in synchrony with dark-light cycles. Ablation of melatonin source (pinealectomy, superior cervical ganglionectomy, and exogenous melatonin) result in changes in the secretion of LH, FSH and PRL (Kennaway *et al.*, 1981).

Other lines of evidence support the view that melatonin effects are mediated through the CNS. Cranial sympathectomy (denervation of the pineal) in rams (Barrell and Lapwood, 1979) renders the animals nonphotoperiodic. In rams with superior cervical ganglia removed during a sexually active period the testes do not decrease in size during short light periods (Lincoln and Short, 1980). In castrated animals the hypothalamic-pituitary axis is more sensitive to gonadal steroid feedback (see Turek *et al.*, 1984).

Melatonin provides cues to photo changes in humans but does not regulate fertility. In jet travelers melatonin concentration in peripheral plasma peaks (a 2.5 fold increase) about 1½ h after onset of sleep. A westward jet lag of 7 h results in a decrease of the circadian cycle; an eastward 7 h jet lag (33 h of sleep deprivation) will result in a total desynchronization (Fevre-Montagne *et al.*, 1981). In this study it took 11 days for the 5 subjects to become fully adopted to local time. The jet lag disturbed adrenal function; ACTH and cortisol circadian rhythm were abnormal, but the adaptation of this function was unrelated to melatonin circadian cycles. In patients who become depressed in winter (in high latitudes), exposure to bright light during the morning (but not during the evening) advances night-time melatonin production and generates an antidepressant response (Lewy *et al.*, 1987).

Various regulatory mechanisms were proposed to explain the connections between photic cues and pineal function. Melatonin may "activate" GnRH producing neurons to increase neurohormone release by several mechanisms:

(i)   by altering the electrical activity of neurons;
(ii) by impairing axonal transport of GnRH-containing granules;
(iii) by changing the synthesis of neuroamines which regulate the release of GnRH.

At the present the mechanism remains to be elucidated, and the above hypotheses are still speculative. Intact retinal photoreceptors are not needed; blinded ewes (see Karsch *et al.*, 1984), or ewes born blind (Legan, 1980), maintain a normal reproductive cycle.

Motor Activity
Kincl *et al.* (1970) and Quay (1972) reported that free circadian running activity in white rats depends on the integrity of the pineal function. Kawashima *et al.* (1981) failed to find such a correlation. The two former groups found that regardless of the qualitative nature of the change rats adjusted their motor activity more slowly to changing photoperiods in the presence of the pineal. Rats without epiphysis cerebri (removed within 24 h of birth) adapted more rapidly to shifts in photoperiod. Both groups suggest that in the absence of pineal homeostasis, rapid light changes could expose animals to asynchrony with the environment.

2.3.4 Olfactory Control

In many mammalian species olfactory cues communicate sexual interest. Chemicals ("pheromone") produced in the vagina, excreted in urine, or secreted from cutaneous glands provide a signal of sexual receptivity to either sex. Pheromones are low molecular weight compounds, often aliphatic hydrocarbon derivatives, volatile, and usually specific for each species. In many invertebrate species (chiefly insects) and some fish species the ability to mate is pheromone dependent (Shorey, 1976).

Male Pheromone
Odors released by males may stimulate (or inhibit) the endocrine function of females. Female mice respond to the odor of male mice of the same strain by an increase in the release of gonadotropins and increased litter size (Beilharz, 1968). In immature females the urine will induce precocious puberty (Castro, 1967). Urine of males of one strain will block pregnancy in females of a different strain (Parkes and Bruce, 1962; Chipman and Fox, 1966; Lott and Hopwood, 1972). The pheromone is androgen dependent; castrated males do not secrete it (Bruce, 1965). Increased prolactin secretion, triggered by the pheromone, mediates most likely the pregnancy block (Bellringer *et al.,* 1980).

Ewes respond to a pheromone produced by rams by an increase in LH pulse frequency, even during the anestrus season (Lincoln *et al.,* 1985). Only rutting males secrete the pheromone. When testosterone secretion is low the production is inconspicuous. Other rutting species produce pheromones; boars are castrated to eliminate odor, associated with 17-deoxy androstane derivatives (5α-16-androsten-3-one and the corresponding 3β-alcohol), eliminated in the urine.

In other species pheromones may serve other purposes. For example, in a black-tailed deer a scent, produced by metatarsal organs, acts as an alarm pheromone. The males use scent produced in the forehead glands and urine to mark their territory (Müller-Schwarze, 1971).

Female Pheromone
Many females produce odors during heat which males find attractive; for example rams show preference in approaching ewes in heat (Kelly, 1937). Olfactory bulb ablation abolishes the response (Lindsay, 1965). During estrus, pheromones are produced in cows (Hart *et al.,* 1946), dogs (Beach and Merari, 1970), hamsters

(Murphy and Schneider, 1970), monkeys (Michael and Keverne, 1968; 1970) and humans (Michael *et al.*, 1971; 1974). Pheromone released by females may result in the suppression or synchronization of estrus in sheep , goats (Shelton, 1960) and mice (Bronson and Marsden, 1964). Inbred male and female mice show preference for each other if they are genetically dissimilar in respect to genes which code for the major histocompatibility complex (Yamazaki *et al.*, 1976; Boyse *et al.*, 1983). The preference is the result of familial imprinting during early postnatal life (Yamazaki *et al.*, 1988). The chemosensory imprinting presumably guides the choice of a mate, helping the male mouse avoid mating with his closest kin.

A "normal" adrenal function and pheromone "make-up" are important. Exposure of juvenile females to urine of adult males causes acceleration of puberty; exposure to urine of females delays the onset (by about 4 days) and decreases reproductive performance. The effect of the delay mimics social stress and is mediated by the function of the adrenals. Adrenalectomized adults produce decreased amounts of pheromones (penten and hepten derivatives), which when made available, accelerate puberty (Novotny *et al.*, 1986).

The importance of olfaction in breeding is less significant in other rodents, rats, rabbits and guinea pigs.

2.3.5 Auditory Cues

Many small rodents vocalize in the ultrasonic range (Sales and Pye, 1974). Male and female rats emit calls between 50 to 60 kilohertz (kHz) during the copulatory sequence; the male produces the calls as he approaches and investigates the female (Sales, 1972) while the call of females is associated with her solicitation behavior (Thomas and Barfield, 1985; Barfield and Thomas, 1985). Males also produce calls around 22 kHz following copulation (Brown, 1979). Ultrasonic communication also takes place between pups and lactating mothers and may play a role in the mother-young interaction (Noirot, 1972; Francis, 1977).

Birds use auditory signals (singing) to defend territory.

2.4 The Hypothalamus-Pituitary-Gonadal Axis

The hypothalamic-pituitary-gonadal axis becomes active at puberty. With age the function begins to decline. Senescent changes include abnormal patterns of gonadotropin secretion, decline in gonadal hormone production, cessation of reproductive functions and infertility. A possible contributory cause is the reduced capability of hypothalamic neurons to respond to stimulatory effects of norepinephrine (Jarjour *et al.*, 1986). The observation that in aged male rats hypothalamic dysfunction can be partially restored by transplantation of hypothalami from younger rats supports the view.

2.4.1 The Hypothalamus

The neurons, located in the hypothalamus (Section 2.3), produce releasing hormones (several RH are known) which stimulate the corresponding tropic hormones of the pituitary. The hormones are specific: gonadotropin releasing hormone (GnRH) stimulates the release and production of FSH and LH but not of thyroid (TSH) or adrenal (ACTH) stimulating hormones.

2.4.1.1 Gonadotropin Releasing Hormone (GnRH)

Nerve terminals activated by α-adrenergic neural pathways elaborate GnRH. The hormone is synthesized as a larger precursor, cleaved during axonal transport into a decapeptide. Castration influences the processing of the prohormone (Culler *et al.*, 1988). The secretory material, synthesized on ribosomes associated with the endoplasmic reticulum, passes into the Golgi apparatus for incorporation into secretory granules. Axoplasmic flow transports the granules to axon terminals from whence they are discharged. Most axons of neurosecretory cells capable of GnRH synthesis terminate on a system of capillary loops (the primary plexus) located in the median eminence, an area of the hypothalamus closest to the pituitary stalk. Blood flows from the brain through the plexus into the portal veins and thence into a second system of capillary loops located within the pituitary gland (the secondary plexus) and then drain into the general circulation (see Figure 2.6).

GnRH is released in episodic events (Section 2.2) which differ from species to species. The release of FSH and LH is governed by pulses controlled by the CNS and modulated by gonadal hormones and other factors. Gonadotropin secretion can be maintained only by intermittent administration of GnRH, but not by a constant infusion. Charlton *et al.* (1983a) demonstrated the need for an intermittent pattern in mice, Karsch *et al.* (1984) in rhesus monkeys and Clarke *et al.* (1986) in ewes. The cyclicity of FSH and LH release had been destroyed by radiofrequency lesions in the arcuate nucleus.

Various neurohormones may interact between themselves. Petraglia *et al.* (1987) reported that corticotropin releasing factor decreased plasma LH concentration by inhibiting the release of GnRH into portal circulation. Hwan and Freeman (1987) purified partially a glycoprotein (mol wt. about 12 kD) which inhibited GnRH induced LH release *in vitro.*

GnRH producing neurons, identified by immunohistochemical methods, are found in the medial preoptic area, accessory olfactory bulbs, anterior hippocampus and septal areas of rats (Samson *et al.*, 1980), cows (Estes *et al.*, 1977) and rabbits (Witkin *et al.*, 1982). In primates the neurons are mainly located in the medial preoptic area, paraventricular region and arcuate nucleus with some found in the mediobasal hypothalamus (Silverman *et al.*, 1982). Placement of radiofrequency lesions in the region of the arcuate nucleus abolishes the episodic release of both GnRH and gonadotropins. The axons project to the median eminence and organum-vasculosum of the lamina terminalis but also to other parts of the brain. This suggests that GnRH may have other, as yet unrecognized functions.

The bulk of the hormone is stored in the arcuate nucleus and median eminence but is also found in other parts of the hypothalamus (Figure 2.9). In genetically hypogonadal mice (GHG), characterized by a deficiency in GnRH production, transplantation of tissue from the fetal preoptic area of normal mice will restore the reproductive function in the GHG strain; the animals will mate, become pregnant and deliver healthy litters (Gibson *et al.*, 1984).

GnRH has been measured in hypothalamic-hypophysial portal blood of rats (Fink and Jamieson, 1975; Eskay *et al.*, 1977) and rhesus monkeys (Carmel *et al.*, 1976; Neill *et al.*, 1977). Evidence for the vital function of the hormone in reproduction has been established in many experiments. For example, section of the pituitary stalk, denervation of the hypothalamus, or immunoneutralization of the hormone will result in cessation of the ovulatory cycle (Fraser and Baker, 1978).

Increased secretion of GnRH occurs during proestrus in rats (Sarkar *et al.*, 1981; Ching, 1982), at the time of LH surge in rabbits (Tsou *et al.*, 1977), rhesus monkeys (Neill *et al.*, 1977) and humans (Miyake *et al.*, 1980; Elkind-Hirsh *et al.*, 1982).

The hormone possesses the dual capacity either to prime (Aiyer *et al.*, 1974) or to depress the gonadotrophs to its own action (see Conn, 1986). The dual response resides in its effect upon the formation of own hormone receptors. Prolonged exposure to GnRH (continuous infusion) will result in downregulation (desensitization) of gonadotrophs due to the reduction of available receptors (Section 2.6).

Gonadotropin releasing hormone is held to stimulate the synthesis and release both of luteinizing (LH) and follicle stimulating (FSH) hormones; some experimental evidence indicates two separate mechanisms:

(i)   in tissue culture of multihormonal gonadotrophs (Section 2.4.2) GnRH stimulation induces preferentially the release of LH. The observation suggests that FSH release may require other, as yet unidentified factor(s) (Lloyd and Childs, 1988);

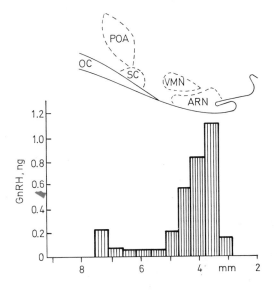

Figure 2.9. Concentration of Gonadotropin Releasing Hormone (GnRH) in Various Areas of Rat Hypothalamus. Average concentrations of GnRH (ng in 400 μm thick sections) determined by radioimmunoassay method. OC, optic chiasma; POA, preoptic area; SC, suprachiasmatic nucleus; VMN, ventromedial nucleus; ARN, arcuate nucleus. Redrawn after Wheaton *et al.* (1975).

(ii) in ovariectomized sheep FSH secretion occurs independently of GnRH administration (Hamernik and Nett, 1988);
(iii) the presence of a suspected FSH-RH is based on the observation that LH pulsatile release can be selectively blocked by specific antisera, or by blocking GnRH receptors. The block does not abolish the pulsatile release of FSH. DePaolo (1985) abolished the release of LH, but not of FSH, in ovariectomized female rats; Culler and Negro-Vilar (1987) reported similar results in male rats.

The differential effect of the neurohormone on pituitary hormones has not been fully elucidated. Differences in clearance rates were thought to be involved (FSH is longer than LH), but it is difficult to explain why LH is released following a single pulse while repetition is needed for the release of FSH (Figure 2.10). A search by several groups for a separate FSH-RH remains fruitless.

The appearance of LH in the peripheral circulation is very fast. It has been

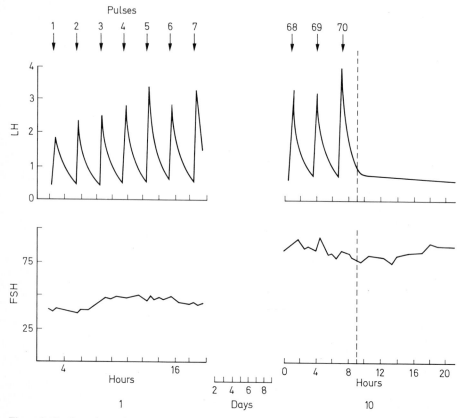

Figure 2.10. Gonadotropin Releasing Hormone Triggers Rapidly the Release of Luteinizing Hormone and Slowly the Release of Follicle Stimulating Hormone. Adult rams were injected (iv) 7 times per day for 10 days 100 ng of GnRH. LH (ng/ml) appeared rapidly in peripheral blood. The response of FSH (ng/ml) was slow, and several days were needed to induce a rise. Drawn with permission from Lincoln and Short (1980).

estimated that gonadotropin secretion takes place within seconds after the neuro-peptide has appeared in portal blood flow (Clarke and Cummins, 1982). The hormone is obligatory in the *de novo* synthesis of LH; it takes about an hour for the production to begin (Starzec *et al.*, 1986). It is possible that FSH synthesis requires a similar stimulus (Clarke *et al.*, 1986).

The degree to which GnRH stimulates gonadotropin release depends upon the enviroment created by gonadal hormones. In female rats LH concentrations are highest during proestrus (estrogen phase) and lowest during diestrus (progesterone phase) (Blake, 1978; Pickering and Fink, 1979). In women the concentration rises during the late follicular phase and is highest prior to the LH surge (Lasley *et al.*, 1975).

Gonadal steroids also provide a negative feedback. There is some evidence that steroid hormone feedback may be sex-linked. Melrose and Gross (1987) demonstrated that neurons, harvested from male rats, will respond *in vitro* to the inhibitory influence of testosterone and dihydrotestosterone but not to estradiol and progesterone. In the male rhesus monkey (Plant, 1982), rats (Steiner *et al.*, 1982) and rams (D'Occhio *et al.*, 1982) bilateral orchidectomy results in an increase of LH frequency (from once every 3 h to once every 1 h). The increase in frequency is prevented by testosterone. Intermittent infusion of GnRH will restore FSH, but not LH secretion. That the increase can be prevented by testosterone indicates a LH block at the level of the hypothalamus while FSH is blocked at the pituitary level (Plant and Dubey, 1984).

It is of interest that the peak concentration in portal blood of humans, 2000 pg/ml (Antunes *et al.*, 1978), is very similar to the concentration found to be effective in restoring pituitary function in lesioned rhesus monkeys (Knobil, 1980).

Extrapituitary Action

The classic endocrine hypothesis stated that hypothalamic hormone regulated only the pituitary function. This view is no longer valid. GnRH influences various extrapituitary functions:

(i) GnRH directly stimulates ovulation in immature rats primed with PMSG thus mimicking the action of LH (Ekholm *et al.*, 1981);

(ii) chronic treatment with pharmacological doses of GnRH or with (1 or more) synthetic analogue inhibits steroidogenesis, the growth of accessory sex organs, folliculogenesis, spermatogenesis, ovulation, ovum transport, implantation and pregnancy (Table 2.4).

Concomitant treatment with GnRH antagonist, or steroid hormones, blocks the effects. Both Leydig and Sertoli cells in males and granulosa cells in the ovaries are particularly sensitive to the paradoxical effects of GnRH. The hormone directly inhibits 17-hydroxylase and 17,20-desmolase, two enzymes essential for the formation of androgens and estrogens (Hsueh and Jones, 1981). Receptors for the neuropeptide are present in rat oocytes (Dekel *et al.*, 1988).

GnRH may act by way of other factors. In the rat ovary the hormone directly stimulates plasminogen activators believed to play an important role in the induction of ovulation (Beers, 1975; Strikland and Beers, 1976; Reich *et al.*, 1985; Hsueh *et al.*, 1988).

Table 2.4 Inhibition of Reproductive Functions by Chronic Administration of GnRH.

Effect	Reference
Males	
Decrease in testis weight	Sandow and Hahn (1978); Cusan *et al.* (1979); Badger *et al.* (1980); Labrine *et al.* (1980).
Decrease in ventral prostate and seminal vesicles growth	Sandow and Hahn (1978); Cusan *et al.* (1979); Oshima *et al.* (1975)
Inhibition of spermatogenesis	Rivier *et al.* (1979); Gore-Langton *et al.* (1981); Knecht *et al.* (1981).
Decrease in testicular androgen production	Haynes *et al.* (1977); Sandow and Hahn (1978); Dufau *et al.* (1979); Badger *et al.* (1980); Hsueh and Erikson (1979); Massicote *et al.* (1980); Wisner and Stalvery (1980).
Females	
Pregnancy termination	Corbin and Beattie (1975); Bowers and Folkers (1976); Hilliard *et al.* (1976); Beattie *et al.* (1977); Beattie and Corbin (1977); Humphrey *et al.* (1977 a, b, 1978); Corbin *et al.* (1978); Beattie (1979); Bex and Corbin (1979); Jones (1979 a, b, c, 1980); Jones *et al.* (1979)
Delay of parturition	Humphrey *et al.* (1978); Berca *et al.* (1980)
Inhibition of folliculogenesis	Amundson and Wheaton (1979)
Decrease in steroid hormone	Hilliard *et al.* (1976); Henderson *et al.* (1976); Beattie *et al.* (1977); Humphrey *et al.* (1977 a, b); Beattie (1979); Bex and Corbin (1979); Jones *et al.* (1979); Cusan *et al.* (1979).
Inhibition of FSH and LH action on the ovary	Rippel and Johnson (1976); Behrman *et al.* (1980); Reddy *et al.* (1980).

2.4.1.2 Peptides Similar to GnRH

The hypothalamus may not be the only tissue producing GnRH. The presence of a GnRH-like peptide has been demonstrated in granulosa cells (Ying *et al.*, 1981), in immature rat ovary (Birnbauer *et al.*, 1985), in bovine and ovine ovaries (Aten *et al.*, 1987), and in Sertoli cells (Sharpe *et al.*, 1981). The placenta produces a peptide identical to GnRH (Khodr and Siler-Khodr, 1980; Gautron *et al.*, 1981). Millar *et al.* (1986) reported that a synthetic peptide, comprising the first 13 amino acids of the GnRH percursor (60 amino acid peptide) stimulated FSH and LH release from human and baboon pituitary cell culture. Other peptides and factors which regulate reproduction are recounted in Section 2.4.4.

Hwan and Freeman (1987) characterized partially a polypeptide, isolated from the hypothalamus, which inhibited the release of GnRH. The peptide is a glycoprotein, apparent molecular weight 12 kD, and the pI 4.1.

2.4.2 The Pituitary Gland

The pituitary gland, located at the base of the brain, is connected to the hypothalamus by a delicate stalk which furnishes a conduit for the hypophyseal portal veins leading into the anterior portion, and axons which project into the posterior part. The gland arises from two different tissues. The anterior and intermediate lobes originate from Rathke's pouch, an outgrowth from the roof of the mouth; the posterior lobe (neurohypophysis) from the infundibulum, an outpocketing of the hypothalamus. Vasopressin and oxytocin are the two main hormones of the posterior lobe. The intermediate section (poorly developed in mammalian species) produces mainly the melanocyte stimulating hormone (MSH). Seven hormones have been obtained from the anterior lobe: growth hormone (somatotropin, GH), adrenocorticotropin stimulating hormone (ACTH), thyrotropin stimulating hormone (TSH), prolactin (PRL), follicle-stimulating hormone (FSH), luteinizing hormone (LH) and lipotropin (LPH). All these hormones are peptides having a molecular weight of about 30,000 daltons; FSH, LH and TSH contain aminosugars.

The gland is an essential link in the neuroendocrine system but has very little capacity to act independently. Many experiments demonstrated the dependence of pituitary function upon uninterrupted flow of blood from the hypothalamus, and its hormone: removal of the gland (hypophysectomy); sectioning of the pituitary stalk (a block of inert material must be inserted between the sections since the portal vessels easily vascularize); *in vitro* experiments; and transplantation procedures.

Based on the affinity to various stains the anterior lobe cell population is classified as acidophil (producing PRL and growth hormone), basophil (producing FSH, LH and TSH) and chromophobe. Following castration the gonadotrophic cells become enlarged, degranulated, and eventually develop large cytoplasmic vacuoles. The "castration" cells persist unless exogenous gonadal hormones are given.

Dada *et al.* (1983) report that in adult rats all FSH-containing cells are capable of LH synthesis while about 25% of gonadotrophs in females and 11% in males contain only LH but not FSH. In females the number of cells that contain LH, or those that contain both hormones, does not change during the estrus cycle, and the cells are smaller than in males. During the periods of elevated FSH in blood (morning of early proestrus) LH concentration does not change in cells from which FSH has been released. Childs *et al.* (1982) classified gonadotrophs in three subtypes: multihormonal cells which contain both hormones; and monohormonal cells which contain either only FSH, or only LH.

Obviously, the release requires the synchronization of a population of cells within the pituitary, but the nature of the synchronization is not known. Perhaps it may be provided by a short loop feedback. Sensory inputs (for example suckling) inhibit the activity of the arcuate oscillator, and pituitary function ceases. When the stimulus is removed, the oscillator becomes reactivated, and FSH is secreted, followed by LH.

The endocrine status influences the molecular form of both hormones. During a cycle the molecular weight (Peckham *et al.*, 1973; Peckham and Knobil, 1975; 1976; Bogdanove *et al.*, 1974 a,b; Blum and Gupta, 1980), biological half-life (Weick, 1977 and above refs.) and isoelectric point (pI) (Robertson *et al.*, 1977; Strollo *et al.*, 1981) change, possibly in response to changes in sialic acid content. Others have noted in rodent pituitary the presence of several FSH hormones (Chappel *et al.*, 1982 a,b; Ulloa-Aguirre and Chappel, 1982) which differ in biological activity, pI and receptor binding(Miller *et al.*, 1983). Further evidence that

Table 2.5 Effect of Aging on the Release of Hormones
from the Hypothalamus, the Pituitary Gland and the
Gonads of Male Rats.

Hormone	Serum Concentration	
	Young[a]	Old[b]
GnRH (pg/ml)[c]	1960	756
FSH (pg/ml)	5.2	1.4
LH (pg/ml)	355	203[d]
Testosterone (mg/ml)	2.6	0.9

[a] 2 months old; [b] 23 months old; [c] total secretion from
hypothalamus *(in vitro)*; [d] serum LH undetectable ($<$
150 pg/ml) in 14/25 animals; data from Jarjour *et al.*
(1986).

the type of gonadotropin produced is influenced by the endocrine status was provided in male and female rats by Robertson *et al.* (1982) and in cynomolgus monkeys by Chappel *et al.* (1984a). The concentrations of both hormones depend on
the time of the cycle and are significantly lower in aged individuals (Table 2.5).

Both hormones are needed for full physiological effects. Concentrations of circulating gonadal hormones mediate the sensitivity of pituitary cells to GnRH stimulation. Ewes respond readily to exogenous GnRH when pretreated with estrogens (Reeves *et al.*, 1971); rats show greatest response during proestrus (Aiyer
et al., 1974). In ovariectomized rats injection of estradiol and progesterone produces an increase of LH stores (Fink and Aiyer, 1974).

The amounts of gonadal hormones are critical; "higher" doses will cause a
negative feedback. The feedback control of gonadotropin release varies from species to species and may be different for the two hormones. Steroid hormones
inhibit hypothalamic function, and pituitary activity ceases (Section 2.5). Polypeptide hormones originating in the gonads either stimulate or inhibit FSH function
(Section 2.4.4).

2.4.2.1 Follicle Stimulating Hormone (FSH)

In the female, FSH promotes the growth and development of follicles, supports
the development of granulosa cells in multiple layers and antrum formation. Interaction with LH is necessary for full folliculogenesis; without LH (in hypophysectomized FSH treated animals) the follicles will not reach full size, and they do not
secrete estrogens. In males, FSH stimulates the growth of seminiferous tubules
and plays a role during the first stages of spermatogenesis. The hormone does not
support the activity of Leydig cells.

The release of FSH during the cycle is episodic, superimposed upon a basal
release (Gay and Sheth, 1972; Chappel *et al.*, 1984; Lumpkin *et al.*, 1984b;
DePaolo, 1985). Amplitude fluctuations are usually small, and the periodicity (in
rats) varies from 50 to 60 min (Condon *et al.*, 1986). In rodents two peaks occur.
The first takes place during the estrus period and the second before the LH surge

(Bast and Greenwald, 1974; Smith *et al.*, 1975). The administration of CNS depressants to hamsters (Siegel *et al.*, 1976; Chappel *et al.*, 1979) and rats (Ashiru and Blake, 1978) abolishes both peaks (and also the LH surge). In hamsters GnRH discharge regulates FSH surge (Coutifaris and Chappel, 1983). In rats (Ashiru and Blake, 1979) and hamsters (Coutifaris and Chappel, 1982) FSH release during estrus can be also induced by systemic FSH or LH. The observations suggest that the estrus release of FSH in rodents is triggered by a synergistic action between GnRH and the gonadotropin(s) (Section 2.5.3). In primates the peak is reached at the same time that the LH surge takes place. The peak is broader than the LH peak and lasts longer.

2.4.2.2 Luteinizing Hormone (LH)

In females LH induces ovulation of graafian follicles and causes a rapid and large increase in blood flow to the ovaries and the formation of the corpus luteum. In males the hormone influences the structure, and function, of Leydig cells. In both sexes LH stimulates the synthesis of respective gonadal hormones (testosterone in males, progesterone and estrogens in females) by stimulating the transport of cholesterol into steroidogenic cells. The manner by which LH facilitates the development of corpora lutea and steroidogenesis differs from species to species (Section 2.4.3). There is some evidence to indicate that LH may directly influence folliculogenesis (Bicsak *et al.*, 1986).

The release of LH from the gonadotrophs in the pituitary gland in response to GnRH stimulus is rapid. It has been estimated that the secretion takes place within seconds after the neuropeptide has appeared in portal blood (Clarke and Cummins, 1982). Wilson *et al.* (1986) state that LH appears in blood within two minutes. Rapid fluctuation of hormone levels in blood does not present a danger that the endocrine tissue could become depleted. The concentration of hormones in the respective endocrine glands usually surpasses manyfold the quantity released. It has been estimated that in the rat less than 5% of the available GnRH is released during each episode. The amount released is in excess of the physiological need. Kalra *et al.* (1971) approximate that only about 15% of LH released during a preovulatory surge is needed for ovulation.

In females the release of LH is characterized by a sudden appearance in blood, the *LH surge*. During the surge, in contrast to FSH, the concentration increases many times over the basal level (Figure 2.11). Increases in the electrical activity in the hypothalamus and increases in the frequency and amplitude of GnRH secretion precede the peak. In about 70% of women the surge takes place in the early hours of the day. In rodents the cascade effect for LH release in the CNS is initiated during the critical period in the afternoon of proestrus; in cyclic, and seasonal, ovulators the surge is caused by a feedback cascade; in induced ovulators in reaction to copulation. The response is modulated by the two gonadal hormones, estradiol and progesterone. Depending on the timing (cycle "status") and the dose, the hormones may stimulate (Caligaris *et al.*, 1971), inhibit (Goodman, 1978), or have no effect on LH release (Section 2.4.5).

Adequate estradiol production is essential to trigger the LH surge. In primates

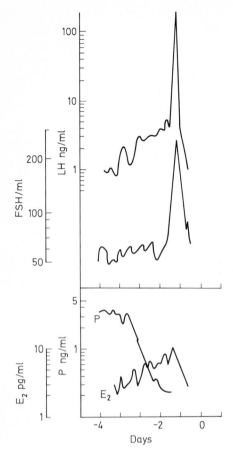

Figure 2.11. Pituitary and Gonadal Hormones in Blood of Ewes Preceding Ovulation. Changes in concentration of LH (ng/ml), FSH (ng/ml), progesterone (ng/ml), and estradiol (pg/ml) prior to ovulation (day 0). Redrawn with permission from Karsch *et al.* (1984).

estradiol influences mainly the pituitary ((Knobil, 1980). In cattle (Kesner *et al.*, 1981; 1982), pigs (Döcke and Bush, 1974; Kesner *et al.*, 1987), rats (Sarkar *et al.*, 1976; Goodman, 1978; Kawakami and Higuchi, 1979) and sheep (Radford *et al.*, 1974; Coppings and Malven, 1976; Clarke and Cummins, 1984) estradiol adjusts both the CNS and the pituitary to LH wave.

Catechol estrogens (2-hydroxyestrone) also induce a rise in serum LH (Naftolin *et al.*, 1975b).

In males testosterone modulates the responsiveness of hypothalamic neurons to neurotransmitting substances, and of pituitary gonadotrophs to the action of GnRH. In castrated male rats low amounts of testosterone stimulate LH release by increasing LH pulse amplitudes and decreasing pulse frequency (see Conn, 1986). Pharmacological amounts will inhibit the release.

2.4.2.3 Prolactin (PRL)

In mammals prolactin is needed to initiate and maintain lactation. In marsupials PRL is essential in the control of reproduction and in rats to maintain functioning of corpora lutea. In those species which do not require PRL as a luteotrophic substance the hormone may exercise a subtle control. Excessive secretion of PRL (during lactation) or induced by pharmacological agents (such as chlorpromazine) may result in the inhibition of ovulation in humans (Rudel and Kincl, 1966; Rolland *et al.*, 1975) and rodents (Lu *et al.*, 1976; Smith, 1978). During hyperprolactinemia it is the release of LH which is disrupted; basal FSH release is not affected (Smith, 1982). Synthetic derivative of an ergot alkaloid, bromocriptine mesylate (Besser *et al.*, 1972; Thorner *et al.*, 1974; 1981), and 2-hydroxyestradiol (Banks *et al.*, 1986) block the hypersecretion.

In rats (and other species) stimuli associated with mating result in the release of prolactin. In some (rabbits) mating has no influence on PRL liberation. In mice PRL synthesis is augmented by estradiol (Charlton *et al.*, 1983b).

Prolactin decreases the sensitivity of the pituitary to the action of GnRH; the gland is less responsive during lactation (Lee *et al.*, 1976; Smith, 1978; Steger and Peluso, 1978). Reduced pituitary reaction, noted also *in vitro* (Smith, 1982; Cheung *et al.*, 1983), is the result of decreased GnRH receptor formation (Clayton *et al.*, 1980; Marian *et al.*, 1981; Marchetti and Labrie, 1982; Garcia *et al.*, 1985).

Placement of pituitary grafts under the kidney capsule increases PRL production. In men (Thorner *et al.*, 1974; Carter *et al.*, 1978) and in male rats (Fang *et al.*, 1974; McNeilly *et al.*, 1978) increased PRL leads to hypogonadism while testosterone production is not influenced. This suggests derangement of the negative feedback on the hypothalamic-pituitary axis.

The concentrations of the hormone in plasma fluctuates hourly, and diurnal peaks are present. There is a progressive increase during the follicular phase with maximum values concomitant with LH peak. Fluctuations also occur circannually with highest levels during the summer months in cattle (Koprowski and Tucker, 1973; Karg and Schams, 1974; Schams and Reinhardt, 1974; Peter and Tucker, 1978), goats (Buttle, 1974; Hart, 1975), sheep (Ravault, 1976; Sanford *et al.*, 1978; Barrell and Lapwood, 1978) and deer (Mirarchi *et al.*, 1978; Schulte *et al.*, 1981). Light cues influence in part the annual cycle in sheep (Pelletier, 1973; Forbes *et al.*, 1975; Lincoln *et al.*, 1978; Barrell and Lapwood, 1979) and cattle (Bourne and Tucker, 1975; Leining *et al.*, 1979).

2.4.2.4 Melanocyte Stimulating Hormone (Melanotropin, MSH)

The concentration of MSH rises and falls during the estrus cycle in patterns similar to those of LH, FSH and PRL (Tomatis and Taleisnik, 1969; Celis, 1975). In rats the hormone may act in synergism with LH and/or progesterone to induce ovulation (Alde and Celis, 1980) or may potentiate the effect of β-endorphin in inhibiting LH secretion (Khorram and McCann, 1986).

In female rats stress increases MSH secretion; the increased production modulates both LH and PRL release (Khorram *et al.*, 1985).

2.4.3 The Gonads

2.4.3.1 Males

The testes are paired structures formed of about 30 coiled seminiferous tubules and interstitial tissue formed of connective tissue and interstitial (Leydig) cells. Two types can be distinguished histologically (Payne *et al.*, 1980). The tubules are filled with spermatozoa in different stages of development, suspended in a fluid. The fluid is a mixture of secretion from the tubules and of secretion formed in the rete testis, a network of anastomosing tubules. Seminiferous tubules connect to the head of a secondary sex organ, the epididymis. Most of the tubule secretion is absorbed here. Spermatozoa mature and become motile in the caput and central portion (corpus region) and are stored in the tail portion. Sperm maturation is regulated by androgens (Orgebin-Crist *et al.*, 1975).

Secondary (accessory) sex organs include: the epididymis; ductus deferens, a conduit through which the sperm is transported; and secretory organs, the seminal vesicles with coagulating glands, the prostate, Cowper's and preputial glands.

Hutson and Donahoe (1986) reviewed hormonal control of testes descent and Ewing and Zirkin (1983) the biochemical aspects of testicular function.

Spermatogenesis

Sperm is formed along the tubule length in a wave-like pattern (spermatogenic wave); a cross section reveals the presence of all the elements, in different proportions. Sperm production requires high concentrations of testosterone which is dependent upon LH and FSH. Sperm maturation is mediated by Sertoli cells influenced by FSH (Fritz, 1978; see also Orth *et al.*, 1988). The youngest germ cells (spermatogonia) rest at the base of the tubule membrane, closely touching the large Sertoli cells. During the process of maturation spermatogonia undergo a series of mitotic divisions. The resulting spermatocytes are transformed by meiotic division (chromosome reduction) into spermatids. Loss of cytoplasm and subsequent changes yield mature spermatozoa which are pushed towards the lumen of the tubules. The process requires about 34 days in mice, 48 days in rats, and 74 days in man. Mammalian spermatozoa leaving the seminiferous tubules are immobile and incapable of fertilization; a further process of maturation takes place in the epididymis. In some species (rat) a further "ripening" (capacitation) must take place in the reproductive tract of females before the sperm becomes fertile.

The hormone produced in the testes by Leydig and Sertoli cells is testosterone. In most target tissues reduction to the 5α-derivative, 5α-dihydrotestosterone (DHT, 17β-hydroxy-5α-androstan-3-one) is a prerequisite for the expression of biological activity (Bruchowsky and Wilson, 1968; Fanf *et al.*, 1969). Failure of conversion (for example in XXY-man with Klinefelter's syndrome) results in inadequate androgenic response (Sulcova *et al.*, 1978).

In rats the hormone controlling spermatogenesis is DHT, produced locally (Purris *et al.*, 1977). Testosterone is the active hormone in men; spermatogenesis proceeds normally in individuals suffering from congenital 5α-reductase deficiency (Imperato-McGinley and Peterson, 1976).

Androgenic hormone(s) support embryonic sex differentiation, postnatal

Table 2.6 Concentration of Androgens in Peripheral Blood in Man and Rat and Relative Androgenic Potency of Each.

	Peripheral plasma concentration ng/nl		Rate of testicular secretion mg/24 h		Androgen potency*
	Man	Rat	Man	Rat	
Testosterone	7	2	6.2	0.075	100
Androstenedione	1	0.3	–	0.01	10
DHA	5	13	0.15	–	3

* measured by seminal vesicles growth; compiled from references in text.

growth of prostate and seminal vesicles, maturation of germ cells and of cells of accessory sex glands; influence the physiological status of accessory sex tissues; condition the appearance of secondary sex characteristics (muscle mass, growth of hair, plumage in birds, horns and antlers) and salivary gland growth (submaxillary glands of male mice differ from females); influence behavior and immunological responses (Castro, 1974).

The accessory sex tissues of males are dependent for growth and physiological activity upon adequate androgen secretion by the testes; all the organs atrophy following castration (orchidectomy). Exogenous androgens restore the growth and secretory function. The production of androgens by the testis depends upon pituitary gonadotropins; the sex organs of hypophysectomized animals resemble those of castrates. The weights of seminal vesicles, of the ventral lobe of the prostate, and of levator ani muscle in immature or castrated rodents (rats) are particularly useful to assay androgenic activity.

The main androgens found in peripheral blood are testosterone, androstenedione (4-androstene-3,17-dione) and dehydroepiandrosterone (DHA, 3β-hydroxy-5-androsten-17-one). DHT, formed *in situ*, is present in peripheral circulation only in small quantities (Table 2.6). The testes of some species (horse) have the capacity to produce significant amounts of estrogens (Beall, 1940).

2.4.3.2 Females

R. de Graaf, a Dutchman, described the gross structure of avian and mammalian ovaries in 1672. The Czech physiologist, J.E. Purkyne, reported the formation of germinal vesicles in the chick in the early 1800s. A German anatomist traced ovum formation to the graafian follicle in 1827. Knauer established in 1900 the endocrine function of the ovary; he showed that ovarian transplants will prevent uterine atrophy in ovariectomized rabbits. In 1917 Stockard and Papanicolaou described cyclic variation in the growth of vaginal epithelium during an estrus phase, and in 1923 Allen and Doisy suggested that cell structure changes (cornification) of the lining of the vagina in spayed mice could be used as a biological test for estrogenic potency. Cohn and Fraenkel (1901) and Fraenkel (1903) observed that the function of the corpus luteum was essential for a normal outcome of pregnancy. Bouin and Ancel (1910) correlated the appearance of secretory changes with the function of corpus luteum, and Clauberg (1930) designed a biological test which is still used today.

I described that in rats a biological clock, located in the suprachiasmatic nucleus of the hypothalamus, dispenses a rhythm which drives the reproductive cycle. In monkeys, the ovary is the zeitgeber which coerces the rhythmicity, and estradiol is the principal modulator to which both the hypothalamus and the pituitary respond. Estradiol action is mediated at the pituitary level by an increase in the sensitivity of gonadotrophs to GnRH whereas progesterone decreases the sensitivity of the gonads to estradiol (Batra and Miller, 1985).

In contrast to the ever present production of sperm in males, females are born with a given number of oocytes. At birth the human ovary may contain about 500,000 oocytes. Assuming an average reproductive span of 30 years, only about 400 ova may mature and ovulate. Thus 99.9% of all oocyte are lost to a degenerative process (follicular atresia).

As the follicle matures, the ova increase in size, the squamous cells lining the primary follicle begin to proliferate, and the follicle becomes encapsulated by several layers of granulosa cells (see Hsueh et al., 1984). The stromal tissue has a predominant enzymatic capacity for the biosynthesis of androgens (see Erickson et al., 1985). Below the granulosa cell layers are located two other structures, the theca interna and theca externa, derived from connective tissue. Blood and lymph vessels penetrate the theca externa and project into the theca interna while the granulosa cells become supplied by vessels only after ovulation when the corpus luteum forms. Luteal cells are present in two sizes: large (about 30% of volume), possibly derived from granulosa cells, and small (about 16% of volume), possibly derived from theca cells; the origin of the remaining 30–40% of luteal cells is not known. The small cells may develop into large cells under stimulation by LH. The large cells are richer in estradiol, progesterone, and prostaglandin receptors while LH receptors are more numerous in the small cells (see Niswender et al., 1985). The cells respond both to FSH and LH stimulation but are more sensitive to LH (Quirk et al., 1986).

Folliculogenesis
The initial growth of the primary follicles is independent of pituitary functions. Follicles will reach preantrum size in hypophysectomized rats and guinea pigs (Dempsey, 1937). FSH (Schwartz, 1974; Hirschfield and Midgley, 1978) and LH (Lostroh and Johnson, 1966; Welschen, 1973) are needed for the growth to the preovulatory stage. The responsiveness of growing follicles to gonadotropic stimulation results from an interaction of steroid hormones and peptide regulators (see Channing et al., 1982). The mature (graafian) follicle features a liquid filled antrum rich in 17-hydroxy-4-pregnene-3,20-dione, 4-androstene-3,17-dione, testosterone, estrone and estradiol. Steroid hormones are not needed for meiotic division, but estradiol is needed for the "fertilizability" of the ova (Yoshimura et al., 1987).

Ovulation
In monotocous species (primates, ungulates) usually only one follicle matures and ovulates. Zeleznik et al. (1985) believe that follicular atresia is the result of rising

production of estradiol which selectively suppresses FSH secretion. In addition to estradiol other agents contribute to the selection. Catechol estrogens, various peptides and 5α-androgens have been proposed as chemical messengers involved in the paracrine control (Hillier, 1987). Preston *et al.* (1987) feel that rising intrafollicular concentration of cAMP, produced by cumulus cells in response to FSH stimulation, prevents maturation of other oocytes.

When steroidogenesis is blocked ovulation (perfused rat ovary) does not take place (Lipner and Wendelken, 1971; Testart *et al.*, 1983). Selective repression of estradiol biosynthesis does not prevent ovulation (Koos *et al.*, 1984).

The ovum is expelled from the follicle (ovulation) in response to a hormonal stimulus (LH). After ovulation the collapsed follicle fills with blood, the granulosa cells become fully vascularized and give rise to a new endocrine structure, the corpus luteum (CL). If fertilization has not taken place the corpus luteum regresses (atretic CL), and the cycle is repeated. The control of CL function is discussed in Section 2.4.3.3.

The ova escapes because of the thinning of the follicle wall caused by lytic enzymes synthesized towards the end of the follicular phase; rising liquid pressure within the follicle does not contribute to the rupture. The eggs, surrounded by a mass of cells (cumulus cells), ooze out through the opening thus created.

Ova Transport

Fertilization takes place in the ampulla, perhaps 8 to 12 h following ovulation. The sojourn of the fertilized and dividing cell mass in the oviduct is about 3 to 4 days for many species, notwithstanding differences in the length of the tube and the duration of gestation. Only in marsupials is the transport rapid; ova reach the uterus in about 24 h. During the period of transport the ovum is exposed to tubal secretion providing a stream of nourishing fluid. Zygote descent, the secretory activity of tubal epithelium and its growth depend on the synergist action of estradiol and progesterone. When one predominates, the orderly transport is disrupted.

The zygote reaches the uterus in the form of a morula (a solid mass of 16 to 32 cells) or as a blastocyst, a hollow sphere with a single peripheral layer of large, flat cells (from which the placenta develops), and a protuberance of smaller cells to one side of the cavity, which will give rise to the embryo. The size of the blastocyst varies from species to species. In mice it contains only about 100 cells while in rabbits the mass may reach several thousand. In sheep and cattle the blastocyst may attain a length of 20 cm, and in the pig it becomes a tread-like tube more than a meter in length. The blastocyst (or morula) is said to implant when it attaches itself to the endometrium, the inner lining of the uterus.

Implantation

The growth and maturation of the uterine lining is essential for proper implantation and embryonic development. During the estrogen dominated phase the endometrial cells begin to grow, and mitotic activity is prominent. During the second half of the cycle the cells mature and acquire a secretory ability necessary for implantation and the competence to nourish the embryo before placentation has taken place.

By the time the blastocyst has entered the uterus the epithelium, influenced by changing hormonal production of the ovaries, has become "prepared". The most visible change is in the permeability of the endometrial capillaries which surround the about-to-implant zygote, termed decidualization. This local vascular change, during which implantation can take place, lasts only a few hours. Thereafter the endometrium begins to lose its capacity for decidua formation, and the uterine environment becomes hostile to the embryo. Estradiol and progesterone are needed to control the process. The presence of both hormones stimulates DNA synthesis in stromal cells and their differentiation into polyploid decidual cells (Moulton and Koenig, 1984).

The manner in which the blastocyst implants varies among species. In rabbits, carnivores, and some monkeys, the cell mass increases in volume and fills the space between the uterine walls. In mice and rats the blastocyst remains the same size and is covered by cell layers originating from the endometrium. In guinea pig and primates the blastocyst embeds itself into the mucosa by active penetration.

In many species, especially in those which breed seasonally, implantation is delayed untill nature provides more clement conditions for offspring survival. Estradiol maintains the blastocyst viable while progesterone induces implantation (Section 2.5).

Patterns of Ovarian Hormones in Plasma

In seasonally ovulating species estradiol peaks at the time of estrus. Progesterone begins to increase at this time. If fertilization does not occur, a new estrus will take place. In cyclically ovulating species the follicular and luteal phases are of approximately the same length. An estradiol peak precedes the LH surge, is followed by a progesterone peak and a second estradiol rise. Concentrations of the two hor-

Table 2.7 Concentrations of Estradiol and Progesterone in Peripheral Plasma During a Menstrual or Estrus Cycle of Various Species.

Species	Estradiol (pg/ml)	Progesterone (ng/ml)	Reference
Primates			
Baboon		5	Stevens *et al.* (1970)
	250	4	Goncharov *et al.* (1976)
Human	125–300	10	Rudel *et al.* (1973)
	125–400	7–12	Brotherton (1976)
Macaca			Hodgen *et al.* (1976)
fascicularis	450	7	
	250	6	Goodman *et al.* (1977)
Macaca mulatta	400	7	Hess and Resko (1973)
Carnivora			
Fox	216	33	Mondain-Monval *et al.* (1977)
Ungulates			
Mare	12	5–15	Noden *et al.* (1975)
Sheep	1000	2	Wheeler *et al.* (1975)
White-tailed deer	30	5	Plotka *et al.* (1977)
Glires			
Guinea pig	12–23	3–5	Sasaki and Hanson (1974)
Rat	88	46	Butcher *et al.* (1974)

mones vary from species to species (Table 2.7). Steroid hormones are also biosynthesized in adipose tissue (Mendelson *et al.*, 1982) and in the brain.

2.4.3.3 The Biosynthesis of Steroid Hormones

Steroid hormones are lipid-like low molecular weight compounds, derived from
cholesterol either accumulated from the systemic circulation or synthesized *de
novo* from acetate. Only cholesterol bound to lipoprotein, or cholesteryl esters, can
be utilized. When the uptake exceeds the rate of steroid synthesis, *de novo* cholesterol formation is suppressed, and the excess of cellular needs is esterified and
stored for future needs (see Gwynne and Strauss III, 1982).

During the process of metabolism (removal from the body) conjugation of
hydroxyl functions with glucuronic, or sulfuric, acid renders steroid metabolites
water soluble, and the sulfates and/or glucuronide are excreted mainly in urine.
The gonads are not the only source of these hormones. The adrenal glands, in
addition to adrenocortical hormones, also produce androgens, progesterone and
estrogens.

Males

In the immature rat the main products of Leydig cells are 5α-reduced androgens,
principally 5α-androstane-3α,17β-diol. During sexual maturation the testes begins
to produce primary testosterone (Payne *et al.*, 1980). In addition to androgens
testes (Sertoli cells) also produce estradiol (Beall, 1940), but 80% or more of aromatization takes place in peripheral tissues (subcutaneous fat and in the brain) and
not in the testes (Table 2.8).

Testicular tissue is capable of *de novo* cholesterol (Srere *et al*, 1950) and testosterone (Brady, 1951), synthesis. Degradation of cholesterol to pregnenolone takes
place in the mitochondria. Pregnenolone then must be transported into the
smooth endoplasmic reticulum where it is converted into testosterone (see Ewing
and Zirkin, 1983). Four different pathways are feasible:

(i) pregnenolone (PR) $\rightarrow$ 3ß,17-dihydroxy-5-pregnene-20-one (17-OHPR) $\rightarrow$ 3β-
     hydroxy-5-androsten-17-one (DHA) $\rightarrow$ 5-androstene-3ß,17β-diol (DIOL) $\rightarrow$
     testosterone (Yanaihara and Troen, 1972);

Table 2.8 Relative Contribution of the Testes, Adrenals and Peripheral Tissue to Circulating Sex
Steroids in Men.

	Testicular Secretion	Adrenal Secretion	Peripheral Conversion
Testosterone	95	<1	<5
Dihydrotestosterone	20	<1	80
Estradiol	20	<1	80
Estrone	2	<1	98
DHA–SO$_4$[a]	10	90	...

[a] dehydroepiandrosterone sulfate; compiled from references in text.

(ii) PR → 17-OHPR → DHA → 4-androstene-3,17-dione (DIONE) → testosterone (Eik-Nes and Kekre, 1963);

(iii) PR → 17-OHPR → 17-hydroxy-4-pregnene-3,20-dione (17-OHP) → DIONE → testosterone (Bedrak and Samuels, 1969);

(iv) PR → progesterone →17-OHP → DIONE → testosterone (Slaunwhite and Samuels, 1956).

In rat testis the main pathway from pregnenolone to testosterone is the 4-ene pathway (Preslock, 1980) while in the human the 5-ene pathway predominates (Yanaihara and Troen, 1972; Hammar and Petersson, 1986). The steroidogenic function is chiefly regulated by LH (Purvis *et al.*, 1981; Dufau *et al.*,1984). LH withdrawal causes a regression of Leydig cell structure and function which is restored if the hormone is resupplied (Schwartz and Merker, 1965; Dym and Raj, 1977). FSH (Chen *et al.*, 1976), growth hormone and prolactin (Swerdloff and Odel, 1977; Bartke *et al.*,1978) and insulin (Adashi *et al.*, 1982) modulate the responsiveness of Leydig cells to LH. Estradiol (Hsueh *et al.*, 1978) and adrenocortical hormones (see Miller, 1988) are inhibitory.

Females
The ovary produces estradiol, progesterone and measurable quantities of androgens. The ability of ovarian tissue to produce such a variety of hormones resides in the different cellular compartments formed by the growing follicle (see Erickson *et al.*, 1985). In the theca cells androgen biosynthesis is stimulated by LH. Testosterone and DIONE cross the basal membrane, enter the granulosa cell layer, and accumulate in the follicular fluid. Depending on the developmental status of the follicle (controlled by FSH) the cells have the capacity to use androgens as substrate for aromatization, or 5α-reduction; 5α-reduced androgens (DHT) are aromatization inhibitors (Hillier, 1987). Granulosa cells, stimulated by FSH and modulated by insulin (Garzo and Dorrington, 1984), have the capacity to produce, in addition to estrogens and androgens, also pregnane derivatives, including progesterone (Armstrong and Dorrington, 1976; Nimrod and Lidner, 1976). The FSH stimulation of progesterone biosynthesis is inhibited by GnRH (Hsueh and Erickson, 1979; Massicotte *et al.*, 1980). GnRH increases the activity of 20α-hydroxy-, and decreases 3β-hydroxysteroid dehydrogenase (Jones and Hsueh, 1982a) and blocks pregnenolone synthesis, probably at the side-chain of cholesterol (Jones and Hsueh, 1982b).

Androgens increase FSH release and stimulate progesterone production and aromatase activity in granulosa cells *in vivo* (Goff *et al.,* 1979; Leung *et al.,* 1979) and *in vitro* (Lucky *et al.,* 1977; Daniel and Armstrong, 1980; Hillier and DeZwart, 1981; Welsh *et al.,* 1982).

Control of Corpus Luteum Function
The control of corpus luteum function varies from species to species (Table 2.9). In rats prolactin augments the action of LH (maintains luteal function; Astwood, 1941; Evans *et al.*, 1941). In the absence of PRL LH becomes luteolytic (see Auletta and Flint, 1988). LH is luteotropic in cow, sheep and primates. In mice and humans low doses of PRL facilitate, but high doses inhibit progesterone pro-

Table 2.9 Control of Luteotrophic and Luteolytic Functions of the Corpus Luteum in Various Species.

Species	Controlling Agent(s)	
	luteotrophic	luteolytic
Rodents (rats, mice)	PRL-LH (estrogens)	$PGF_{2\alpha}$
Ferrets	PRL-LH (estrogens)	
Hamsters	PRL + FSH	
Guinea Pigs		$PGF_{2\alpha}$
Ruminants (cow, sheep)	LH	$PGF_{2\alpha}$
Primates	LH	estradiol
Rabbits	estrogens	

Data from references in text.

duction (McNatty *et al.*, 1974; 1976). Possibly, PRL may also inhibit aromatase activity.

Luteolysis (regression of the corpus luteum) is induced in guinea pigs (Blatchley and Donovan, 1969), ewes, sows, and ruminants (Niswender *et al.*, 1985) by prostaglandins whereas estradiol is luteolytic in monkeys (Auletta *et al.*, 1972; Karsch *et al.*, 1973) and women (Hoffman, 1960; Gore *et al.*, 1973); see also Auletta and Flint (1988). The responsiveness to the luteolytic effect of estradiol increases as the corpus luteum ages. In monkeys estradiol induces the regression of the CL only in intact, spontaneously ovulating females. It is only during the midluteal phase that the CL becomes sensitive to estradiol effect; exogenous estradiol given during the early luteal phase does not shorten the cycle. Luteolysis does not take place if gonadotropin secretion is maintained by exogenous GnRH (Hutchison *et al.*, 1987a). This indicates that the hypothalamus is the major site at which estrogens initiate luteal regression in primates. For reviews see Nalbandov (1974), Niswender *et al.* (1985), and Auletta and Flint (1988).

Steroidogenesis

An outline of biosynthetic pathways in the gonads (and also in the adrenals) is shown in Figure 2.12. A complete description of control biosynthetic and metabolic pathways is found in Pasqualini and Kincl (1986) and Miller (1988). The origin of carbon atoms in the cholesterol molecule has been mainly elucidated by Bloch (1960).

The biosynthesis proceeds from acetate via mevalonate and squalene intermediates. Cleavage of the cholesterol side chain by the action of mitochondrial enzymes 20α-hydroxylase, 22-hydroxylase, and 20–22 desmolase yields the key intermediate pregnenolone (3β-hydroxy-5-pregnen-20-one, I).

Pregnanes: Following the formation of I (21 carbons) two different steroidogenic pathways are possible. In one, the intermediates retain the double bond between C5 and C6, the 5-ene pathway. In the other pathway the double bond is shifted with a concomitant oxidation of the 3ß alcohol to a 3-ketone (4-ene pathway). The oxidation and isomerization involve the action of two enzymes (delta-5-3β-hydroxysteroid dehydrogenase and an isomerase which shifts the double

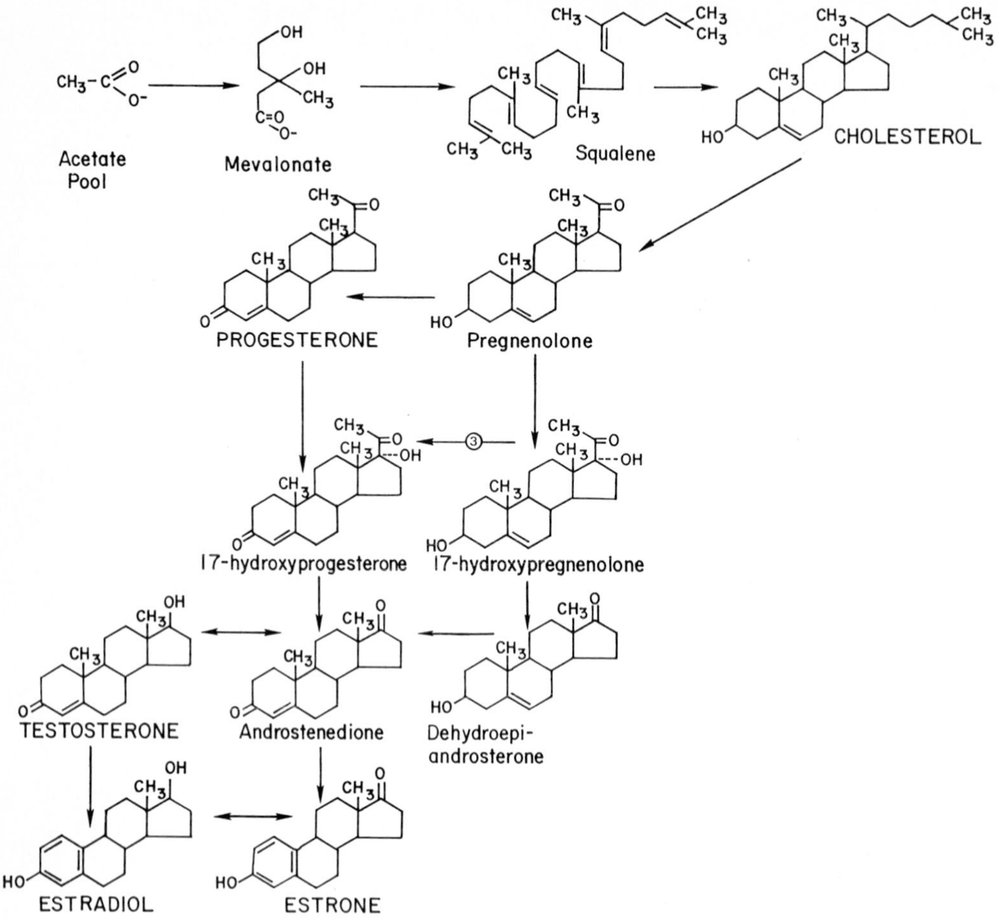

Figure 2.12. Steroidogenesis Pathways. See text for details.

bond between C-5＝C-6 to C-3＝C-4). Both enzymes are located in the micro-somal fraction and require NAD as cofactor. The resulting hormone, progesterone (II), is the main product of the corpus luteum.

Hydroxylation at C-17 position in II is requisite to the cleavage of the preg-nene side chain. The enzyme, C21,17-hydroxylase, is located in the microsomal fraction. The same enzyme can utilize either I or II as a substrate.

Androstanes: Side chain cleavage of either 17-hydroxy pregnenolone or 17-hydroxy progesterone by 17–20 desmolase yields C_{19} steroids (andro-gens). Depending on the precursor two key intermediates are formed: dehydro-epiandrosterone (DHA, 3β-hydroxy-5-androsten-3-one, III) or androstenedione (4-androstene-3,17-dione, IV); III is readily converted to testosterone (17β-hydroxy-4-androsten-3-one, V) by the action 3β-hydroxysteroid dehydrogenese and 5-ene-4-ene isomerase. Dehydroepiandrosterone sulfate present in high concentrations in peripheral blood is mainly of adrenal origin (Table 2.7). Reduction of the keto function in IV at C-17 is catalyzed by 17β-hydroxysteroid dehydrogenase which requires NADPH as a cofactor. The reaction is reversible and may play an important overall role in the physiologic effect of androstanes; the 17β-hydroxy derivatives are more potent than the 17-ketones.

Aromatization: The sequence of reactions by which III and IV are converted into C-18 steroids (estrogens) involves the oxidation and elimination of the C-19 carbon atom with a formation of an unstable intermediate which then rearranges to yield an aromatic ring A nucleus. Estradiol (1,3,5(10)-estratriene-3,17β-diol) is a more potent estrogen than the corresponding 17-ketone (estrone) which is only about 1/10th as active (Tables 2.14 and 2.15).

Aromatization also take place peripherally; in postmenopausal women adipose tissue is a major site of estrogen formation (Grodin *et al.*, 1973). The pituitary (Callard *et al.*, 1981) and the brain also possess the enzymatic capacity to convert androgens to estrogens (Naftolin *et al.*, 1975a). Estrone and estradiol are formed from IV and V in various tissues isolated from human fetus, rhesus monkey, rabbits, rats and mice (Ryan *et al.*, 1972; Table 2.10). Extragonadal conversion in humans is higher in females than in males. In women estrone production is about 44µg/24 h; 20%-50% is derived from IV. In men only 18 µg/24 h is produced, and only 10%-25% is derived from IV. In females castration, or pretreatment with either estradiol or testosterone, increases aromatization (Ryan *et al.*, 1972; Naftolin *et al.*, 1975a). Aromatization of IV to estrone in hypothalamic nuclei (hamsters' brain) has been shown *in vitro* by Callard *et al.* (1979) who failed to detect conversion to estradiol. Henderson *et al.* (1979) found a preferential concentration of estradiol at the expense of testosterone in the brain of male rats. The group speculated that the permeability of the blood-brain barrier may be different for the two classes of steroid hormones.

Catechol Estrogens: 2-Hydroxylation yields catechol estrogens believed to play a role in CNS regulatory processes (Fishman, 1976; Fishman *et al.*, 1976). 2-Hydroxy estrone is the most abundant catechol estrogen circulating in peripheral plasma (Yoshizawa and Fishman, 1971). Catechol estrogens induce serum LH

Table 2.10 Conversion of 4-Androstene-3,17-dione to Estrone in Tissues of Various Species.

Tissue used	Estrone formation proved	
	Yes	No
Hypothalamus	hf, rh, rb, r, m	
Pituitary gland	hf	rb, r
Cortex	hf	rb, r
Skin		hf
Liver	hf	

hf, human fetus; rh, rhesus monkey; rb, rabbit; r, rat; m, mouse; after Naftolin *et al.* (1971 a, b; 1975 a).

rise in immature male rats (Naftolin *et al.*, 1975b) and inhibit LH in mature animals (Pfaff, 1968; Okatani and Fishman, 1984a,b; 1986). 2-Hydroxy estradiol, synthesized by granulosa cells (Hammond *et al.*, 1986), augments the production of progesterone. Possibly, a synergistic interaction with catecholamines is involved, but the mechanism of action remains obscure (Spicer and Hammond, 1987).

Control of Steroidogenesis

Cholesterol formation is lower in females due to higher metabolic inactivation of mevalonate (Hughes *et al.*, 1979; see also Miller, 1988). Males possess an increased synthetic ability of peripheral organs (skin) to produce the precursors (Feingold *et al.*, 1983). The rate-limiting step in cholesterol biosynthesis is mevalonate formation. The transport of cholesterol from the outer to the inner mitochondrial membrane is regulated by luteinizing hormone (Ghosh *et al.*, 1987). During the cleavage of the cholesterol side chain the 20α-hydroxylation step is rate limiting (Sulimovici and Boyd, 1969; McIntosh *et al.*, 1971; Arthur and Boyd, 1974). The formation of the key intermediate, pregnenolone, is regulated by several hormones. FSH facilitates the action of LH by increasing the activity of 3β-hydroxy steroid dehydrogenase (van Beurden *et al.*, 1978) and the cleavage of the side chain (Jones and Hsueh, 1982b). In the granulosa cells pregnenolone production is stimulated by a synergistic action of estradiol and FSH (Veldhuis *et al.*, 1982a; Goldring *et al.*, 1987) and LH (Veldhuis *et al.*, 1982b).

In males progesterone secretion is augmented by DHT and inhibited by estradiol (Hanley and Schomberg, 1978). The action of androgens to increase progesterone biosynthesis is magnified by FSH (Armstrong and Dorrington, 1976; Nimrod and Lindner, 1976) and LH (Quirk *et al.*, 1986).

The cleavage of the C-21 side chain is stimulated by LH but not by progesterone. The stimulation is the result of increased activity of 17-hydroxylase 17,20 lyase enzyme(s) controlled by changes in the fluctuation of LH (Bogovich and Richards, 1982). Estradiol provides a negative feedback by inhibiting the activity of this enzyme (van Beurden *et al.*, 1978; Magoffin and Erickson, 1981). Progesterone metabolism (reduction to 20α-hydroxyl) is increased by GnRH (Jones and Hsueh, 1981).

Androgen production is stimulated by PRL, by increased steroidogenesis (Bartke, 1971; Johnson, 1974), and by increased activity of 17β-hydroxysteroid dehydrogenase (Musto *et al.*, 1972). The production of testosterone is lower during stress in humans (Kreuz *et al.*, 1972; Aono *et al.*, 1976; Morville *et al.*, 1979) and animals (Bliss *et al.*, 1972; Nakashima *et al.*, 1975; Gray *et al.*, 1978; Goncharov *et al.*, 1979), often in the presence of a sufficient concentration of LH in the blood. The decrease is brought about by hyposensitivity of Leydig cells to gonadotropins and post cAMP blockade (Charpenet *et al.*, 1981; 1982) caused by a direct inhibition by arginine-vasopressin (Collu *et al.*, 1984).

CNS depressing drugs depress 5α-reductase activity in the brain (Kaneyuki *et al.*, 1979).

Aromatase activity in ovarian granulosa cells (Dorrington *et al.*, 1975; Erickson and Hsueh, 1978; Norton *et al.*, 1988) and Sertoli cells from testis of immature rats (Dorrington and Armstrong, 1975) is stimulated by FSH (Hsueh *et al.*, 1983).

In theca cells (Fortune and Armstrong, 1977) and Leydig cells obtained from mature male rats (Fritz *et al.*, 1976) aromatase activity is increased by LH. In man prolactin suppresses estrogen production by stimulating LH induced 5α-reductase activity in testes (Takeyama *et al.*, 1986). Lactorrhea in man is associated with delayed sexual development, hypogonadism and infertility (Carter *et al.*, 1978).

The paracrine control of biosynthesis by polypeptide hormones is discussed below. For the discussion of ACTH regulatory function during steroidogenesis see Hall (1985).

2.4.4 Nonsteroid Hormones

The gonads produce, in addition to steroid hormones, polypeptide hormones whose existence was suggested some 60 years ago by Martins and Rocha (1931) and Moore and Price (1932). Both groups suspected the presence of a substance in the seminal fluid which directly regulated pituitary function. Several hormones,

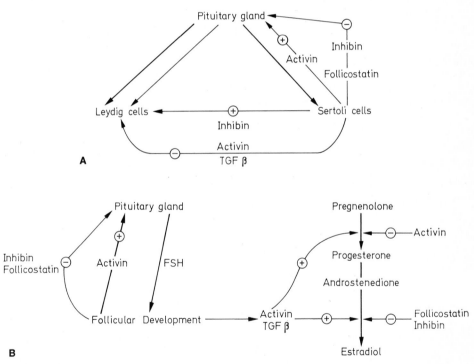

Figure 2.13. Action of Inhibins on the Gonads. In males (A) inhibin (and follicostatin) suppress FSH release by the pituitary and decrease testosterone production by Leydig cells; activin enhances FSH release and increases testosterone production by Leydig cells; TGFβ (a protein widely distributed in the body) also stimulates testosterone production. In females (B) inhibins and follicostatins suppress whereas activin and TGFβ enhance the secretion of FSH by the pituitary; in the developed follicle the polypeptides augment or restrict progesterone and estradiol biosynthesis.

which influence only the function of FSH, termed inhibins, activins, and follicostatins (see Ying, 1988), have been identified. In addition to those, prostaglandins
and insulin also modify reproductive functions. The action of the various polypeptides is summarized in Figure 2.13.

2.4.4.1 Inhibin

In 1932 McCullagh isolated from testes a water-soluble, nonsteroidal hormone
that induced atrophy of the gonads and accessory sex organs and suggested that
the substance inhibited FSH action. Vidgoff and coworkers (Vidgoff et al., 1939;
Vidgoff and Vehrs, 1941) confirmed the presence of such a substance and named
the hormone inhibin. The hormone, whose existence was often disputed, is present
in the granulosa cells of the ovary (Erikson and Hsueh, 1978) and Sertoli cells in
males (Steinberger and Steinberger, 1976). The pure hormone was obtained from
bovine follicular fluid (see Ying, 1988).

More than one hormone of similar biological activity exists; an antiserum to inhibin neutralizes activity in the ovaries but not in the testes (van Dijk et al., 1986).
The principal hormone (molecular weight 58 kD) is a polypeptide composed of
two subunits, joined by disulfide bonds. The sequence of amino acids is known
(see Ying, 1988). Literature prior to 1981 has been reviewed by Grady et al. (1982).

Inhibin(s) block FSH secretion by a direct action upon the pituitary gland (Lee
et al., 1976; DeJong, 1979; Williams and Lipner, 1982) by blocking the response of
gonadotropin to GnRH (Martin et al., 1986; Campen and Vale, 1988). In ewes
inhibin causes increases in LH pulses and delays the onset of estrus. This results
from reduced negative feedback of estradiol brought about by reduction of FSH
secretion (Wallace and McNeilly, 1986). There is some evidence to indicate that
the hormone(s) may also interfere with some functions of the hypothalamus
(Lumpkin et al., 1981). The suppression of FSH activity results in the decrease in
the number of Sertoli cells and stem cells in rat, and ram, testis (Cahoreau et al.,
1979).

Inhibin also influences directly steroidogenesis in the gonads. In granulosa
cells it decreases progesterone secretion and diminishes FSH stimulated aromatase activity (Ying et al., 1986). In Leydig cells inhibin may potentiate testosterone
production (Hsueh et al., 1987).

Inhibin disappears rapidly from the blood after ovariectomy and its levels
increase following FSH administration (Lee et al., 1981; Sander et al., 1986;
Zhiven et al., 1987). In males damage to the seminiferous epithelium results in
increase of serum FSH concentration (Au et al., 1984). Both observations suggest a
feedback relationship.

2.4.4.2 Activins

Activins (two forms are known) were isolated from porcine follicular fluid (Ling
et al., 1985; Rivier et al., 1985). The hormones are dimeric polypeptides (see
Appendix) which stimulate FSH secretion from the pituitary gland (Adashi and

Resnick, 1986). The hormones are more potent stimulators than GnRH (Ling
et al., 1986a,b). The secretion of LH is not affected.

Activins regulate locally granulosa cell functions. Both hormones enhance
FSH induced aromatase activity and inhibit progesterone synthesis (Adashi and
Resnick, 1986; Ying *et al.*, 1987). In Leydig cells activins suppress androgen bio-
synthesis stimulated by LH (Hsueh *et al.*, 1987).

2.4.4.3 Follicostatin

Follicostatin is a single chain polypeptide present in follicular fluid (Ueno *et al.*,
1987). The hormone is a inhibitor of FSH release from the pituitary (Ying *et al.*,
1987).

2.4.4.4 Prostaglandins

Prostaglandins (PG) are lipids that possess a common C_{-20} structure related to
prostanoic acid (see Pasqualini and Kincl, 1985); the lipids are found in almost all
cells. Prostaglandins arise from biologically potent, but very short lived thromboxanes ($t_{1/2}$ about 30 seconds) and endoperoxide ($t_{1/2}$ few minutes).

Prostaglandins are luteolytic; PG receptors are present in the corpus luteum of
several species (Kirton *et al.*, 1976).

Prostaglandins produced within the ovary may contribute to ovulation. Their
concentration rises during ovulation (Armstrong, 1981). The administration of an
inhibitor (indomethacin) of PG synthesis blocks it (Espey, 1982). The effect of the
inhibitor is not clear. In the rabbit stimulated to ovulate with hCG high doses
of indomethacin (10 mg/kg) block ovulation but not the synthesis of PGs in the
follicles (Espey, 1986). The contradictory results leave open the question whether
prostaglandins participate in the ovulatory process, and if so, how important the
contribution may be.

Prostaglandins cause a release of ACTH and LH from the pituitary, may interact with estrogens during decidualization (Kennedy, 1980) and are needed for LH-
induced ovulation in rats (Tsafriri *et al.*, 1973; Bauminger and Lindner, 1975), rabbits (Diaz-Infante *et al.*, 1974; Hamada *et al.*, 1977), guinea pigs and other species
(see LeMaire *et al.*, 1978). In seasonal breeders PGs synthesized in the hypothalamus are thought to be needed to modulate the effect of estradiol leading to LH
surge (McCraken, 1974). Granulosa cells and thecal tissue produce prostaglandins
(Erikson *et al.*, 1977; Clark *et al.*, 1978, Ainsworth *et al.*, 1984). The synthesis can
be induced (*in vitro*) by hCG (Hedin *et al.*, 1987) and by FSH (Shaykh *et al.*, 1985).

2.4.4.5 Other Factors

Insulin
Clinical observations and *in vitro* studies indicate that insulin and insulin-like
growth factors may contribute to the regulation of gonadal function, especially in

females. Diabetes mellitus is associated with ovarian hypofunction: late menarche, low pregnancy rate, high rate of anovulatory cycles (Berqvist, 1954; Djursing *et al.*, 1982), and other endocrine abnormalities (Distiller *et al.*, 1975; Walton *et al.*, 1983). Diabetes, induced experimentally in rats, leads to decreases in the weights of ovaries in the presence of "normal" circulating concentrations of LH and FSH (Liu *et al.*, 1972). An inadequate response to GnRH stimulation, possibly due to the degeneration of neurons in the arcuate nucleus, has been reported (Bestetti *et al.*, 1985).

Insulin may stimulate ovarian functions by influencing the growth and function of granulosa cells (Channing *et al.*, 1976; May and Schomberg, 1981; Savion *et al.*, 1981) and by influencing the aromatization steps, probably in synergism with FSH and LH (Garzo and Dorrington, 1984; Adashi *et al.*, 1985). Indirect evidence for the role of insulin is provided by the observation that receptors to insulin, and insulin-like growth factors, are present in human stromal, follicular and granulosa cells (Poretsky *et al.*, 1984; 1985) and in rat granulosa cells (Davoren *et al.*, 1986).

Transforming Growth Factor (TGFβ)
TGFβ, a dimeric polypeptide (derived from human platelets), stimulates FSH-induced aromatase activity and blocks the effect of inhibin (Hsueh *et al.*, 1987; Hutchinson *et al.*, 1987b).

Other Peptides
A variety of neuropeptides related to vasopressin (Kasson *et al.*, 1986) act as testicular paracrine regulators which modulate the LH stimulated androgen production and the function of Sertoli cells. Stimulatory peptide(s), termed gonadocrinins (Ying *et al.*, 1981; Sluss *et al.*, 1987), are present in ovarian follicular fluid. The peptides may be identical to activins and follicostatins (see above). Testicular peritubular cells secrete an androgen dependent protein(s) of unknown structure (mol.wt. between 50 and 100 kD) which stimulates the secretion of androgen binding protein and inhibits the aromatase activity of Sertoli cells (Verhoven and Cailleau, 1988).

Brain neurons synthesize atrial natriuretic factor (ANF), and receptors for ANF are present in the hypothalamus, the median eminence, and the anterior pituitary. A continuous infusion of ANF (0.1 μg/min) leads to an inhibition of LH release (in castrated rats) possibly by a direct inhibitory effect on the median eminence (Samson *et al.*, 1988).

2.5 Activity of Steroid Hormones

Hormones are released into the systemic circulation in periodic bursts of various duration and intensity (Section 2.2). The biological half-life of steroid hormones is short (see Appendix). In pharmacological tests hormones are given usually in a single dose. This means that sufficient amounts must be given to mimic endoge-

Table 2.11 Androgenic-Anabolic Activity of Testosterone Propionate Compared with Testosterone in 21-Day-Old Castrated Male Rats (test compounds injected subcutaneously daily for 10 days).

	Total dose mg	Organ Weights, mg		
		Seminal vesicles	Ventral prostate	Levator ani
Control	0	8.0	9.5	23.0
Testosterone	0.3	8.0*	15.2	19.6*
	0.6	10.8	20.2	19.3*
	1.2	27.7	47.7	30.5
	2.4	55.4	65.7	36.7
Testosterone propionate	0.3	10.6	26.0	21.5
	0.6	15.9	34.0	26.7
	1.2	37.2	59.3	33.7
	2.4	82.7	82.1	45.3

Data by Cancer Chemotherapy National Service Center, Endocrine Evaluation Branch; * not statistically different from control.

nous plasma concentrations during the time before the next dose is given. Dividing the dose decreases the need for higher amounts and improves drug efficiency. A sustained release form which releases continuously the active principle is more efficient. This has been shown by Kincl and Rudel (1971); see also Rudel and Kincl (1971) who found that by using a sustained release form biological effectiveness increased from 10 to 30 fold.

Ester formation is another method which will improve the activity and duration. Junkmann studied extensively the effect of ester formation on the biological activity of 17-esters of androgens (1952), estrogens (1953), gestagens (1954a) and 21-esters of corticoids (1954b). The studies established a relation between activity and the nature of the esterifying agent. The longer the fatty acid chain the longer is the duration. The esters are absorbed from the injection site, deposited in body fat and cell lipids from which they slowly diffuse. During this step saponification takes place, and the free hormone produces its physiological effect (Junkmann, 1957).

A single injection of testosterone acetate produces a maximal response after one week; of the propionate, butyrate and valerate forms after 2 weeks; and of the caproate and enanthate forms after 3 weeks. The most commonly used ester of testosterone, the propionate, is about 1.5 times more active (Table 2.11) when tested in standard assays (for details of biological tests see Appendix).

2.5.1 Stimulatory and Inhibitory Effects

All the three major gonadal hormones, estradiol, progesterone and testosterone, given in sufficient doses and for a sufficient time will inhibit the hypothalamic-pituitary function and thus the growth, and function, of the gonads. Accessory sex organs will continue to be stimulated. Interference with fertility in mature animals is reversible; depending upon the dose and duration of treatment, the ability to reproduce will be restored after varying periods of time (Burrows, 1937). The inhibitory action is only incompletely specific: adrenocortical hormones (cortisol and corticosterone) also interfere with reproductive function (Section 2.7).

Of the three hormones estradiol is pharmacologically the most active; only microgram quantities are needed to stimulate or inhibit various activities. Progesterone and testosterone must be given in milligram quantities to achieve their respective effects.

2.5.1.1 Androgens

In males testosterone promotes the growth and function of the epididymis, vas deferens, prostate, seminal vesicles, penis, and secondary sex characteristics and behavior, contributes to the muscle and skeletal growth (sex dimorphism in body size), the anabolic effect (nitrogen retaining property) and other functions (Section 2.4.3.1). In the ovary increased androgen biosynthesis (testosterone, 4-androstene-3,17-dione) in theca cells (stimulated by LH) provides the granulosa cells with the necessary substrate for aromatization (Section 2.4.3.3).

Testosterone (and other androgens) inhibit the function of the hypothalamic-pituitary axis. The brain mediates the suppression. In rats joined in parabiosis testosterone and a CNS active drug (chlorpromazine) inhibit ovulation in the intact female partner (Table 2.12). Decreases in ovarian response (number of ovulations) and in the weight of the ovaries show the suppressive effect. Increases in the weight of the ventral prostate in the castrated partner reveal the androgenic influence. It should be noted the chlorpromazine was not androgenic.

Requirements for Activity

An intact 17β-hydroxyl function is needed for androgenic activity. The double bond between C-3 and C-4 may or may not be needed; in some tissues reduction to the 5α-dihydro product is essential for the expression of action. An introduction of a 17α-alkyl-, or allyl-, group (methyl, ethyl) renders the derivative orally active, presumably by protecting the 17β-hydroxyl from oxidation. The derivatives, if used for prolonged periods of time, may cause liver damage.

Table 2.12 Gonadotrophin Inhibitory Activity of Testosterone and Chlorpromazine Assayed in Intact Female-Castrate Male Parabiotic Rats.

Compound	Daily Dose mg	Body wt. g+SE (males)	Ovarian response*	Ovaries mg±SE	Ventral prostate (castrate male), mg
0	0	103±1.8	8/8	186.8±17.9	9.4±0.6
Testosterone	0.05	102±2.9	5/8	93.8±18.6	47.2±4.5
	0.15	106±2.8	4/8	28.7± 3.1	116.1±7.4
	0.45	102±2.9	3/8	32.3± 4.4	174.8±7.5
Chlorpromazine	0.1	109±1.8	8/8	180.2±25.3	11.9±0.6
	0.3	101±1.2	7/7	192.1±14.6	9.7±0.7
	0.9	96±3.2	5/8	94.4±24.4	12.7±0.2

* Ovarian response, number of rats with CL/total number of rats; from Rudel and Kincl (1966).

2.5.1.2 Anabolic Agents

A great research effort took place in the 1950s and 1960s to create "anabolic" agents possessing minimal androgenic and high anabolic (nitrogen retaining) properties (Kochakian and Murlin, 1935; 1936). Hundreds of synthetic steroids were made, screened in animal tests and many assayed in humans. Increases in anabolic potency at the expense of androgenicity were found by introducing alkyl-, or halo-, groups at C-1, C-2, C-6, introducing 9α-halo-11β-hydroxy moiety and by removing the C-19 angular methyl group (19-nor steroids). The studies revealed that the presence of a 3-ketone is not needed; 3-deoxy compounds are active hormones. Kincl (1965) discussed the relationship between structure and activity.

Anabolic steroids aid in muscle build-up but are not recommended. In men, taken in high amounts, many are liver toxic and may increase blood pressure, cause azoospermia, testicular atrophy and mood shifts to aggressiveness. In women, anabolic agents may cause androgenization (acne, growth of facial hair and deep voice) (see Wilson, 1988).

Since androgen receptors in muscles are similar to those found in sex organs a purely myotropic agent may not be possible (Mooradian et al., 1987).

2.5.1.3 Estrogens

Estrogens induce the cornification of vaginal epithelium, stimulate the growth of the vagina and of the uterus, promote mitosis in the endometrium and increase tonicity and blood supply (hyperemia) to the myometrium and the permeability of uterine capillaries. Stimulation of cell division in deeper layers causes a more rapid replacement of the cornified stratum which is seen also in peripheral tissue (skin). Within a follicle estrogens act as autocrine regulators which stimulate mitosis in granulosa cells and amplify the action of FSH and LH by increasing the number of receptors (Section 2.6). Estrogens are strong inhibitors of the hypothalamic-pituitary axis in both sexes, more potent than androgens (Schoeller et al., 1936). Estradiol modulates both the activity and release of GnRH and blocks directly the transcription (especially of LHß) of the gonadotropin gene (Shupnik et al., 1988). Estrogens inhibit ova maturation and, in sufficient doses, arrest embryonic development and cause fetal death.

Estrogens are also needed to promote mammary gland development; synergistic action with other hormones (adrenocorticoids, progesterone, thyroxine and prolactin) is essential for full development and lactation. Calcium metabolism is estrogen dependent; significant loss of calcium from bones may develop when estrogen production declines in menopausal women. The effect of estradiol on the formation of progesterone receptors is discussed in Section 2.6.

Females are more sensitive to hormonal feedback on the pituitary than are males. The data of Meyer and Herz (1937) and Biddulph et al. (1940) demonstrate this in the parabiotic rats test (Table 2.13). The possibility that the differences could be attributed to varying biological half-time or to the development of receptor molecules cannot be excluded.

Table 2.13 Minimum Daily Doses of Estrogens Needed to Inhibit Castration-Induced Hypersecretion of Gonadotropins.

Steroid used	Males (μg needed)	Females (μg needed)
Estradiol	0.15	0.025
Estrone	1.0	0.20
Estriol	10	1.5

Data from Meyer and Herz (1937) and Biddulph *et al.* (1940).

Effects on Neurons

Estradiol stimulates neuron growth and may modify levels of synaptic constituents
and neuron sensitivity to afferent neuronal inputs (McEwen *et al.*, 1982). Estrogens also increase electrical activity in the basomedial hypothalamus which facilitates mating (Pfaff, 1983).

In high doses estradiol causes degeneration of neural processes associated with
microglial and astrocyte actions in the arcuate nucleus of the hypothalamus
(Brawer and Sonnenschein, 1975; Brawer *et al.*, 1978; 1980). Damage induced by
constant illumination is less severe in ovariectomized animals (Schipper *et al.*,
1981). Estradiol induced degeneration is decreased by concomitant administration
of 5α-dihydrotestosterone. It has been proposed that reproductive aging in female
rats may be the consequence of estradiol-dependent hypothalamic impairment
(Aschheim, 1976) and that DHT may provide a "protective" action to prevent
hypothalamic "aging" in male rats (Brawer *et al.*, 1983).

Inhibition of Gonadotropin Secretion

Prolonged treatment with estradiol will cause changes in the morphology in the
pituitary gland; basophil, acidophil and chromophobe cells become enlarged and
degranulated (Hohlweg, 1934). The gland will increase in size; that the gland is
always larger in females than in males was noted in 1929 by Evans and Simpson.
If the treatment is long (several months in rats) the pituitary may lose its distinc-
tive morphology, and the growth of connective tissue and hyalin formation
increase at the expense of other elements. In extreme cases the treatment may lead
to the formation of benign, or even malignant, tumors.

Males: In males exogenous estrogens will cause azoospermia, destruction of the
maturation wave of the seminiferous epithelium, atrophy of the testis and accessory sex organs; prolonged therapy may be associated with loss of body weight.
Leydig cells become atrophic and may revert to fibroblast. The atrophy may proceed so that only Sertoli cells remain in the tubules, but even these may disappear
with the induction of peritubular hyalinization and sclerosis. Simultaneous administration of exogenous gonadotropin, or androgens, prevents the atrophy (Moore
and Price, 1932). Prolonged treatment may also cause multiplication and swelling
of Leydig cells, deposition of lipid material in germinal epithelium, decrease in
glycogen content, and testicular hyperplasia (Burrows, 1937). In young rats estradiol prevents the descent of testes, produces atrophy, and inhibits spermatogenesis
(Pallos, 1941; Gardner, 1949).

The inhibitory effect, if not too severe, is reversible. In rats treated with estradiol regeneration begins in about 2 weeks (Bourg *et al.*, 1952) and is complete within 6 weeks (Lynch, 1952).

Females: If estrogens are given to adolescent animals, the development of the ovaries is arrested, and the follicles do not ripen. When the treatment is stopped, the ovary becomes functional, estrus cycles appear, and the animals are able to reproduce (Selye, 1939). A direct effect on the ovary appears excluded. In 22- or 30-day-old hypophysectomized rats follicle formation can be stimulated with FSH or PMSG and ovulation induced two days later with LH or hCG. In this preparation ovulation is not inhibited by 20 to 200 µg of estradiol or 20 mg/kg of estrone (Krahenbühl and Desaulles, 1964; Rudel and Kincl, 1966).

Transitory inhibitory action and ovarian atrophy caused in adults have been reported in numerous publications (see Kincl, 1972). Low doses of estrogens will cause an irregular, accelerated pattern of ova movement in the oviduct; high doses will arrest the movement (tube locking) and cause ova disintegration (Whitney and Burdick, 1936; 1938; Ketchel and Pincus, 1964). Forcelledo *et al.* (1986) correlated zygote transport with the velocity of estradiol pulse after iv infusion and found that short, high amplitude oscillations (which produced > 10 ng/ml in plasma) were not effective in accelerating the transport. Longer infusion which produced sustained elevations, comparable to those obtained following sc injection (about 1 ng/ml), slowed tubal transport.

Other Effects on Fertility
In pregnant animals estrogens may produce placental separation, loss of decidua, degeneration of the endometrium and fetal death. In rats and mice a single injection of 53 µg of estradiol on day 4 after insemination is sufficient to interrupt pregnancy in one-half of the animals (Emmens and Finn, 1962).

Estrone is usually less active than estradiol while the esters are significantly more active (Tables 2.14 and 2.15).

Requirement for Activity
Hydroxy groups at C-3 and C-17 are essential (Hähnel *et al.*, 1973; Chernayaev *et al.*, 1975) for the expression of biological activity – the binding to receptor molecules. Raynaud *et al.* (1985) defined the function of the two groups by correlating

Table 2.14 Comparative Activity of Various Estrogens in Inhibiting Nidation, Ovulation and Gonadotropin Release in the Rat (by subcutaneous injection).

Steroid	Relative potency		
	Anti-nidation	Anti-ovulation	Anti-gonadotropin release
Estradiol	100	100	100
Estrone	70	150	30
Estriol	12	15	10

Data from Kincl (1972).

Table 2.15 Doses of Various Estrogens Needed to Suppress Follicle Development in Hamsters (a single dose injected at diestrus; atresia of developing follicles studied 5 days later).

Steroid	Dose range used (µg)	Effective dose (µg)
Estradiol	10–250	250
Estradiol benzoate	10– 25	10
Estradiol cyclopentyl propionate	10– 25	10
Estrone	100–500	500

Data after Greenwald (1965).

crystalline conformation, binding affinity and biological potency. According to this group the hydroxyl on C-3 is involved in the first recognition step and acts as a H-donor. Introduction of a 17α-alkyl-, or allyl- (ethinyl), group imparts oral activity. Potency is not increased if the compounds are injected. Substitution on other portions of the molecule decreases the potency (see Emmens, 1962; Emmens and Martin, 1964). Introduction of a 11β-methoxy moiety may increase potency (Raynaud *et al.*, 1973). Nonsteroidal triphenylethylene derivatives (diethylstilbestrol), some occurring in plants, mimic some physiological actions of estrogens (Emmens and Martin, 1964).

2.5.1.4 Progesterone

Progesterone is essential in support of gestation, produces secretory glandular changes in an estrogen-primed endometrium, influences the structure and function of fallopian tubes, implantation and decidua formation, sperm capacitation and transportation in the female reproductive tract and is essential in mammary gland development. Many of the above functions are achieved more fully in synergism with estrogens and pituitary hormones. In many species rise in plasma progesterone during estrus is associated with ovulation and mating receptivity, but in one rodent species, the Djungarian hamster (*Phodopus campbelli*) the peak is absent (Wynne-Edwards *et al.*, 1987).

Progesterone inhibits the function of the hypothalamic-pituitary axis and can act to inhibit estrogen (anti-estrogenic) and androgen (anti-androgenic) action. Synthetic progestogens (compounds that mimic progesterone function) are being used, in combination with a synthetic, orally active estrogen, to inhibit fertility in humans. This is "the Pill" which has found a wide use for family planning purposes.

Progesterone does not induce fetal or neonatal abnormalities, and a full discussion of its pharmacological properties is outside the scope of this chapter. For a full description of the physiology and pharmacology of the hormone see Tausk (1971).

Effect on Neurons

A metabolite of progesterone (3α-hydroxy-5α-pregnan-20-one) is a potent barbiturate-like agent which inhibits the activity of the neurotransmitter γ-aminobutyric acid (Majewska *et al.*, 1986). The ability of pregnane derivatives to act as anesthetic agents and CNS depressants has been recognized many years ago (see P'an and Laubach, 1964).

Inhibition of Ovulation

Observations that progesterone blocks ovulation were made many years ago. Animals with an active corpus luteum do not ovulate (Makepeace *et al.*, 1937; Boyarsky *et al.*, 1947). Progesterone treatment inhibits ovulation in induced ovulators. Rabbits are particularly sensitive to the test (Astwood and Fevold, 1939; Pincus and Chang, 1953; Barnes *et al.*, 1959). An injection of 0.1 to 0.3 mg of progesterone is partially effective; a dose of 1 mg is inhibitory. In contrast 30 mg are needed to produce a significant effect in the parabiotic rat (Kincl and Dorfman, 1963).

The inhibiting effects of progesterone are mediated through the central nervous system (Section 2.3.1). Progestational agents do not act directly on the ovaries (Kincl, 1966). In rabbits, ovulation is not inhibited by progestational agents when ovaries are directly stimulated by hCG (human chorionic gonadotropin) or by copper acetate (Kincl, 1963). Ovulation is also not inhibited in immature hypophysectomized rats in which follicle formation has been induced by the administration of PMSG and ovulation by hCG (Rudel and Kincl, 1966).

Antiandrogenic Activity

Progesterone (and other progestational agents) inhibit testosterone from expressing its activity at the target sites (Kincl, 1971a). Mice and rats are the test animals of choice (Dorfman, 1963a,b). Inhibition of 5α-reductase activity of binding to cytosol and nuclear receptors has been shown to be the steps at which antiandrogens express their activity (Neumann and Steinbeck, 1974). Relatively high amounts are needed to achieve a significant effect (Table 2.16).

A group from Schering AG in Berlin led by Neumann (Hamada *et al.*, 1963) found that a synthetic progestational agent cyproterone acetate (1,2-methylene-6-chloro-17-acetoxy-4,6-pregnadiene-3,20-dione) is a potent antiandrogen. The ste-

Table 2.16 Antiandrogenic Activity of Progesterone in Various Tests.

Test animal	Androgen used (stimulatory dose)	Minimum inhibitory dose (mg)	Mean inhibition %	
			Prostate	Seminal vesicles
Rats	Testosterone	50	0	21
	(2 mg)	250	26	52
Mice	Testosterone			
	(0.8 mg)	40	0	42
	Methyl testosterone	10	0	24
	(0.8 mg)	20	37	36

Data from Dorfman (1963 a, b).

Table 2.17 Antiestrogenic Action of Progesterone Assayed in Several Tests.

Animals	End point	Estrogen used	Dose μg	Progesterone dose, μg	References
Rats	Decidual development	Estrone	1.6	1500	Velardo and Hisaw (1951)
		Estradiol	0.1	1500	Velardo and Hisaw (1951)
		Estradiol benzoate	0.3	1500	Velardo and Hisaw (1951)
		Estriol	20	1500	Velardo and Hisaw (1951)
			0.3	8000	Edgren et al. (1961)
Rats	Uterine growth	Estrone	0.1	1000	Huggins (1956)
		Estriol	50	1000	Huggins (1956)
			100	1000	Edgren et al. (1961)
Mice	Uterine growth	Estrone	0.3	100	Edgren and Calhoun (1957)
		Estrone	0.4	500	Dorfman et al. (1961)
		Estriol	100	1000	Edgren et al. (1961)
Rabbits	Uterine histology	Estriol	128	50	Edgren et al. (1961)
	Carbonic anhydrase	Estrone	0.0025	2000	Miyake and Pincus (1958)
		Estradiol	0.0025	500	Miyake and Pincus (1958)
		Estriol	0.0025	2000	Miyake and Pincus (1958)

roid blocks testosterone induced stimulation in castrated rats and androgen action in intact animals. Cyproterone acetate inhibits the uptake of testosterone in the CNS and conversion of testosterone to 5α-dihydrotestosterone in the CNS and accessory sex tissues, impairs spermatogenesis, libido and aggressive behavior, and induces feminization of male fetuses (Neumann and Steinbeck, 1974).

Antiestrogenic Activity
Progesterone inhibits estrogen induced activity (see Dorfman, 1962; Kincl, 1971b) in mice, rats, or rabbits (Table 2.17). In smaller amounts progesterone synergies with estrogens (Section 2.5.3).

2.5.2 Feedback

Gonadal inhibition of the hypothalamic-pituitary axis which leads to a decrease of gonadal function is referred to as *negative* feedback. Stimulatory activity of gonadal hormones is described as *positive* feedback. Sex linked differences exist: the male of the laboratory rat is more sensitive to testosterone feedback, which regulates the 5α-reductase activity in the pituitary, than is the female (Kniewald and Milkovic, 1973).

The dual action of steroid hormones was recognized as early as in 1932 by Moore and Price in the USA and Hohlweg and Junkmann in Germany. The dual role is best described as "small doses stimulate large doses inhibit" rule.

The conditions of the animal (or test) will determine whether stimulation or inhibition will take place. For example Clarke and Cummins (1985) reported that in ovariectomized sheep estradiol benzoate will elicit first a decrease and then an increase in LH secretion. Figure 2.14 illustrates the inhibitory action of estradiol on LH release in sheep.

The hypothalamus is sensitive to a positive estradiol feedback at a relatively early stage. In cycling species estradiol induces LH release prior to the initiation of first ovulation. In seasonally ovulating species, sheep (Foster, 1983) and monkeys (Wilson *et al.*, 1986), estradiol feedback is modulated by photoperiods. Early exposure to short days (a reversal of the "natural" state) disturbs estradiol reaction and LH secretion. In primates (also women) estradiol elicits a LH surge only after menarche. Decrease in the sensitivity prior to puberty may involve alterations in hypothalamic neural pathways (Plant and Zorub, 1982; Terasawa *et al.*, 1984). Age dependent increase in growth hormone production may provide the necessary impetus (Wilson *et al.*, 1986).

Other species display an intermediate pattern. In the guinea pig ovarian steroids can stimulate LH release in the immature females but not until an age approaching the normal time of the first ovulation (Nass *et al.*, 1984a,b).

That the brain is the main site of the inhibitory action has been documented in Section 2.3.1 (see also Everett, 1952; Kincl, 1972). The site of action is not uniform

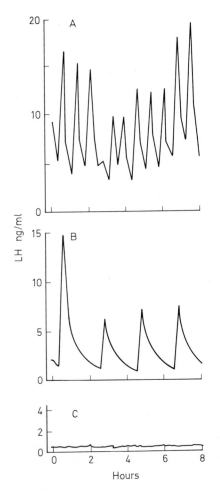

Figure 2.14. Reduction of LH Pulse Frequency by Negative Estradiol Feedback in Ewes. LH concentration (ng/ml) pulses are shown in control anestrus ewes (A), or in animals implanted subcutaneously with dimethylpolysiloxane tubes containing estradiol 6 days prior to blood sampling; group B, 1 cm long implant (estimated release 2.5 μg); group C, 2 cm implant releasing about 5 μg of estradiol in 24 h. Redrawn with permission from Lincoln *et al.* (1985).

and may vary from species to species and be sex dependent. In the hypogonadal mouse, a mutant strain, estradiol provides the main negative feedback in the pituitary, whereas in the male testosterone may act on the hypothalamic nuclei or brain. The sensitivity of the hypothalamus to the negative feedback is often modulated by opioid peptides of the brain (Gabriel et al., 1986) and by prolactin (Charlton et al., 1983; McNeilly et al., 1983). During the cycle the changing hormonal milieu within the gonad modulates the action of GnRH and gonadotropins on the gonad itself (Section 2.4.4).

In addition to the "long loop" (steroid) feedback a "short loop" feedback (a direct inhibitory action of gonadotropins on the hypothalamus) was postulated more than 20 years ago by David et al. (1966). The existence of a short loop feedback mechanism for the control of LH secretion has been proposed for the rat (Ojeda and Ramirez, 1969; Hirone et al., 1970; Döcke and Glaser, 1971), rabbit (Molitch et al., 1976; Patritti-Laborde and Odell, 1978; 1982; Odell et al., 1979), pig (Ziecik et al., 1988) and human (Miyake et al., 1977). A short loop for FSH has been also proposed (Laborde et al., 1981).

Anatomical evidence favors this possibility. The anterior and posterior sections are joined by a common capillary bed, which permits recirculation within the gland and the return of blood to the infundibular process and upward to the median eminence (Bergland and Page, 1978). This makes it possible for the endocrine information to be carried back to the brain (Bergland and Page, 1979). Tanycytes lining internal plexus vessels within the median eminence may support an additional route capable of transporting hormones into the ventricular system. Retrograde axonal flow from the neurohypophysis could create yet another exit. High concentrations of biologically active material and immunologically active pituitary hormones in portal blood provide additional evidence for the possibility of retrograde flow (see Pasqualini and Kincl, 1985).

2.5.2.1 Males

In males testosterone, DHT and estradiol suppress LH secretion (Drouin and Labrie, 1976; Denef et al., 1980). The negative feedback operates both by altering LH pulse characteristics (Section 2.2) and by modulating the responsiveness of pituitary cells to GnRH (Section 2.4).

Endogenous opioid peptides modulate the sensitivity of the hypothalamus to gonadal hormones (Gabriel et al., 1986). Prolactin inhibits the effectiveness of testosterone to inhibit LH and FSH (Bartke et al., 1984).

2.5.2.2 Females

The sensitivity to the feedback varies during the cycle. In rats LH pulse amplitude and frequency increase on day 1(D1), or 2, of diestrus in response to ovariectomy. This demonstrates that estrogens exert a negative feedback on the LH release during this period (Gallo, 1981). In ovariectomized animals estradiol has no effect on LH pulses if castration is carried out within a 24-h interval between estrus and D1 when the concentration of estradiol and of progesterone is low (Gallo and Bona-Gallo, 1985). During estrus estradiol contributes to the induction of high LH frequency pulses (Karsch *et al.*, 1984). Others have described that *in vitro* testosterone inhibits but estradiol stimulates LH secretion from pituitary cells (Tang, 1980; Kamel and Krey, 1982; Kamel *et al.*, 1987b). A positive effect of estradiol on LH secretion has also been described for pituitary cells obtained from cows (Padmanabhan *et al.*, 1978) and sheep (Huang and Miller, 1980; Miller and Huang, 1985). These results show that LH pulse characteristics are influenced by the prevailing steroid milieu.

2.5.3 Synergism

Synergistic action (augmentation) between various hormones operates for such diverse functions as the formation of receptor molecules, induction of ovulation, ova transport and nidation, embryonic growth, and mating behavior. Synergistic action also controls some steps during steroid biosynthesis.

2.5.3.1 Gonadotropin Releasing Hormone

The action of GnRH is facilitated by gonadal hormones, by interaction with neuropeptide Y (Crowley *et al.*, 1987), and by opiate-serotonin synergism (Lenahan *et al.*, 1987). Steroid hormones act either:

(i) on the hypothalamus by modifying GnRH pulse characteristics, or
(ii) at the pituitary level by modulating pituitary responsiveness to GnRH.

In seasonally ovulating species (ewe) small doses of estradiol (large doses are inhibitory) modulate the performance of GnRH neurons leading to increased LH pulse frequency (Karsch *et al.*, 1984). Progesterone decreases the sensitivity of gonadotrophs to GnRH (Batra and Miller, 1985).

Hamsters exhibit two FSH surges during a cycle: a proestrus surge and a second one during estrus (Bast and Greenwald, 1974). The estrus surge is a neural event. It can be extinguished by transection of the neural connections between the rostral hypothalamus and the arcuate-median eminence or blocked by phenobarbital. Phenobarbital block can be overridden by FSH infusion into the third ventricle (Coutifaris and Chappel, 1982). Intraventricular injection elicits a selective increase in serum FSH by an increase in hypothalamic GnRH content. The results indicate that in the hamster, during proestrus, FSH stimulates the neural mechanism necessary for GnRH release (Coutifaris and Chappel, 1983).

2.5.3.2 Pituitary Hormones

Luteinizing hormone, possibly in synergism with estrogens, has been credited with influencing the secretion of relaxin by granulosa cells from preovulatory follicles (Loeken *et al.*, 1983). Progesterone interacts with estradiol in facilitating increases in both frequency and amplitude of LH (Miyake *et al.*, 1972) and FSH (Drouin and Labrie, 1981) in rats and in monkeys (Teresawa, 1987) regardless of the estradiol environment. Stimulatory action of progesterone may also involve a synergistic action with prostaglandin E_2 (Ramirez *et al.*, 1985).

The stimulatory action of estradiol is in part the result of increased sensitivity of gonadotrophs to GnRH noted *in vitro* by Drouin *et al.* (1976). Increased sensitivity of the pituitary prior to ovulation has been observed in women (Jaffe and Keye, 1974; Nillius and Wide, 1972) and in rats (Arimura and Schally, 1971; Gordon and Reichlin, 1974; Ferland *et al.*, 1975).

During the early part of the estrus cycle in rats, progesterone multiplies the effect of LH and supports the growth of small antral follicles to the preovulatory stage (Richards and Bogovich, 1982).

A direct effect of some hormones is postulated from *in vitro* studies. Pituitary cell cultures obtained from rats (Drouin *et al.*, 1976; Kamel *et al.*, 1987), cows (Padmanabhan *et al.*, 1978) and sheep (Huang and Miller, 1980) are sensitive to testosterone, which inhibits LH secretion in the presence of sufficient amounts of GnRH.

Prolactin interacts with several hormones. PRL synthesis is stimulated by an increased concentration of estradiol (Charlton *et al.*, 1983b). The hormone interacts with LH to stimulate 5α-reductase activity in the testes.

2.5.3.3 Gonadal Hormones

Androgens
Synthetic progestational agents potentiate the action of testosterone stimulating to greater activity β-glucuronidase in mouse kidney (Mowszowicz *et al.*, 1974), submaxillary gland epidermal growth factor (Bullock *et al.*, 1975) and alcohol dehydrogenase (Gupta *et al.*, 1978).

Estradiol-Progesterone
The synergism between the two hormones is manifested in many physiological functions (see Kincl, 1971b).

Ovulation: Administration of small doses of estrogens to immature female rats advances the onset of puberty and causes a rise in LH plasma levels. McCormack and Meyer (1963) treated 22-day-old females with suboptimal doses of pregnant's mare serum gonadotropin (PMSG) to induce follicle maturation and found that ovulation was facilitated by small amounts of progesterone while larger amounts blocked the response. Döcke and Dörner (1965) induced ovulation in immature rats by injecting estradiol benzoate (1 µg). When progesterone (0.1–5 mg) was added 3 days later the percentage of females that ovulated increased threefold. Small amounts of progesterone synergise with estradiol to facilitate ovulation in rabbits (Sawyer, 1952) while larger doses of progesterone are inhibitory. In sheep both estradiol and progesterone are involved in triggering the LH surge (Hauger *et al.*, 1977). Individually estradiol and progesterone reduce LH pulse frequency; acting together both hormones restrain any changes in LH pulse amplitude (Leipheimer *et al.*, 1986).

Ova Transport: The transport of fertilized eggs through the oviduct depends on the relative amounts of estrogen and progesterone. In ovariectomized animals ova movement is slow. Estrogen dominance (induced by superovulation) may accelerate the movement in domestic animals and mice. Progesterone dominance decreases the rate in rabbits (see Rudel and Kincl, 1966). Quantitative changes in the movement in response to changing hormonal milieu in rats were described by Rudel and Kincl (1966).

Implantation: The process of implantation is controlled by estrogen-progesterone synergism. If the ovaries are removed from mated rats during the time the zygotes are travelling through the oviduct, implantation will not take place, and the zygote will degenerate. If progesterone is injected at the time of ovariectomy, the blastocyst will remain viable for as long as progesterone is available but will not implant (delayed implantation). The process occurs naturally in many species. The dose of progesterone needed for delayed implantation is between 0.5 and 4 mg per day in rats ovariectomized on the second day after fertilization. Injection of estrone (0.3 µg) will induce the blastocyst to grow and implant. When the estrone dose is increased to 10 µg, nidation is prevented (see Psychoyos, 1973; 1974).

Progesterone (4 mg daily) will maintain pregnancy in ovariectomized (day 14 of gestation) rats, but many fetuses will bear hematomata on their extremities. The addition of estradiol (0.2 µg, two injections) will prevent the injury (Tamada and Ichikawa, 1980).

Some rodent species are less sensitive to the synergistic influence of the two hormones: In hamsters implantation will take place with progesterone alone, but addition of estrone will increase the number of surviving implants. In guinea pigs, ovarian or exogenous progesterone is not required for implantation but is needed for proper embryonic growth and development (see Psychoyos 1973; 1974).

Effects on the Uterus: There is considerable evidence that both hormones are needed to stimulate uterine growth; the cell size, cell number and nucleic acid synthesis reach an optimal state if both hormones act synergistically. High amounts of estrogen inhibit uterine growth and endometrial proliferation (see Kincl,1971b). Edgren *et al.* (1961) used 10 mg of progesterone in addition to 1 µg of estriol in spayed rats. The same dose potentiated also the action of estrone.

Pregnancy: During pregnancy both estrogens and progesterone appear in the blood in ever increasing amounts, and appropriate concentrations are essential for embryonic growth and development. As term nears, the relative amounts of both hormones change (see Pasqualini and Kincl, 1985).

In cows, goats, and rats both hormones are necessary for normal mammary development. Synergism with other hormones (adrenal steroids, thyroid hormones, neurohormone of the posterior pituitary) may be needed in other species to initiate and maintain lactation (see Kincl, 1971b).

2.5.3.4 Other Hormones and Factors

During the synthesis of prostaglandins estradiol synergizes with arachidonic acid to enhance PG production (Schatz *et al.*, 1987). Androgens and FSH synergize to induce the synthesis of androgen binding protein in Sertoli cells (Fritz *et al.*, 1979).

2.6 Hormone Receptors

Hormones are carried by blood, and all cells are exposed to them, but only the "target" organs respond. The selective receptiveness resides in the ability of cells to recognize hormones by means of high affinity binding proteins, termed "receptors". Receptor molecules are proteins, possibly lipo- or glycoproteins. Those that bind high molecular weight hormones (polypeptides and glycoproteins) are found on cell surfaces. To express its physiological activity a hormone bound to a cell surface stimulates a secondary messenger, such as cAMP. Receptor molecules (HR) for low molecular weight hormones (steroids) are found within the cytoplasm and the nucleus. Cell functions thought to be regulated by steroid HR include synthesis of specific polypeptides and proteins (GnRH, gonadotropins, PRL, ovalbumin) and enzymes important for cell proliferation (kinases, DNA polymerase, ornithine decarboxylase, etc.).

Low molecular weight hormones are thought to diffuse across cell membranes and bind to cytosol receptors; the complex is translocated to the nucleus, where a series of molecular events results in the formation of a protein(s) which brings on the biological response. The hypothesis rests on several assumptions:

(i)   diffusion across the cell membrane is accomplished with little resistance;
(ii) association of the hormone with cytosol receptors alters the nature of the protein, rendering the complex nucleotropic;
(iii) the response of the cell is the result of protein(s) production which was either absent or present in different measure.

A cell responds to hormonal stimulation by translocating cytosol HR complex to the nucleus. This means that cytosol receptors become depleted and must be replenished. The dynamics of the process (uterine estradiol receptors) was studied by Jensen and coworkers (Jensen and Jacobson, 1962; Jensen *et al.*, 1982) and groups led by Gorski (Shyamala and Gorski, 1969; Sarff and Gorski, 1971). The binding and translocation to the nucleus are complete within an hour; by this time cytosol-R reach a nadir and nuclear-R are at a peak. The *de novo* synthesis of cytosol receptors is estradiol dependent, depletion is needed to initiate replenishment, and the pituitary seems to elaborate an inhibitor to renewal (see Walters, 1985).

Ligand and receptor binding data are routinely analyzed by constructing Scatchard graphs. It has been questioned whether this analysis yields acceptable data relating to the number of receptor sites (Klotz, 1982; Light, 1984). Thus, most numbers purportedly concerning this aspect of receptor research may represent only approximations.

While undoubtedly intracellular HR play an important role in the stimulation of cell proliferation, some experimental evidence contests the role of HR as direct, obligatory mediators of cell-type specific protein synthesis and cell growth. The dissention is based on the activity of estradiol in castrated female rats and rests on the following observations:

(i) cells which contain significant concentrations of estrogen receptors (ER) do not proliferate when exposed to estradiol (Strobl and Thompson, 1985);

(ii) organs not considered to be estrogen target tissue contain cell populations with ER (Jensen and Desombre, 1972; Strobl and Thompson, 1985). In this alternate view the control of hormone (estrogen) action depends upon estrogens interacting with cell membrane inhibitors of proliferation (colyones). The interaction then permits the cell to grow (see Soto and Sonnenschein, 1987).

2.6.1 Nature of Receptor Molecules

The steroid hormone receptor activity is a dynamic process understood only incompletely. The main difficulty in unraveling the process resides in the nature of the problem: trying to adapt *in vitro* data to an *in vivo* process.

Not all proteins that bind steroids are receptors. A number of proteins, capable of binding the hormones with high affinity, low capacity in a manner similar to that seen in receptors, are present in blood. In addition there are proteins (albumins) which display a low affinity for many steroids. These steroid-protein complexes exist in equilibrium with free steroids and provide a reservoir readily available to prevent extreme fluctuations of free steroid concentration. This assures a constant supply for cell uptake at any given time (Westphal, 1971). Receptor proteins must meet several criteria.

Finite Binding Capacity
If hormone-receptor (HR) formation is required for biological response then the quantity of HR (binding sites) should be limited.

High Affinity
The concentration of hormones is 10^{-10} to 10^{-8} M. If a tissue is to respond to a hormone via the receptor mechanism then the receptor must react to the hormone in a range which corresponds to blood concentrations. Otherwise, no response would take place. The affinity is expressed either as an association (affinity) constant (K_a) or the dissociation constant (K_d), calculated from Scatchard graphs.

Tissue Specificity
Only those tissues which respond to the hormonal stimulus (target tissue) should react without interference from other signals. Hormones of the same class should compete for the same receptors while not affecting others. Table 2.18 illustrates tissue specificity of various tissues to estrogen uptake. In a target tissue not all the cells react. In neuroendocrine tissues only between ¼ to ½ may bind the hormone (Table 2.19).

Table 2.18 Estrogen Receptor Content in Target and Nontarget Tissue.

Tissue	ER content[a]
Uterus	5.92
Vagina	2.15
Pituitary	1.43
Kidney	0.20
Diaphragm	0.06
Spleen	0.02

[a] ER = estrogen receptor content, pico-moles per 100 mg tissue; data from Clark *et al.* (1980).

Table 2.19 Estrogen Binding at Peak Estrogen Concentration in Hypothalamic Nuclei.

Area	Number of cells	Percent concentration[a]
Preoptic	6626	23
Ventromedial n	4481	37
Arcuate n	3148	28
Amygdala	4885	48

[a] Percent of estrogen concentrating cells; n nucleus; data from Pfaff (1983).

Table 2.20 Specificity of Cytosol Progesterone Receptor[a] to Various Steroid Hormones.

Steroid	Relative binding affinity (%)
Progesterone	100
5α-Pregnane-3,20-dione	50
Cortisol	0
Estradiol	1
Testosterone	8

[a] Receptor isolated from chick oviduct; data from Clark *et al.* (1976).

Specificity of response is illustrated in Table 2.20. Compounds which share similarity of structure (progesterone, 5α-pregnane-3,20-dione) compete for binding sites whereas other steroids (testosterone, estradiol and cortisol) do not.

Nature of the Receptor

Most receptor molecules are a single polypeptide chain composed of two subunits sometimes of similar length. Each subunit binds one molecule of hormone. The molecular weight of the dimer may be 220 to 250 kDa. During *in vitro* studies changes in temperature, pH, or ionic strength will alter the receptors' activity. The effects are prevented by prior binding of a respective hormone.

The subunits may differ in their ability to bind to nuclear material. Progesterone receptor isolated from chick oviduct separates readily into two subunits designated A and B. The A subunit (mol. wt. 110 kDa) binds only to free DNA while

Table 2.21 Molecular Weight of Cytosol and Nuclear Receptors in the Neuroendocrine Tissues and the Pituitary Gland.

Hormone receptor	Estimated sedimentation coefficient(s)	
	Cytosol	Nuclear
Androgen	~7	3–4
Estrogen	~8	4–5
Progesterone	6–9	5

Data from McEwen *et al.* (1982).

the subunit B (mol. wt. 117 kDa) only to chromatin. The intact dimer binds readily to nuclear material, and both subunits are present. The number of receptor molecules has been estimated to be about 20,000 per cell (Clark *et al.*, 1976).

Cytosol receptors are higher molecular entities than nuclear receptors. Table 2.21 lists sedimentation coefficients (S) found for receptors in neuroendocrine tissue. The nature of the receptor molecule is believed to be the same regardless of the tissue of origin.

Receptor Transformation
Cytoplasmic HR are unable to be stored in nuclei or bind to DNA unless they develop a structural modification. The event is believed to be associated with changes in quaternary structure and possibly molecular weight. Walters (1985) summarized the nature of the changes.

2.6.1.1 Receptor Interaction

Receptor molecules are hormone specific, but situations exist in which a hormone influences a function of a different HR. Androgens retard the action of estrogens (7 c), are antiestrogenic, and progesterone is antiandrogenic. Other hormones stimulate unrelated functions (Section 2.4.5).

Androgens stimulate the growth of the uterus. The stimulation takes place because the androgen-estrogen receptor complex induces estrogen-dependent protein synthesis (Ruh *et al.*, 1974; Garcia and Rochefort, 1977), nuclear translocation of the estrogen-receptor complex is influenced by androgens (Ruh and Ruh, 1975), and androgens (DHT) bind to estrogen receptors (Rochefort and Garcia, 1976).

Hormone-hormone interactions which induce receptor formation are discussed below.

2.6.1.2 Sex Differences

Sex dimorphism is present in respect to HR in sex organs. In others, differences may exist. Estrogen receptor concentration in the thymus is higher in female mice; females are more immunocompetent. Orchidectomy raises estrogen receptors in males to female levels (Gillette and Gillette, 1979).

There are some sex related differences in the neuroendocrine tissues. In male and female rats the concentration and properties of estrogen receptors are the same in the hypothalamus and pituitary (Korach and Muldoon, 1973; 1974a,b). The number of serotonin receptors in females fluctuates during the cycle; it increases in ovariectomized females in response to estrogens (Fischette et al., 1983). The development of pituitary dopamine (DA) receptors is dependent upon the presence of gonads in both male and female rats (Ojeda et al., 1977b). Gonadal secretion appears to be needed to confer either a male (Watanobe and Takebe, 1987a) or female (Watanobe and Takebe, 1987b) formation of pituitary DA receptors in the adult.

2.6.1.3 Receptors in Neonatal and Aged Animals

Steroid HR begin to appear during fetal life. Indeed, one study states that an implanted blastocyst influences the formation of estradiol and progesterone cytosol and nuclear receptors in the endometrium (Logeat et al., 1980).

Androgenic Receptors (AR)
AR are present in the neuroendocrine tissues of neonatal female rats (Kato, 1976). Testosterone and DHT receptors in the urogenital tract of rabbits are not different from AR receptors isolated from the fetus (George and Noble, 1984).

Estrogen Receptors (ER)
ER develop in target organs early in life. Estradiol binds to cells in the hypothalamus after birth, and the uptake increases six-fold between days 20–26 (Plapinger and McEwen, 1973). MacLusky et al. (1976), Westley et al. (1976) and Vito and Fox (1979) all demonstrated the presence of ER in the neonatal brain. Pasqualini et al. (1976) reported the presence of ER in the brain and uterus of fetal guinea pigs and purified the complex (Pasqualini and Cosquer-Clavreul, 1978). The nucleus arcuatus and the preoptic area are the main sites of ER localization (Tardy and Pasqualini, 1983). The same group also reported on the occurrence of ER in the thymus of fetal guinea pigs (Screpanti et al., 1982).

The estrus cycle transforms in aged rats. Females 12–15 months old are usually in constant estrus; later (24–30 months) the estrus is replaced by diestrus interspersed with intermittent periods of estrus. The changes are accompanied by decreased ER in the brain (Kanungo et al., 1975) and in the uterus (Hsueh et al., 1979).

Progesterone Receptors (PR)
PR are present in the fetal uterus and ovary of guinea pigs (Pasqualini and Nguyen, 1980) and cerebral cortex of neonatal female rats (Kato and Onouchi, 1981). The amount of PR declines in aged hamsters (Blake and Leavitt, 1978).

2.6.2 Gonadotropin Releasing Hormone

Gonadotropin releasing hormone autoregulates (up-regulation) the number of its own receptors in the pituitary. In females the number increases after the administration of exogenous GnRH in a dose related manner (Bourne *et al.*, 1980; 1982) or when the concentration of GnRH in portal blood is high (Sherwood and Fink, 1980). Autoregulation has been also reported in hamsters (Adams and Spies, 1981), sheep (Wagner *et al.*, 1979), and primates (Adams *et al.*, 1981). In mice, unlike in rats, pituitary GnRH receptor content decreases after castration indicating either persistent receptor occupancy by endogenous GnRH secretion, GnRH induced receptor loss, or a species difference to the removal of negative steroid feedback (Naik *et al.*, 1984a). The number of receptors is not affected by GH, FSH or PRL, whereas LH and estradiol reduce the posthypophysectomy rise in receptor concentration (Bourne and Marshall, 1984). Increased activity in receptor synthesis results from enhanced transcription of RNA; actinomycin D will block increases in receptor formation (Schally *et al.*, 1969).

The regulatory function is established early in life; in rats it begins from about day 10 of life (Duncan *et al.*, 1983).

After castration receptors for both GnRH and gonadotropin increase, and the rise can be prevented by exogenous gonadal steroids (Clayton and Catt, 1981; Frager *et al.*, 1981; Marian *et al.*, 1981) and by lesion in the medial basal hypothalamus (Clayton *et al.*, 1982; Pieper *et al.*, 1982). Testosterone, or DHT, causes a reduction in the number of pituitary GnRH-binding sites (Giguere *et al.*, 1981; Zolman, 1983).

Prolactin contributes to the regulatory action. During hyperprolactinemia gonadotropin and gonadal function are suppressed. Gonadotropin response to castration is abolished by high PRL concentrations in the blood (Grandison *et al.*, 1977), and the number of pituitary receptors for GnRH is diminished by a transient elevation of PRL (Garcia *et al.*, 1985). Inhibin appears to decrease the number of GnRH receptor sites on pituitary cell surfaces (Wang *et al.*, 1988).

Gonadotropin releasing hormone binds to cells in the gonads and directly inhibits the secretion of steroid hormones. After hypophysectomy decreased secretion of gonadal hormones is restored by replacement therapy with LH (Section 2.5). In males LH inhibits the number of GnRH testicular receptors (Bourne and Marshall, 1984). Since GnRH inhibits directly androgen synthesis, the LH inhibitory action may be one of the mechanisms by which gonadotropin enhances androgen production.

2.6.3 Pituitary Hormones

Glycoprotein hormones of the pituitary bind to cell surfaces of the gonads as the first step in the expression of biological activity. Hypophysectomy usually results in a decrease of receptor sites suggesting an antagonistic-facilitative effect between pituitary gonadotropins and steroid hormones of the gonads.

2.6.3.1 Follicle Stimulating Hormone

The activity of Sertoli cells depends not only on the amount of FSH circulating in the blood but also on the number of FSH receptor sites on the cell surface. An increase in the number of receptors translates into higher biochemical activity of the cells (Welsh and Wiebe, 1978; Salhanick and Wiebe, 1980; Van Sickle *et al.*, 1981). In rodents (rats, mice) the number of binding sites decreases if the testes are exposed to high doses of FSH; hypophysectomy decreases the number of sites in rats, but the reverse happens in mice (O'Shaughnessy and Brown, 1978; Francis *et al.*, 1981; Tsutsui *et al.*, 1985). The binding of FSH to Sertoli cells (Desjardins *et al.*, 1974; Steinberger *et al.*, 1974; Ketelslegers *et al.*, 1978; Barenton *et al.*, 1983) is followed by the synthesis of androgen binding protein (Sanborn *et al.*, 1975; Means and Tindal, 1975) and estradiol biosynthesis (Armstrong *et al.*, 1975; Dorrington *et al.*, 1976).

2.6.3.2 Luteinizing Hormone

In males the stimulatory effect of LH is mediated by a specific receptor located on the plasma membrane of Leydig cells. As is the case with FSH receptors the numbers increase during sexual maturation. The increase is modulated by a stimulatory action of FSH in rats (Chen *et al.*, 1976; Odell and Swerdloff, 1976; Ketelslegers *et al.*, 1978) and mice (Mori *et al.*, 1985).

The hormone autoregulates itself. LH (or hCG) induce loss of testicular LH receptors (Hsueh *et al.*, 1976; Sharpe and Fraser, 1980; Purvis *et al.*, 1977; Sharpe and McNeilly, 1978).

The number of LH receptors in cultured granulosa cells increases by treatment with FSH (Erickson *et al.*, 1979; Rani *et al.*, 1981).

Females

In females both FSH and LH are needed during follicular maturation for the completion of steroidogenesis. The theca cells, stimulated by LH, produce androgens which are utilized by the granulosa cells, stimulated by FSH, as precursors in the aromatization process (Section 2.4.3.3). During the process the number of FSH receptors remains constant, while the number of LH receptors augments during the late stages of follicular maturation (Uilenbroek and Richards, 1979; Uilenbroek and van der Linden, 1983). LH receptor formation is induced by FSH and augmented by estrogens (Zeleznik *et al.*, 1974). The FSH induction is potentiated by estrogens (Jia *et al.*, 1985) and is inhibited by 5α-dihydrotestosterone (Farookhi, 1980) and two other nonaromatizable steroids, 5α-androstane-$3\beta,17\beta$-diol and 17α-methyl-17β-hydroxy-4,9,11-estratrien-3-one (Jia *et al.*, 1985).

2.6.4 Androgen Receptors

Radioactivity after injection of labeled testosterone is found in the neural and genital tissue of rats (Roy and Laumas, 1969), both in the cytosol and the nucleus (Sar and Stumpf, 1973). Rezek (1975) described the uptake of labeled testosterone in various tissues. Receptors for testosterone and DHT are found in the cytosol (Samperez *et al.*, 1974; Clark and Nowell, 1979) and the nucleus (Lieberburg *et al.*, 1977) in the hypothalamus and the pituitary, androgen sensitive tissues, seminiferous tubules (Mulder *et al.*, 1975), Sertoli cells (Tindall *et al.*, 1977) and also in granulosa cells (Schreiber *et al.*, 1976). The nature of the receptors is similar in fetal and adult rabbits (George and Noble, 1984). The complementary DNA for human androgen receptor has been cloned (Lubahn *et al.*, 1988). Androgen receptors interact with estrogen receptors (Rochefort and Garcia, 1976; Garcia and Rochefort, 1977).

2.6.4.1 Regulation of Formation

Receptor formation is androgen sensitive. The number decreases after orchidectomy in male rats within days (Baulieu and Jung, 1970; Mainwaring and Mangan, 1973). In Sertoli cells both FSH and testosterone are needed to stimulate the formation of AR (Verhoven and Cailleau, 1988). Receptor formation is increased by prolactin (Prins, 1987). Progesterone suppresses the formation of AR in males (Connolly *et al.*, 1988).

2.6.5 Estrogen Receptors

In 1956 Sulman and in 1959 Glascock and Hoekstra implied the existence of estrogen binding sites in the uterus of rats. A group led by Jensen (Jensen and Jacobson, 1962; see also Jensen *et al.*, 1982) provided definitive proof that estradiol binds to proteins in the uterus. ER are present in the target (vagina, uterus) tissues of rodents (Payne and Katzenellenbogen, 1980; Cunha *et al.*, 1983), in the urinary tract of both sexes in carnivores (Schulze and Barrack, 1987) and humans (Wiegerinck *et al.*, 1980), in neuroendocrine tissues of adults of both sexes (Vreeburg *et al.*, 1975) and in fetal and neonatal tissue (Nguyen *et al.*, 1986).

Cells in the hypothalamus and the pituitary display also an ability to bind catechol estrogens (Davies *et al.*, 1975). The activity indicates a possibility of interaction between the two estrogens. When both testosterone and estradiol are injected into adult male rats the uptake and the retention of estradiol in the hypothalamus and the pituitary are significantly higher than that of testosterone. The difference indicates the importance of estrogens to the hypothalamus-pituitary-gonadal axis (Henderson *et al.*, 1979).

The distribution of cells with ER receptors in the neuroendocrine tissue is not uniform. Only between 20%–50% of the nuclei possesses the ability to capture estrogens. The nature and function of the remainder of the cells is not understood (Table 2.21).

2.6.5.1 Structure

ER isolated from calf uterus are single polypeptides with a molecular weight of 70 kDa, which sediments at 8S (Pucca *et al.*, 1980). Receptor molecules isolated from rat uteri may undergo dimer formation during translocation, but this has not been clarified (see Walters, 1982). The structure of the receptor molecule isolated from human endometrium was described by Notides *et al.* (1976).

2.6.5.2 Specificity of Estrogen Receptors

The presence of free phenolic hydroxyl (C-3) and 17β-hydroxyl function on C-17 are essential for highest affinity binding. The presence or absence of an angular methyl group on C-13 has no influence on the affinity. Additional oxygen substitution on ring D, on ring A, and unsaturation on ring B decrease steroid affinity for the receptor (Hähnel *et al.*, 1973; Raynaud *et al.*, 1985). These findings agree closely with effects of substitution on biological activity. The distance between C-3 and C-17 and spatial configuration are of importance. l-Estradiol, the optical isomer which has the opposite configuration of asymmetric centers, binds with low affinity to a cytosol receptor isolated from rabbit uterus (Chernayaev *et al.*, 1975).

Androgens (Eisenfeld and Axelrod, 1967; Rochefort and Garcia, 1976; Garcia and Rochefort, 1977) interfere with estrogen binding in the CNS *in vivo*, most likely due to *in situ* aromatization (Ogren *et al.*, 1976). Nonsteroidal estrogen (clomiphene) competes with estradiol for binding sites in the uterus (Roy *et al.*, 1964) and the brain (Kato *et al.*, 1968). A CNS active drug, phenothiazine, also competes with estradiol for binding sites (Shani *et al.*, 1971).

2.6.5.3 Regulation of Formation

Estradiol provides the main regulation of receptor synthesis, reutilization and recycling (see Muldoon, 1980).

Progesterone inhibits estradiol receptor binding in the pituitary but not in the hypothalamus. The transfer of cytosol ER to the nucleus is influenced by estradiol and estrone. Androgens influence ER transfer to the nucleus, possibly by modifying nuclear membrane permeability (see Walters, 1985). Cortisol and progesterone have no effect on the transfer (Rochefort *et al.*, 1972).

2.6.6 Progesterone Receptors

Progesterone receptors have been detected in the cytosol and nuclei in rats (Davies and Ryan, 1972), mice (Feil *et al.*, 1972), guinea pigs (Kontula *et al.*, 1972), rabbits (McGuire and Bariso, 1972) and humans (MacLaughlin and Richardson, 1976). Chick oviduct PR have been studied in great detail (see O'Malley and Means, 1976). Nuclear receptor numbers rise between the noon of diestrus and the noon of proestrus in rats and guinea pigs (Wilcox and Feder, 1983).

2.6.6.1 Structure

The kinetics of PR in human tissue was described by Jänne *et al.* (1976). Progesterone receptors, isolated from endometrial carcinoma grown in nude mice but similar to PR from normal human uterus, are two proteins, 116 kDa and 81 kDa. The 116 kDa protein forms triplet or doublet isoforms depending upon the stage of the menstrual cycle. The formation of the protein is induced by a synthetic progestational agent, medroxyprogesterone acetate (Feil *et al.*, 1988). The nature of rat uterine PR was reported by Ilenchuk and Walters (1987).

2.6.6.2 Regulation of Formation

The formation of PR in uterine tissue (Toft and O'Malley, 1972; Faber *et al.*, 1972) and in the pituitary and hypothalamus of immature, castrated rats is induced by estradiol (Kato and Onouchi, 1977; MacClusky and McEwen, 1978; 1980). The formation peaks 12 h after estradiol treatment then falls to a plateau which is maintained for an additional 20 h. If estradiol is given during this period, progesterone receptors are again formed in 12 h. During the peak formation progesterone induces the development of its own receptors within 1–2 h (Calderon *et al.*, 1987).

2.7 Interactions with Other Hormonal Systems

Normal development of reproductive organs depends, in addition to the influences of the hypothalamic-pituitary axis, also on the proper functioning of other endocrine systems and nutrition (Leathem, 1961). The influences of adrenals and thyroid hormones on reproduction were recognized more than 50 years ago. Yet, despite this long time, many questions still remain unanswered. One of the problems in separating the effects of these two endocrine glands lies in the multiple effects both exert upon general body functions. The thyroid affects the general metabolism and the adrenals, both protein and vitamin metabolism (Villee, 1961).

2.7.1 Adrenal Function

In 1939 Selye observed that reproductive function in rodents decreased in response to stress. Others confirmed the observation (Collu *et al.*, 1979; Tache *et al.*, 1980; Welsh and Johnson, 1981); stress also lowers reproductive ability in humans (Baker *et al.*, 1981). Reproductive function is inhibited by an increase in the secretion of ACTH in response to stress and by an direct inhibitory effect of glucocorticoids on gonadotropin secretion. Kamel and Kubajal (1987) described direct inhibition of LH secretion by corticosterone *in vitro*.

The effects are less pronounced in males than in females. The influence of corticotropin releasing factor on reproduction is covered in Section 2.3.1.4.

2.7.1.1 Males

In males adrenalectomy has no effect on the histological appearance of the testis, and there is no change in the structure and function of the accessory sex organs (Albert, 1961), even in animals dying of adrenal insufficiency (Gaunt and Parkins, 1933).

Chronic stress (Krulich *et al.*, 1974; Du Ruisseau *et al.*, 1978; Gray *et al.*, 1978) or chronic ACTH infusion (Baker *et al.*, 1950; Asling *et al.*, 1951; Saez *et al.*, 1977; Mann *et al.*, 1985; Vreeburg *et al.*, 1984) will result in a decreased secretion of LH and of testosterone (Tache *et al.*, 1980; Charpenet *et al* 1981; Ringstrom and Schwartz, 1985). In rats the decrease in the LH release is mediated by an increase in the discharge of the main adrenocortical hormone, corticosterone; it is not mediated by the action on the CNS since central nervous depressants do not prevent the effect (Charpenet *et al.*, 1982). Diminished testosterone production results from a decreased sensitivity of the testis to LH and reduced testosterone response to GnRH, mediated by corticosterone (Tibolt and Childs, 1985). An increase in the sensitivity of the hypothalamus to the negative feedback of testosterone appears to be excluded (Mann *et al.*, 1987).

2.7.1.2 Females

In adrenalectomized female rats the ovaries atrophy, and the females exhibit an irregular cycle (Chester and Jones, 1957; Zanisi *et al.*, 1983). The effect can be prevented by administration of cortisol (Davis and Plotz, 1954; Mandl, 1954). In intact animals both ACTH (Hagino *et al.*, 1969; Ogle, 1977) and glucocorticoid (Baldwin and Sawyer, 1974; Padmanabhan *et al.*, 1983) inhibit ovulation, an action mediated by a decrease in LH release. The effect of adrenal steroids on FSH secretion is more variable. Pituitary cells in culture respond to corticosterone by a small decrease in LH production and an increase in FSH synthesis (Suter and Schwartz, 1985).

In the human female suffering from the adrenocortical syndrome deficiency (decreased cortisol production) the ovaries are infantile, and menstrual cycles are absent. The adrenals of these patients produce, however, large quantities of androgens which block the normal production of gonadotropins.

2.7.2 Thyroid

Thyroid and gonadal hormones interact with each other closely. In women the thyroid enlarges at puberty, at menstruation and during pregnancy. Estrogens cause increased accumulation of iodine and its incorporation into thyroid hormones (T_4). The gonads of male and female offspring of cretin rats are subnormal. The testes may contain only a few spermatocytes and no spermatozoa; Leydig cells produce few or no androgens.

2.7.2.1 Hypothyroidism

The reproductive system of the adult is less affected by a decrease in thyroid function than that of immature males (Chandrasekhar et al., 1985a). Hypothyroid male rats have smaller testes and accessory sex organs, and spermatogenesis is decreased (Smelser, 1939), but the males are capable of siring litters (Jones et al., 1949). Young rats made hypothyroid at birth may show a delay in sexual maturation (Scow and Marks, 1945; Scow and Simpson, 1945) or may have a normal reproductive history (Goddard, 1948). In adult hypothyroid rats the response of the testes and accessory sex organs to exogenous LH and testosterone is normal (Kalland et al., 1978). In growing mice decreased thyroid function retards sexual maturation (Maqsood and Reineke, 1950).

In adult thyroidectomized rams sperm maturation but not spermatogenesis is inhibited (Berliner and Warbritton, 1937), and testosterone concentrations in plasma are low (Chandrasekhar et al., 1986). Decreased thyroid hormone production also affects spermatogenesis in rabbits (Maqsood, 1955).

In the human male myxedema is associated with disturbed androgen metabolism (Gordon et al., 1969). The endogenous production is low, and abnormal amounts of testosterone are converted to androsterone (Hellman et al., 1959).

In the females of many species hypothyroidism decreases fecundity. Corpora lutea are absent, and ovarian lipid and cholesterol concentrations are low (Leathem, 1961).

Feeding an antithyroid drug (thiouracil) will decrease litter size in guinea pigs (Hoar et al., 1957) and rats (Leathem, 1961). Long term treatment (8 months) with antithyroid drugs of rats results in the formation of cystic ovaries (Janes, 1944; Leathem 1961). The ovaries of myxedema patients are cystic.

2.7.2.2 Hyperthyroidism

An excess of thyroid hormones increases appetite and absorption of food from the intestinal tract, results in the loss of body fat, increased protein metabolism, and, if not checked, eventually in death.

Testes are degenerated in hyperthyroid rats, testosterone production is decreased, and accessory sex organs are atrophied (Smelser, 1939). The number and volume of Leydig cells are decreased (Lenzi and Marino, 1947).

In the male thyroxine augments the effect of small doses of testosterone. Amounts not active per se will produce atrophy in rats (Masson and Romanchuck, 1945). Large doses of testosterone will prevent the damage caused by thyroxine in rats (Roy et al., 1955) and in guinea pigs (Richter, 1944; Richter and Winter, 1947).

Excessive thyroid function in adolescent female rats prevents ovarian maturation and in the adult causes atrophy of the ovaries and cessation of the estrus cycles (Drill, 1943).

2.8 Reproduction and the Immune Response

The function of the thymus has been linked to reproductive functions for many years, and yet the nature of the dependence remains elusive. The difficulty arises mainly from separating the influence of the thymus on immunological competence and control of the reproductive processes.

Two systems constitute the entire immune response. The nonspecific (innate) system comprises nonantigen dependent reactions, such as phagocytosis and inflammation. The acquired (specific) system involves the antigen-dependent reactions of lymphocytes (T-cells and B-cells): T-cells are regulators of cell-mediated immune responses (B-cell function) and phagocytosis; B-cells produce antibodies (immunoglobulins).

2.8.1 Sex Dimorphism

Some autoimmune diseases are more prevalent in women (rheumatoid arthritis, systemic lupus, Graves' disease, or autoimmune thyroiditis) than in men (see Dörner et al., 1980; Grossman, 1985). Wallotton and Forbes (1966), deMoraes-Ruehsen et al. (1972), Vazquez and Kenny (1973), and Sikorski et al. (1975) all suspect in women that ovarian failure is an autoimmune disorder.

In many rodent species females generate a higher titer of immunoglobulin and demonstrate a higher ratio of T to B-cells (see Grossman, 1985). Immunoglobulins fluctuate during the estrus cycle, and their formation may be estradiol dependent (Wira and Sandoe, 1980). Estradiol inhibits suppressor T-cell activity. The activity is translated into increased antibody production since suppressor T-cells restrain B-cells from manufacturing immunoglobulin. Androgens affect immune responses in males (Castro, 1974).

2.8.2 Thymus

The thymus fulfills a critical role during neonatal life in establishing immunological competence. Removal of the gland causes loss of resistance to infectious agents; the animals do not grow and remain stunted (see Metcalf, 1966). The incidence of the wasting disease can be as high as 75%. The consequence of thymectomy can be prevented, or reversed, by injecting a suspension of thymic cells or of thymosin (Asanuma et al., 1970).

Two polypeptide hormones were isolated from the gland. Thymosin, a heat stable glycoprotein, mol. wt. 12–15 kDa, which regulates lymphoid structure and function (Goldstein et al., 1972) and a smaller polypeptide (mol. wt. 3 kDa) by Bezssonoff and Comsa (1958); see also Bernardi and Comsa (1965). In rats thymosin stimulates the release of GnRH and LH (Rebar et al., 1981; Hall et al., 1982), corticosterone (McGillis et al., 1985), and prolactin and growth hormone (Spangelo et al., 1987). In prepubertal monkeys the hormone elevates concentrations of β-endorphin, ACTH and cortisol (Healy et al., 1983).

2.8.2.1 Effects of Thymectomy

Ablation of the thymus in female mice shortly after birth produces ovarian dysgenesis characterized by the absence of follicles and corpora lutea and hypertrophy of interstitial tissue (Nishizuka and Sakakura, 1971). The ovaries retain steroidogenic capability to utilize cholesterol, pregnenolone, progesterone, and 17-hydroxy progesterone but produce abnormally large quantities of androgens (Nishizuka et al., 1973).

In male rats neonatal thymectomy causes a transitory increase in concentrations of LH and testosterone while the concentration of ACTH decreases (Deschaux et al., 1979). See also Chapter 4.

2.8.2.2 Effect of Hormones

Androgens and estrogens induce regression of the thymus gland, influence and attenuate antibody formation, may facilitate the development of autoimmune diseases and exhibit antiinflammatory activity (see Mooradian et al., 1987). Immune responses during pregnancy, when estrogen concentrations are high, are depressed in most mammalian species (see Grossman, 1985). Stimson and Hunter (1976) characterized the immunosuppressive properties of estradiol.

Siiteri et al. (1977) repressed granuloma formation (a measure of antiinflammatory, or immune, response) by using sustained release progesterone implant. Kincl and Ciaccio (1980) maintained allograft skin viability with biodegradable progesterone preparation as long as progesterone was available. Similar preparations made with pregnenolone, estradiol, or cortisol were not effective.

Estradiol receptors are present in the cytosol of thymus cells. Corticosterone (and estradiol) compete for the binding sites (Reichman and Villee, 1978; Brodie et al., 1980). Su et al. (1988) suggested a link between endocrine, nervous and immune systems.

References

Adams TE, Spies HG (1981) GnRH-induced regulation of GnRH receptor concentration in the phenobarbital-blocked hamster. Biol Reprod 25: 298–302

Adams TE, Norman RL, Spies HG (1981) Gonadotrophin-releasing hormone receptor binding and pituitary responsiveness in estradiol-primed monkeys. Science 213: 1388–1390

Adashi EY, Resnick CE (1986) Antagonistic interactions of transforming growth factors in the regulation of granulosa cell differentiation. Endocrinology 119: 1879–1881

Adashi EY, Fabics C, Hsueh AJW (1982) Insulin augementation of testosterone production in a primary culture of rat testicular cells. Biol Reprod 26: 270–280

Adashi EY, Resnick CE, Brodie AMH, Svoboda ME, Van Wyk JJ (1985) Somatomedin-C-mediated potentiation of follicle-stimulating hormone-induced aromatase activity of cultured rat granulosa cells. Endocrinology 117: 2313–2320

Adler BA, Johnson MD, Lynch CO, Crowley WR (1983) Evidence that norepinephrine and epinephrine systems mediate the stimulatory effects of ovarian hormones on luteinizing hormone and luteinizing hormone-releasing hormone. Endocrinology 113: 1431–1438

Ainsworth L, Tsang BK, Marcus GJ, Downey BR (1984) Prostaglandin production by dispersed granulosa and theca interna cells from porcine preovulatory follicles. Biol Reprod 31: 115–121

Aiyer MS, Fink G, Greig FA (1974) Changes in the sensitivity of the pituitary gland to luteinizing hormone releasing factor during the oestrus cycle of the rat. J Endocrinol 60: 47–64

Albert A (1961) The mammalian testis. In: Young WC (ed) Sex and internal secretions, 3rd edn. The Williams and Wilkins Co, Baltimore, pp 305–365

Alde S, Celis ME (1980) Influence of α-melanotropin on LH release in the rat. Neuroendocrinology 31: 116–120

Allen E, Doisy EA (1923) An ovarian hormone. Preliminary report on its localization, extraction and partial purification, and action in test animals. J Am Med Assoc 81: 819–821

Allen LG, Kalra SP (1986) Evidence that a decrease in opioid tone may evoke preovulatory luteinizing hormone release in the rat. Endocrinology 118: 1275–1279

Allen LG, Hahn E, Caton D, Kalra SP (1988) Evidence that a decrease in opioid tone on proestrus changes the episodic pattern of luteinizing hormone (LH) secretion: implications in the preovulatory LH hypersecretion. Endocrinology 122: 1004–1013

Almeida OFX, Nikolarakis KE, Herz A (1988) Evidence for the involvement of endogenous opioids in the inhibition of luteinizing hormone by corticotropin-releasing factor. Endocrinology 122: 1034–1041

Amoroso EC, Marshall FHA (1960) External factors in sexual periodicity. In: Parkes AS (ed) Marshall's physiology of reproduction, vol 1. Longmans, Green, New York, pp707–831

Amundson BC, Wheaton JE (1979) Effects of chronic LHRH treatment on brain LHRH content, pituitary and plasma LH and ovarian follicular activity in the anestrous ewe. Biol Reprod 20: 633–638

Anton-Tay F, Anton SM, Wurtman RJ (1970) Mechanism of changes in brain norepinephrine metabolism after ovariectomy. Neuroendocrinology 6: 265–273

Antunes JL, Carmel PW, Housepian EM, Ferin M (1978) Luteinizing hormone – releasing hormone in human pituitary blood. J Neurosurg 49: 382–386

Aono T, Kurachi K, Miyata M, Nakasima A, Koshiyama K, Uozumi T, Matusmo K (1976) Influence of surgical stress under general anesthesia on serum gonadotropin levels in male and female patients. J Clin Endocrinol Metab 42: 144–149

Arimura A, Schally AV (1971) Augmentation of pituitary responsiveness to LH-releasing hormone (LHRH) by estrogen. Proc Soc Exp Biol Med 136: 290–293

Armstrong DT (1981) Prostaglandins and follicular function. J Reprod Fertil 62: 283–294

Armstrong DT, Dorrington JH (1976) Androgens augment FSH-induced progesterone secretion by cultured rat granulosa cells. Endocrinology 99: 1411–1414

Armstrong DT, Moon YS, Fritz IB, Dorrington JH (1975) Synthesis of estradiol-17β by Sertoli cells in culture: stimulation by FSH and dibutyryl cyclic AMP. In: French FS, Hansson V, Ritzen EM, Nayfeh SN (eds) Current topics in molecular endocrinology, vol 2. Plenum Press, New York, p 117

Arthur JR, Boyd GS (1974) The effect of inhibitors of protein synthesis on cholesterol side chain cleavage in the mitochondria of luteinized rat ovaries. Eur J Biochem 49: 117–422

Asanuma Y, Goldstein AL, White A (1970) Reduction in the incidence of wasting disease in neonatally thymectomized CBA/w mice by the injection of thymosin. Endocrinology 86: 600–610

Aschheim S, Zondek B (1927) Hypophysenvorderlappenhormon und Ovarialhormon im Harn von Schwangeren. Klin Wochenschr 6: 1322–1329

Aschhein P (1976) Aging in the hypothalamic-hypophyseal ovarian axis in the rat. In: Everett AV, Burgess JA (eds) Hypothalamus, pituitary and aging. Charles C Thomas, Springfield, p 396

Ashiru OA, Blake CA (1978) Restoration of periovulatory FSH surges in sera by LHRH in phenobarbital-blocked rats. Life Sci 23: 1507–1513

Ashiru OA, Blake CA (1979) Stimulation of endogeneous FSH release during estrus by exogenous follicle-stimulating hormone or luteinizing hormone at proestrus in the phenobarbital-blocked rat. Endocrinology 105: 1162–1167

Asling CW, Reinhardt WO, Li CH (1951) Effects of adrenocorticotropic hormone on body growth, visceral proportions, and white blood cell counts of normal and hypophysectomized male rats. Endocrinology 48: 534–541

Astwood EB (1941) The regulation of corpus luteum function by hypophyseal luteotrophin. Endocrinology 28: 309–315

Astwood EB, Fevold HL (1939) Action of progesterone on the gonadotropic activity of the pituitary. Am J Physiol 127: 192–198

Aten RF, Ireland JJ, Weems CW, Behrman HR (1987) Presence of gonadotropin-releasing hormone-like proteins in bovine and ovine ovaries. Endocrinology 120: 1717–1733

Au CL, Robertson DM, de Kretser DM (1984) Relationship between testicular inhibin content and serum FSH concentrations in rats after bilateral efferent duct ligation. J Reprod Fertil 72: 351–357

Audy M-C, Bonnin M, Souloumiac J, Ribes C, Kerdelhue B, MondainMonval M, Scholler R, Canivenc R (1985) Seasonal variations in plasma luteinizing hormone and testosterone levels in the European badger Meles meles. Gen Comp Endocrinol 57: 445–453

Auletta FJ, Flint APF (1988) Mechanisms controlling corpus luteum function in sheep, cow, non-human primates, and women especially in relation to the time of luteolysis. Endocr Rev 9: 88–105

Auletta FJ, Caldwell BV, Van Wagenen G, Morris JM (1972) Effects of postovulatory estrogen on progesterone and PGF levels in the monkey. Contraception 6: 411–418

Backström CT, McNeilly AS, Leask RM, Baird DT (1982) Pulsatile secretion of luteinizing hormone, FSH, prolactin,estradiol and progesterone during the human menstrual cycle. Clin Endocrinol 17: 29–42

Badger TM, Rosenblum PM, Clement RE, Loughlin JJ (1980) Effects of chronic luteinizing hormone-releasing hormone administration on gonadotropin dynamics of adult male rats. Proc Soc Exp Biol Med 165: 253–259

Baird DT, Swanston I, Scaramuzzi RJ (1976) Pulsatile release of LH and secretion of ovarian steroids in sheep during the luteal phase of the estrous cycle. Endocrinology 98: 1490–1496

Baker BL, Schairer MA, Ingle DJ, Li CH (1950) The induction of involution in the male reproductive system by treatment with adrenocorticotropin. Anat Rec 106: 345–356

Baker ER, Mathur RS, Kirk RF, Williamson HO (1981) Female runners and secondary amenorrhea: correlation with age, parity, mileage, and plasma hormonal and sex-hormone-binding globulin concentrations. Fertil Steril 36: 183–187

Baldwin DM, Sawyer CH (1974) Effects of dexamethasone on LH release and ovulation in the cyclic rat. Endocrinology 94: 1397–1403

Banks PK, Inkster SE, White N, Jeffcoate SL (1986) 2Hydroxyoestradiol acutely inhibits prolactin secretion from the superfused pituitary glands of normal female rats: evidence for a cyclical effect. J Endocrinol 111: 199–204

Barenton B, Hochereau-de Reviers MT, Perreau C, Saumande J (1983) Changes in testicular gonodotropin receptors and steroid content through postnatal development until puberty in the lamb. Endocrinology 112: 1447–1453

Barfield RJ, Thomas DA (1985) The role of ultrasonic vocalizations in the regulation of reproduction in rats. Ann NY Acad Sci 474: 33–43

Barnes LE, Schmidt FL, Dulin WE (1959) Progestational activity of 6 α-methyl-17 α-acetoxyprogesterone. Proc Soc Exp Biol Med 100: 820–822

Barraclough CA, Sawyer CH (1955) Inhibition of the release of pituitary ovulatory hormone in the rat by morphine. Endocrinology 57: 329–335

Barraclough CA, Sawyer CH (1957) Blockade of the release of pituitary ovulating hormone in the rat by chlorpromazine and reserpine: possible mechanisms of action. Endocrinology 61: 341–351

Barraclough CA, Sawyer CH (1959) Induction of pseudopregnancy in the rat by reserpine and chlorpromazine. Endocrinology 65: 563–571

Barraclough CA, Wise PM, Selmanoff MK (1984) A role for hypothalamic catecholamines in the regulation of gonadotropin secretion. Recent Progr Horm Res 40: 487–529

Barrell GK, Lapwood KR (1978) Seasonality of semen production and plasma luteinizing hormone, testosterone, and prolactin levels in romney, merino and polled dorset rams. Anim Reprod Sci 1: 213–219

Barrell GK, Lapwood KR (1979) Effects of pinealectomy on the secretion of luteinizing hormone, testosterone and prolactin in rams exposed to various lighting regimes. J Endocrinol 80: 397–405

Bartke A (1971) Effects of prolactin on spermatogenesis in hypophysectomized mice. J Endocrinol 49: 311–316

Bartke A, Hafiez AA, Bex FJ, Dalterio S (1978) Hormonal interactions in regulation of androgen secretion. Biol Reprod 18: 44–54

Bartke A, Matt KS, Siler-Khodr TM, Soares MJ, Talamantes F, Goldman BD, Hogan MP, Hebert A (1984) Does prolactin modify testosterone feedback in the hamster? Pituitary grafts alter the ability of testosterone to suppress luteinizing hormone and follicle-stimulating hormone release in castrated male hamsters. Endocrinology 115: 2311–2317

Bast JD, Greenwald GS (1974) Serum profiles of follicle stimulating hormone, luteinizing hormone and prolactin during the estrous cycle of the hamster. Endocrinology 94: 1295–1299

Batra SK, Miller WL (1985) Progesterone decreases the responsiveness of ovine pituitary cultures to luteinizing hormone-releasing hormone. Endocrinology 117: 1436–1440

Baulieu E-E, Jung I (1970) A prostatic cytosol receptor. Biochem Biophys Res Commun 38: 599–604

Bauminger S, Lidner HR (1975) Periovulatory changes in ovarian prostaglandin formation and their hormone control in the rat. Prostaglandins 9: 737–751

Beall C (1940) The isolation of β-estradiol and estrone from horse testis. Biochem J 34: 1293–1298

Beattie CW (1979) Serum hormone levels during a post-implantation LH-RH induced luteolysis in the rat. Steroids 34: 365–380

Beattie CW, Corbin A (1977) Pre- and postcoital contraceptive activity of LH-RH in the rat. Biol Reprod 16: 333–339

Beattie CW, Corbin A, Cole G, Corry S, Jones RC, Koch K, Tracy J (1977) Mechanism of the postcoital contraceptive effect of LH-RH in the rat. I. Serum hormone levels during chronic LH-RH administration. Biol Reprod 16: 322–331

Bedrak E, Samuels LT (1969) Steroid biosynthesis by the equine tests. Endocrinology 85: 1186–1195

Beers WH (1975) Follicular plasminogen and plasminogen activator and the effect of plasmin on ovarian follicle wall. Cell 6: 379–384

Behrman HR, Preston SL, Hall AK (1980) Cellular mechanism of the antigonadotropic action of luteinizing hormone-releasing hormone in the corpus luteum. Endocrinology 107: 656–664

Beilharz RG (1968) Effect of stimuli associated with the male on litter size in mice. Aust J Biol Sci 21: 583–585

Bellringer JF, Pratt HPM, Keverne EB (1980) Involvement of the vomeronasal organ and prolactin in pheromonal induction of delayed implantation in mice. J Reprod Fertil 59: 223–228

Ben-Jonathan N, Oliver C, Weiner HJ, Mical RS, Porter JC (1977) Dopamine in hypophysial portal plasma of the rat during the estrous cycle and throughout pregnancy. Endocrinology 100: 452–458

Bercu BB, Hyashi A, Poth M, Alexandrova M, Soloff MS, Donahoe PK (1980) Luteinizing hormone-releasing hormone-induced delay of parturition. Endocrinology 107: 504–508

Berge PJ (1966) Eleven-month "embryonic diapause" in a marsupial. Nature 211: 435–436

Bergland RM, Page RB (1978) Can the pituitary secrete directly to the brain? (affirmative anatomical evidence). Endocrinology 102: 1325–1338

Bergland RM, Page RB (1979) Pituitary-brain vascular relations: a new paradigm. Science 204: 18–24

Bergqvist N (1954) The gonadal function in female diabetics. Acta Endocrinol [Suppl] (Copenh) 19: 3–49

Bernardi G, Comsa J (1965) Purification chromatographique d'une preparation de thymus douée d'une activité hormonale. Experientia 21: 416–418

Bestetti G, Locatelli V, Tirone F, Rossi GL, Muller EE (1985) One month of streptozotocin-diabetes induces different neuroendocrine and morphological alterations in the hypothalamo-pituitary axis of male and female rats. Endocrinology 117: 208–216

Bex FJ, Corbin A (1979) Mechanism of the postcoital contraceptive effect of luteinizing hormone-releasing hormone: ovarian luteinizing hormone receptor interactions. Endocrinology 105: 139–145

Bezssonoff NA, Comsa J (1958) Preparation d'un extrait purifié de thymus. Ann Endocrinol (Paris) 19: 222–227

Bicsak TA, Tucker EM, Cappel S, Vaughan J, Rivier J, Vale W, Hsueh AJW (1986) Hormonal regulation of granulosa cell inhibin biosynthesis. Endocrinology 119: 2711-2719

Biddulph C, Meyer RK, Gumbreck LG (1940) The influence of estriol, estradiol and progesterone on the secretion of gonadotropic hormones in parabiotic rats. Endocrinology 26: 280-284

Birnbaumer L, Shahabi N, Rivier J, Vale W (1985) Evidence for a physiological role of gonadotropin-releasing hormone (GnRH) or GnRH-like material in the ovary. Endocrinology 116: 1367-1370

Blaha GC, Leavitt WW (1978) Uterine progesterone receptors in the aged golden hamster. J Gerontol 33: 810-815

Blake CA (1975) Effects of "stress" on pulsatile luteinizing hormone release in ovariectomized rats. Proc Soc Exp Biol Med 148: 813-815

Blake CA (1976) A detailed characterization of the proestrus luteinizing hormone surge. Endocrinology 98: 445-450

Blake CA (1978) Changes in plasma luteinizing hormone-releasing hormone and gonadotropin concentrations during constant rate intravenous infusion of luteinizing hormone-releasing hormone in cyclic rats. Endocrinology 102: 1043-1052

Blatchley FR, Donovan BT (1969) Luteolytic effect of prostaglandin in the guinea pig. Nature 221: 1065-1066

Bliss E, Frischat A, Samuels L (1972) Brain and testicular function. Life Sci 11: 231-235

Bloch KE (1960) Lipid metabolism. J Wiley & Sons, New York

Blum W, Gupta D (1980) Age and sex-dependent nature of the polymorphic forms of rat pituitary FSH: the role of glycosylation. Neuroendocrinol Lett 6: 357-360

Bogdanove EM, Campbell GT, Blair ED, Mula ME, Miller AE, Grossman GH (1974a) Gonad-pituitary feedback involves qualitative change: androgens alter the type of FSH secreted by the rat pituitary. Endocrinology 95: 219-228

Bogdanove EM, Campbell GT, Peckham WD (1974b) FSH pleomorphism in the rat - regulation by gonadal steroids. Endocr Res Commun 1: 87-100

Bogovich K, Richards J-AS (1982) Androgen biosynthesis in developing ovarian follicles: evidence that luteinizing hormone regulates thecal 17α-hydroxylase and C_{17-20}-lyase activities. Endocrinology 111: 1201-1208

Bouin P, Ancel P (1910) Recherches sur les fonctions du corps jaune gestatif. J Physiol Path Gen 12: 1-16

Bourg R, Van Meensel F, Gompel G (1952) Action des doses massives de benzoate d'oestradiol au niveau des testicules du rat adulte. Ann Endocrinol (Paris) 13: 195-199

Bourne GA, Marshall JC (1984) Anterior pituitary hormonal regulation of testicular gonadotropin-releasing hormone receptors. Endocrinology 115: 723-727

Bourne GA, Regiani S, Payne AH, Marshall JC (1980) Testicular GnRH receptors: characterization and localization on interstitial tissue. J Clin Endocrinol Metab 51: 407-409

Bourne GA, Dockrill MR, Regiani S, Marshall JC, Payne AH (1982) Induction of testicular gonadotropin-releasing hormone (GnRH) receptors by GnRH: effects of pituitary hormones and relationship to inhibition of testosterone production. Endocrinology 110: 727-733

Bourne RA, Tucker HA (1975) Serum prolactin and LH responses to photoperiod in bull calves. Endocrinology 97: 473-474

Bowers CY, Folkers K (1976) Contraception and inhibition of ovulation by minipump infusion of the luteinizing hormone releasing hormone, active analogs and antagonists. Biochem Biophys Res Commun 72: 1003-1005

Boyarsky LH, Baylies H, Casida LE, Meyer RK (1947) Influence of progesterone upon the fertility of gonadotrophin-treated female rabbits. Endocrinology 41: 312-319

Boyse EA, Beauchamp GK, Yamazaki K (1983) The sensory perception of genotypic polymorphism of the major histocompatibility complex and other genes: some physiological and phylogenetic implications. Hum Immunol 6: 177-183

Brady RO (1951) Biosynthesis of radioactive testosterone in vitro. J Biol Chem 193: 145-153

Brawer JR, Sonnenschein C (1975) Cytopathological effects of estradiol on the arcuate nucleus of the female rat: a possible mechanism for pituitary tumorigenesis. Am J Anat 144: 57-88

Brawer JR, Naftolin F, Martin J, Sonnenschein C (1978) Effects of a single injection of estradiol valerate on the hypothalamic arcuate nucleus and on reproductive function in the female rat. Endocrinology 103: 501-512

Brawer JR, Schipper H, Naftolin F (1980) Ovary-dependent degeneration in the hypothalamic arcuate nucleus. Endocrinology 107: 274–279

Brawer JR, Schipper H, Robaire B (1983) Effects of long term androgen and estradiol exposure on the hypothalamus. Endocrinology 112: 194–198

Bresser GM, Parke L, Edward CRW, Forsyth IA, McNeilly AS (1972a) Galactorrhea: successful treatment with reduction of plasma prolactin levels by bromergocryptine. Br Med J 3: 669–672

Bresser GM, Parke L, Edward CRW, Forsyth IA, McNeilly AS (1972b) Galactorrhea: successful treatment with reduction of plasma progesterone upon the fertility of gonadotrophin-treated female rabbits. Endocrinology 41: 312–319

Brodie JY, Hunter IC, Stimson WH, Green B (1980) Specific estradiol binding in cytosols from the thymus glands of normal and hormone-treated male rats. Thymus 1: 337–345

Bronson FH, Marsden HM (1964) Male-induced synchrony of estrus in deermice. Gen Comp Endocrinol 4: 634–637

Brooks CMc (1937) The role of the cerebral cortex and of various sense organs in the exitation and execution of mating activity in the rabbit. Am J Physiol 120: 544–553

Brooks CMc (1938) A study of the mechanism whereby coitus excites the ovulation producing activity of the rabbits' pituitary. Am J Physiol 121: 157–177

Brown RE (1979) The 22-kHz pre-ejaculatory vocalization of the male rat. Physiol Behav 22: 483–489

Bruce HM (1965)Effect of castration on the reproductive pheromones of male mice. J Reprod Fertil 10: 141–143

Bruchovsky N, Wilson JD (1968) Conversion of testosterone to 5α-androstan-17β-ol-3-one by rat prostate in vivo and in vitro. J Biol Chem 243: 2012–2021

Bullock LP, Barthe PL, Mowszowicz I, Orth DN, Bardin CW (1975) The effect of progestins on submaxillary gland epidermal growth factor demonstration of androgenic, synandrogenic and antiandrogenic actions. Endocrinology 97: 189–195

Burrows H (1937) Oestrogens. In: Biological action of sex hormones, Chapter 4. Cambridge University Press, Cambridge

Butler WR, Krey LC, Lu KH, Peckham WD, Knobil E (1975) Surgical disconnection of the medial basal hypothalamus and pituitary function in the rhesus monkey. IV. Prolactin secretion. Endocrinology 96: 1099–1105

Buttle HL (1974) Seasonal variation of prolactin in plasma of male goats. J Reprod Fertil 37: 95–99

Cahoreau C, Blanc MR, Dacheux JL, Pisselet C, Courot M (1979) Inhibin activity in ram rete testis fluid: depression of plasma FSH and LH in the castrated and cryptorchid ram. J Reprod Fertil 26: 97–102

Calderon J-J, Muldoon TG, Mahesh VB (1987) Receptor-mediated interrelationships between progesterone and estradiol action on the anterior pituitary-hypothalamic axis of the ovariectomized immature rat. Endocrinology 120: 2428–2435

Caligaris L, Astrada JJ, Taleisnik S (1971) Biphasic effect of progesterone on the release of gonadotropin in rats. Endocrinology 89: 331–337

Callard GV, Hoffman RA, Petro Z, Ryan KJ (1979) In vitro aromatization and other androgen transformations in the brain of the hamster (Mesocricetus auratus). Biol Reprod 21: 33–38

Callard GV, Petro Z, Ryan KJ (1981) Biochemical evidence for aromatization of androgen to estrogen in the pituitary. Gen Comp Endocrinol 44: 359–364

Campen CA, Vale W (1988) Interaction between purified ovine inhibin and steroids on the release of gonadotropins from cultured rat pituitary cells. Endocrinology 123: 1320–1328

Carmel PW, Araki S, Ferin M (1976) Pituitary stalk portal blood collection in rhesus monkeys: evidence for pulsatile release of gonadotropin-releasing hormone (GnRH). Endocrinology 99: 243–248

Carter JN, Tyson JE, Tolis G, van Vliet S, Faiman C, Friesen HG (1978) Prolactin-secreting tumors and hypogonadism in 22 men. N Engl J Med 299: 847–852

Castro BM (1967) Age of puberty in female mice: relationship to population density and the presence of adult males. Ann Acad Brasil Sci 39: 289–291

Celis ME (1975) Serum MSH levels and the hypothalamic enzymes involved in the formation of MSH-RF during the estrus cycle in the rat. Neuroendocrinology 18: 256–262

Chandrasekhar Y, D'Occhio MJ, Holland MK, Setchell BP (1985a) Activity of the hypothalamo-pituitary axis and testicular development in prepubertal ram lambs with induced hypothyroidism or hyperthyroidism. J Endocrinol 117: 1645-1651

Chandrasekhar Y, Holland MK, D'Occhio MJ, Setchell BP (1985b) Spermatogenesis, seminal characteristics and reproductive hormone levels in mature rams with induced hypothyroidism and hyperthyroidism. J Endocrinol 105: 39-46

Chandrasekhar Y, D'Occhio MJ, Setchell BP (1986) Reproductive hormone secretion and spermatogenic function in thyroidectomized rams receiving graded doses of exogenous thyroxine. J Endocrinol 111: 245-253

Channing CP, Tsai V, Sachs D (1976) Role of insulin, thyroxin and cortisol in luteinization of porcine granulosa cells grown in chemically defined media. Biol Reprod 15: 235-247

Channing CP, Anderson LD, Hoover DJ, Kolena J, Osteen KG, Seymour H, Pomerantz SH, Tanabe K (1982) The role of nonsteroidal regulators in control of oocyte and follicular maturation. Recent Prog Horm Res 38: 331-408

Chappel SC, Norman RL, Spies HG (1977) Regulation of the second (estrous) release of FSH in hamsters by the medial basal hypothalamus. Endocrinology 101: 1339-1342

Chappel SC, Norman RL, Spies HG (1979) Evidence for a specific neural event that controls the estrous release of FSH in golden hamsters. Endocrinology 104: 169

Chappel SC, Coutifaris C, Jacobs SJ (1982a) Studies on the microheterogeneity of FSH present within the anterior pituitary gland of ovariectomized hamsters. Endocrinology 110: 847-854

Chappel SC, Ulloa-Aguirre A, Ramaley JA (1982b) Sexual maturation in female rats: time course of the appearance of multiple species of anterior pituitary FSH. Biol Reprod 28: 196-205

Chappel SC, Bethea CL, Spies HG (1984a) Existence of multiple forms of follicle-stimulating hormone within the anterior pituitaries of cynomolgus monkeys. Endocrinology 115: 452-461

Chappel SC, Miller C, Hyland L (1984b) Regulation of the pulsatile releases of luteinizing and follicle-stimulating hormones in ovariectomized hamsters. Biol Reprod 30: 628-636

Charlton HM, Halpin DMG, Iddoon C, Rosie R, Levy G, McDowell IFW, Megson A, Morris JF, Bramwell A, Speight A, Ward BJ, Broadhead J, Davey-Smith G, Fink G (1983a) The effects of daily administration of single and multiple injections of gonadotropin-releasing hormone on pituitary and gonadal function in the hypogonadal (hpg) mouse. Endocrinology 113: 535-544

Charlton HM, Speight A, Halpin DMG, Bramwell A, Sheward WJ, Fink G (1983b) Prolactin measurements in normal and hypogonadal (hpg) mice: developmental and experimental studies. Endocrinology 113: 545-551

Charpenet G, Tache Y, Forest M, Haour F, Saez J, Bernier M, Ducharme J, Collu R (1981) Effects of chronic intermittent immobilization stress on rat testicular androgen function. Endocrinology 109: 1254-1258

Charpenet G, Tache Y, Bernier M, Ducharme JR, Collu R (1982) Stress-induced testicular hyposensitivity to gonadotropins in rats. Role of the pituitary gland. Biol Reprod 27: 616-623

Chen Y-DI, Payne AH, Kelch RP (1976) FSH stimulation of Leydig cell function in the hypophysectomized immature rat. Proc Soc Exp Biol Med 153: 473-475

Chernayaev GA, Barkova TI, Egorova VV, Sorokina IB, Ananchenko SN, Mataradze GD, Sokolova NA, Rozen VB (1975) A series of optical, structural and isomeric analogs of estradiol: a comparative study of the biological activity and affinity to cytosol receptor of rabbit uterus. J Steroid Biochem 6: 1483-1488

Chester Jones I (1957) The adrenal cortex. Cambridge University Press, Cambridge

Cheung CY (1983) Prolactin suppresses luteinizing hormone secretion and pituitary responsiveness to luteinizing hormone releasing hormone by a direct action at the anterior pituitary. Endocrinology 113: 632-638

Childs GV, Ellison DG, Lorenzen JR, Collins TS, Schwartz NB (1982) Immunocytochemical studies of gonadotropin storage in developing castration cells. Endocrinology 111: 1318-1328

Ching M (1982) Correlative surges of LHRH, LH and FSH in pituitary stalk plasma and systemic plasma of rat during proestrus. Effect of anesthetics. Neuroendocrinology 34: 279-285

Chipman RK, Fox KA (1966) Factors in pregnancy blocking: age and reproductive background of females: numbers of strange males. J Reprod Fertil 12: 399-403

Chu EW, Wurtman RJ, Axelrod J (1964) An inhibitory effect of melatonin on the estrous phase of the estrous cycle of the rodent. Endocrinology 75: 238-242

Cicero TJ, Bell RD, Weist WG, Allison JH, Polakoski K, Robins E (1975) Function of the male sex organs in heroin and methadone users. N Engl J Med 292: 882–885

Cicero TJ, Schainker BA, Meyer ER (1979) Endogenous opioids participate in the regulation of the hypothalamic-pituitary luteinizing hormone axis and testosterone's negative feedback control of luteinizing hormone. Endocrinology 104: 1286–1291

Clark CR, Nowell NW (1979) Binding properties of testosterone receptors in the hypothalamic-preoptic area of the adult male mouse brain. Steroids 33: 407–426

Clark JH, Peck EJ Jr., Schrader WT, O'Malley BW (1976) Estrogen and progesterone receptors: methods for characterization, quantification and purification. Methods Can Res XII: 367–418

Clark JH, Markaverich B, Upchurch S, Eriksson H, Hardin JW, Peck EJ Jr. (1980) Heterogeneity of estrogen binding sites: relationship to estrogen receptors and estrogen responses. Recent Prog Horm Res 36: 89–134

Clark MR, Marsh JM, Le Maire WJ (1978) Mechanism of luteinizing hormone regulation of prostaglandin synthesis in rat granulosa cells. J Biol Chem 253: 7757–7761

Clarke IJ, Cummins JT (1982) The temporal relationship between gonadotropin releasing hormone (GnRH) and luteinizing hormone (LH) secretion in ovariectomized ewes. Endocrinology 111: 1737–1739

Clarke IJ, Cummins JT (1984) Direct pituitary effects of estrogen and progesterone on gonadotropin secretion in the ovariectomized ewe. Neuroendocrinology 39: 267–274

Clarke IJ, Cummins JT (1985) Increased gonadotropin-releasing hormone pulse frequency associated with estrogen-induced luteinizing hormone surges in ovariectomized ewes. Endocrinology 116: 2376–2383

Clarke IJ, Burman KS, Doughton BW, Cummins JT (1986) Effects of constant infusion of gonadotrophin-releasing hormone in ovariectomized ewes with hypothalamo-pituitary disconnection: further evidence for differential control of LH and FSH secretion and the lack of a priming effect. J Endocrinol 111: 43–49

Clauberg C (1930) Der biologische Test für das Corpus luteum Hormon. Klin Wochenschr 9: 2004–2005

Clayton RN, Catt KJ (1981) Regulation of pituitary gonadotropin releasing hormone receptors by gonadal steroids. Endocrinology 108: 887–895

Clayton RN, Solano AR, Garcia-Vila A, Dufau ML, Catt KJ (1980) Regulation of pituitary receptors for gonadotropin-releasing hormone during the rat estrous cycle. Endocrinology 107: 699–706

Clayton RN, Channabasavaiah K, Stewart JM, Catt KJ (1982) Hypothalamic regulation of pituitary gonadotropin releasing hormone receptors: effects of hypothalamic lesions and a GnRH antagonist. Endocrinology 110: 1108–1115

Clifton DK, Sawyer CH (1979) LH release and ovulation in the rat following depletion of hypothalamic norepinephrine: chronic vs. acute effects. Neuroendocrinology 28: 442–449

Collu R, Tache Y, Ducharme JR (1979) Hormonal modifications induced by chronic stress in rats. J Steroid Biochem 11: 989–995

Collu R, Gibb W, Bichet DG, Ducharme JR (1984) Role of arginine vasopressin (AVP) in stress-induced inhibition of testicular steroidogenesis in normal and in AVP-deficient rats. Endocrinology 115: 1609–1615

Condon TP, Sawyer CH, Whitmoyer DI (1986) Episodic patterns of luteinizing hormone and follicle-stimulating hormone release: differential secretory dynamics and adrenergic control in ovariectomized rats. Endocrinology 118: 2525–2533

Conn PM (1986) The molecular basis of gonadotropin-releasing hormone action. Endocr Rev 7: 3–10

Connolly PB, Handa RJ, Resko JA (1988) Progesterone modulation of androgen receptors in the brain and pituitary of male guinea pigs. Endocrinology 122: 2547–2553

Coppings RJ, Malven PV (1976) Biphasic effect of estradiol on mechanisms regulating LH release in ovariectomized sheep. Neuroendocrinology 21: 146–156

Corbin A, Beattie CW (1975) Post-coital contraceptive and uterotrophic effects of luteinizing hormone releasing hormone. Endocr Res Commun 2: 445–458

Corbin A, Beattie CW, Tracy J, Jones R, Foell TJ, Yardley J, Rees RWA (1978) The anti-reproductive pharmacology of LH-RH and agonistic analogues. Int J Fertil 23: 81–92

Coutifaris C, Chappel SC (1982) Intraventricular injection of follicle-stimulating hormone (FSH) during proestrus stimulates the rise in serum FSH on estrus in phenobarbital-treated hamsters through a central nervous system-dependent mechanism. Endocrinology 110: 105–109

Coutifaris C, Chappel SC (1983) Involvement of hypothalamic luteinizing hormone-releasing hormone in the regulation of the estrous follicle-stimulating hormone surge in the female golden hamster. Endocrinology 113: 563–569

Crowley WF Jr., Filicori M, Spratt DI, Santoro NF (1985) The physiology of gonadotropin-releasing hormone (GnRH) secretion in men and women. Recent Prog Horm Res 41: 473–531

Crowley WR, Hassid A, Kalra SP (1987) Neuropeptide Y enhances the release of luteinizing hormone (LH) induced by LH-releasing hormone. Endocrinology 120: 941–945

Culler MD, Negro-Vilar A (1986) Evidence that pulsatile follicle stimulating hormone secretion is independent of endogenous luteinizing hormone-releasing hormone. Endocrinology 118: 609–612

Culler MD, Valenca MM, Merchenthaler I, Flerko B, Negro-Vilar A (1988) Orchidectomy induces temporal and regional changes in the processing of the luteinizing hormone-releasing hormone prohormone in the rat brain. Endocrinology 122: 1968–1976

Cunha GR, Shannon JM, Vanderslice KD, McCormick K, Bigsby RM (1983) Autoradiographic demonstration of high affinity nuclear binding and finite binding capacity of [^{3}H]estradiol in mouse vaginal cells. Endocrinology 113: 1427–1430

Cusan L, Auclair C, Belanger A, Ferland L, Kelly PA, Seguin C, Labrie F (1979) Inhibitory effects of long term treatment with a luteinizing hormone-releasing hormone agonist on the pituitary gonadal axis in male and female rats. Endocrinology 104: 1369–1376

Dada MO, Campbell GT, Blake CA (1983) A quantitative immunocytochemical study of the luteinizing hormone and follicle stimulating hormone cells in the adenohypophysis of adult male rats and adult female rats throughout the estrous cycle. Endocrinology 113: 970–984

Daniel SAJ, Armstrong DT (1980) Enhancement of follicle-stimulating hormone-induced aromatase activity by androgens in cultured rat granulosa cells. Endocrinology 107: 1027–1033

David MA, Fraschini F, Martini C (1966) Control of LH secretion: role of a "short" feedback mechanism. Endocrinology 78: 55–60

Davies IJ, Ryan KJ (1972) The uptake of progesterone by the uterus of the pregnant rat in vivo and its relationship to cytoplasmic progesterone-binding protein. Endocrinology 90: 507–515

Davies IJ, Naftolin F, Ryan KJ, Fishman J, Siu J (1975) The affinity of catechol estrogens for estrogen receptors in the pituitary and anterior hypothalamus of the rat. Endocrinology 97: 554–557

Davis FC, Gorski RA (1985) Development of hamster circadian rhythms: prenatal entrainment of the pacemaker. J Biol Rhythms 1: 77–84

Davis GJ, Meyer RK (1973) Seasonal variation in LH and FSH of bilaterally castrated snowshoe hares. Gen Comp Endocrinol 20: 61–68

Davis ME, Plotz EJ (1954) The effects of cortisone acetate on intact and adrenalectomized rats during pregnancy. Endocrinology 54: 384–389

Davoren JB, Kasson BG, Li Ch, Hsueh AJW (1986) Specific insulinlike growth factor (IGF)I- and II-binding sites on rat granulosa cells: relation to IGF action. Endocrinology 119: 2155–2162

Deanesly R (1963) Further observations on the effects of oestradiol on tubal eggs and implantation in the guinea pig. J Reprod Fertil 5: 49–57

de Greef WJ, Voogt JL, Visser TJ, Lamberts SWJ, van der Schoot P (1987) Control of prolactin release induced by suckling. Endocrinology 121: 316–322

de Jong FH (1979) Inhibin - fact or artifact. Mol Cell Endocrinol 13: 1–10

Dekel N, Lewysohn O, Ayalon D, Hazum E (1988) Receptors for gonadotropin releasing hormone are present in rat oocytes. Endocrinology 123: 1205–1207

DeMoraes-Ruehsen M, Blizzard RM, Garcia-Bunuel R, Jones GS (1972) Autoimmunity and ovarian failure. Am J Obstet Gynecol 112: 693–699

Dempsey EW (1937) Follicular growth and ovulation after various experimental procedures in the guinea pig. Am J Physiol 120: 126–132

Dempsey EW, Hertz R, Young WC (1936) The experimental induction of oestrus (sexual receptivity) in the normal and ovariectomized guinea-pig. Am J Physiol 116: 201–209

Denef C, Hautekeete E, Dwals R, de Wolf A (1980) Differential control of luteinizing hormone and follicle-stimulating hormone secretion by androgens in rat pituitary cells in culture: functional diversity of subpopulations separated by unit gravity sedimentation. Endocrinology 106: 724–729

De Paolo LV (1985) Differential regulation of pulsatile luteinizing hormone (LH) and follicle-stimulating hormone secretion in ovariectomized rats disclosed by treatment with a LH-releasing hormone antagonist and phenobarbital. Endocrinology 117: 1826–1833

De Paolo LV, Ojeda SR, Negro-Vilar A, McCann SM (1982) Alterations in the responsiveness of median eminence luteinizing hormone-releasing hormone nerve terminals to norepinephrine and prostaglandin E_2 in vitro during the rat estrous cycle. Endocrinology 110: 1999–2005

Deschaux P, Massengo B, Fontanges R (1979) Endocrine interaction of the thymus with the hypophysis, adrenals and testes: effects of two thymic extracts. Thymus 1: 95–108

Desjardins C (1981) Endocrine signaling and male reproduction. Biol Reprod 24: 1–16

Desjardins C, Lopez MJ (1983) Environmental cues evoke differential responses in pituitary-testicular function in deer mice. Endocrinology 112: 1398–1406

Desjardins C, Zeleznik AJ, Midgley AR Jr (1974) In vitro binding and autoradiographic localization of human chorionic gonadotropin and follicle stimulating hormone in rat testes during development. In: Dufau ML, Means AR (eds) Current topics in molecular endocrinology, vol 1. Plenum Press, New York, 221

Diaz-Infante A, Wright KH, Wallach EE (1974) Effects of indomethacin and prostaglandin $F_{2\alpha}$ on ovulation and ovarian contractility in the rabbit. Prostaglandins 5: 567–579

Diefenbach WP, Carmel PW, Frantz AG, Ferin M (1976) Suppression of prolactin secretion by L-DOPA in the stalk-sectioned rhesus monkey. J Clin Endocrinol Metab 43: 638–642

Dierschke DJ, Yamaji T, Karsch FJ, Weick, RF, Weiss G, Knobil E (1973) Blockade by progesterone of estrogen-induced LH and FSH release in the rhesus monkey. Endocrinology 92: 1496–1501

Distiller LA, Sagel J, Morley JE, Joffe BI, Seftel HC (1975) Pituitary responsiveness to luteinizing hormone-releasing hormone in insulin-dependent diabetes mellitus. Diabetes 24: 378–380

Djursing H, Nyholm HC, Hagen C, Carstensen L, Pedersen LM (1982) Clinical and hormonal characteristics in women with anovulation and insulin-treated diabetes mellitus. Am J Obstet Gynecol 143: 876–882

D'Occhio MJ, Schanbacher BD, Kinder JE (1982) Relationship between serum testosterone concentration and patterns of luteinizing hormone secretion in male sheep. Endocrinology 110: 1547–1552

Döcke F, Busch W (1974) Evidence for anterior hypothalamic control of cyclic gonadotrophin secretion in female pigs. Endokrinologie 63: 415–421

Döcke F, Dörner G (1965) The mechanism of the induction of ovulation by estrogens. J Endocrinol 63: 491–499

Döcke F, Glaser D (1971) Internal feedback of luteinizing hormone in cyclic female rats. J Endocrinol 51: 403–408

Dorfman RI (1962) Anti-estrogenic compounds. In: Dorfman RI (ed) Methods in hormone research, vol II. Academic Press, New York, pp 113–126

Dorfman RI (1963a) Anti-androgenic compounds. In: Cori CF, Foglia VG, Leloir LF, Ochoa S (eds) Perspectives in biology. Elsevier, The Netherlands, pp 43–55

Dorfman RI (1963b) Anti-androgens in a castrated mouse test. Steroids 2: 185–193

Dorfman RI, Kincl FA, Ringold HJ (1961) Anti-estrogen assay of neutral steroids administered by subcutaneous injection. Endocrinology 68: 17–24

Dorner G, Eckert R, Hinz G (1980) Androgen-dependent sexual dimorphism of the immune system. Endokrinologie 76: 112–114

Dorrington JH, Armstrong DT (1975) Follicle-stimulating hormone stimulates estradiol-17 synthesis in cultured Sertoli cells. Proc Natl Acad Sci USA 72: 2677–2681

Dorrington JH, Moon YS, Armstrong DT (1975) Estradiol-17β biosynthesis in cultured granulosa cells from hypophysectomized immature rats; stimulation by follicle-stimulating hormone. Endocrinology 97: 1328–1331

Dorrington JH, Fritz IB, Armstrong DT (1976) Site at which FSH regulates estradiol-17β synthesis in Sertoli cell preparations in culture. Mol Cell Endocrinol 6: 117–122

Drouin J, Labrie F (1976) Selective effect of androgens on LH and FSH release in anterior pituitary cells in culture. Endocrinology 98: 1528-1534

Drouin J, Labrie F (1981) Interactions between 17β-estradiol and progesterone in the control of luteinizing hormone and follicle stimulating hormone release in rat anterior pituitary cells in culture. Endocrinology 108: 52-57

Drouin J, Lagacé L, Labrie F (1976) Estradiol-induced increase of the LH responsiveness to LH releasing hormone (LHRH) in rat anterior pituitary cells in culture. Endocrinology 99: 1477-1481

Drouva SV, Gallo RV (1976) Catecholamine involvement in episodic lutenizing hormone release in adult ovariectomized rates. Endocrinology 99: 651-658

Drouva SV, Laplante E, Kordon C (1982) α_1- Adrenergic receptor involvement in the LH surge in ovariectomized estrogen-primed rats. Eur J Pharmacol 81: 341-346

Dufau ML, Cigorraga S, Baukal AJ, Sorrell S, Bator JM, Neubauer JF, Catt KJ (1979) Androgen biosynthesis in Leydig cells after testicular desensitization by luteinizing hormone-releasing hormone and human chorionic gonadotropin. Endocrinology 105: 1314-1321

Dufau ML, Winters CA, Hattori M, Aquillano D, Barano JLS, Nozu K, Baukal A, Catt KJ (1984) Hormonal regulation of androgen production by the Leydig call. J Steroid Biochem 20: 161-167

Duncan JA, Dalkin AC, Barkan A, Regiani A, Marchall JC (1983). Gonadal regulation of pituitary gonadotropin-releasing hormone receptors during sexual maturation in the rat. Endocrinology 113: 2238-2246

Du Ruisseau P, Tache Y, Brazeau P, Collu R (1978) Pattern of adenohypophyseal hormone changes induced by various stressors in female and male rats. Neuroendocrinology 27: 257-271

Dym M, Raj HGM (1977) Response of adult Sertoli cells and Leydig cells to depletion of luteinizing hormone and testosterone. Biol Reprod 17: 676-696

Eckstein P, Zuckerman S (1962) The oestrus cycle in the mammalia. In: Parkes AS (ed) Marshall's physiology of reproduction, vol 1. Longmans, Green, London, pp 226-396

Edgren RA, Calhoun DW (1957) Estrogen antagonism. Inhibition of oestrone-induced uterine growth by testosterone propionate, progesterone, and 17α-ethyl-19-nortestosterone. Proc Soc Exp Biol Med 94: 537-539

Edgren RA, Elton RL, Calhoun DW (1961) Studies on the interactions of oestriol and progestrone. J Reprod Fertil 2: 98-105

Eik-Nes KB, Kekre M (1963) Metabolism *in vivo* of steroids by the canine testes. Biochim Biophys Acta 78: 449-456

Eisenfeld AJ, Axelrod J (1967) Evidence for oestradiol binding sites in the hypothalamus, effect of drugs. Biochem Pharmacol 16: 1781-1785

Ekholm C, Hillensjö T, Isaksson O (1981) Gonadotropin-releasing hormone agonists stimulate oocyte meiosis and ovulation in hypophysectomized rats. Endocrinology 108: 2022-2024

Elkind-Hirsch K, Ravnikar V, Schiff I, Tulchinsky D, Ryan KJ (1982) Determinations of endogenous immunoreactive luteinizing hormone-releasing hormone in human plasma. J Clin Endocrinol Metab 54: 602-607

Ellis GB, Desjardins C (1982) Male rats release LH and testosterone episodically. Endocrinology 110: 1618-1627

Ellis GB, Desjardinis C (1984) Orchidectomy unleashes pulsatile luteinizing hormone secretion in the rat. Biol Reprod 30: 619-624

Emmens CW (1962) Estrogens. In: Dorfman RI (ed) Methods in hormone research, vol 2. Academic Press, New York, pp 59-111

Emmens CW, Finn CA (1962) Local and parental action of oestrogens and anti-oestrogens on early pregnancy in the rat and mouse. J Reprod Fertil 3: 239-245

Emmens CW, Martin L (1964) Estrogens. In: Dorfman RI (ed) Methods in hormone research, vol IIIA. Academic Press, New York, pp 1-80

Erickson GF, Hsueh AJW (1978) Stimulation of aromatase activity by follicle stimulating hormone in rat granulosa cells *in vivo* and *in vitro*. Endocrinology 102: 1275-1282

Erickson GF, Challis JRG, Ryan KJ (1977) Production of prostaglandin F by rabbit granulosa cells and thecal tissue. J Reprod Fertil 49: 133-134

Erickson GF, Wang C, Hsueh AJW (1979) FSH induction of functional LH receptors in granulosa cells cultured in chemically defined medium. Nature 279: 336-338

Erickson GF, Magoffin DA, Dyer CA, Hofeditz C (1985) The ovarian androgen producing cells: a review of structure/function relationships. Endocr Rev 6: 371-399

Eskay RL, Mical RS, Porter JC (1977) Relationship between luteinizing hormone releasing hormone concentration in hypophysial portal blood and luteinizing hormone release in intact, castrated, and electrochemically-stimulated rats. Endocrinology 100: 263-270

Espey LL (1982) Optimum time for administration of indomethacin to inhibit ovulation in the rabbit. Prostaglandins 23: 329-334

Espey LL, Norris C, Saphire D (1986) Effect of time and dose of indomethacin on follicular prostaglandins and ovulation in the rabbit. Endocrinology 119: 746-754

Estes KS, Simpkins JW (1982) Resumption of pulsatile luteinizing hormone release after α-adrenergic stimulation in aging constant estrous rats. Endocrinology 111: 1778-1784

Estes KS, Padmanabhan V, Convey EM (1977) Localization of gonadotropin releasing hormone (GnRH) within the bovine hypothalamus. Biol Reprod 17: 706-711

Estes KS, Simpkins JW, Kalra SP (1982) Resumption with clonidine of pulsatile LH release following acute norepinephrine depletion in ovariectomized rats. Neuroendocrinology 35: 56-62

Evans AJ, Simpson ME, Lyons WR (1941) Influence of lactogenic preparations on production of traumatic placentoma in the rat. Prog Soc Exp Biol Med 46: 586-590

Evans HM, Simpson ME (1929) A sex difference in the hormone content of the anterior hypophysis of the rat. Am J Physiol 89: 375-378

Everett JW (1952) Presumptive hypothalamic control of spontaneous ovulation. Ciba Found Coll Endocrinol 4: 167-176

Everett JW (1961) The mammalian female reproductive cycle and its controlling mechanisms. In: Young WC (ed) Sex and internal secretion, vol 1. Williams & Wilkins, Baltimore, pp 497-555

Everett JW, Sawyer CH (1950) A 24-hour periodcity in the "LH release apparatus" of female rats, disclosed by barbiturate sedation. Endocrinology 46: 198-218

Everett JW, Sawyer CH, Markee JE (1944) A neurogenic timing factor in control of the ovulatory discharge of luteinizing hormone in the cyclic rat. Endocrinology 44: 234-250

Ewing LL, Zirkin B (1983) Leydig cell structure and steroidogenic function. Recent Prog Horm Res 39: 599-635

Faber LE, Sandmann ML, Stavely HE (1972) Progesterone-binding proteins of the rat and rabbit uterus. J Biol Chem 247: 5648-5649

Fang S, Anderson KM, Liao S (1969) Receptor proteins for androgens. J Biol Chem 244: 6584-6595

Fang VS, Refetoff S, Rosenfield RL (1974) Hypogonadism induced by a transplantable prolactin-producing tumor in male rats: hormonal and morphological studies. Endocrinology 95: 991-996

Farookhi R (1980) Effects of androgen on induction of gonadotropin receptors and gonadotropin-stimulated adenosine 3': 5' monophosphate production in rat ovarian granulosa cells. Endocrinology 106: 1216-1223

Feil PD, Glasser SR, Toft DO, O'Malley BW (1972) Progesterone binding in the mouse and rat uterus. Endocrinology 91: 738-746

Feil PD, Clarke CL, Satyaswaroop PG (1988) Progestin-mediated changes in progesterone receptor forms in the normal human endometrium. Endocrinology 123: 2506-2513

Feingold KR, MacRae G, Moser AH, Wu J, Siperstein MD, Wiley MH (1983) Differences in de novo cholesterol synthesis between the intact male and female rat. Endocrinology 112: 96-103

Ferland L, Borgeat P, Labrie F, Bernard J, De Lean A, Raynaud JP (1975) Changes of pituitary sensitivity to LHRH during the rat estrous cycle. J Mol Cell Endocrinol 2: 107-111

Fevre-Montange M, Van Cauter E, Refetoff S, Desir D, Tourniaire J, Copinschi G (1981) Effects of jet lag on hormonal patterns 2. Adaptation of melatonin circadian periodicity. J Clin Endocrinol Metab 52: 642-649

Fink G, Aiyer MS (1974) Gonadotrophin secretion after electrical stimulation of the preoptic area during the oestrous cycle of the rat. J Endocrinol 62: 589-604

Fink G, Jamieson MG (1975) Immunoreactive luteinizing hormone releasing factor in rat pituitary stalk blood: effects of electrical stimulation of the medial preoptic area. J Endocrinol 68: 71-87

Fischette CT, Biegon A, McEwen BS (1983) Sex differences in serotonin 1 receptor binding in rat brain. Science 222: 333–335

Fishman J (1976) The catechol estrogens. Neuroendocrinology 22: 363–374

Fishman J, Naftolin F, Davies IJ, Ryan KJ, Petro Z (1976) Catechol estrogen formation by the human fetal brain and pituitary. J Clin Endocrinol Metab 42: 177–180

Forbes JM, Driver PM, El Shabat AA, Boaz TG, Scanes CG (1975) The effect of daylength and level of feeding on serum prolactin in growing lambs. J Endocrinol 64: 549–554

Forcelledo ML, de la Cerda ML, Croxatto HB (1986) Effectiveness of different estrogen pulses in plasma for accelerating ovum transport and their relation to estradiol levels in the rat oviduct. Endocrinology 119: 1189–1194

Fortune JE, Armstrong DT (1977) Androgen production by theca and granulosa isolated from proestrous rat follicles. Endocrinology 100: 1341–1347

Foster DL (1983) Photoperiod and sexual maturation of the female lamb: early exposure to short days perturbs estradiol feedback inhibition of luteinizing hormone secretion and produces abnormal ovarian cycles. Endocrinology 112: 11–17

Fox SR, Smith MS (1985) Changes in the pulsatile pattern of luteinizing hormone secretion during the rat estrous cycle. Endocrinology 116: 1485–1492

Fraenkel L (1903) Die Funktion des Corpus-luteum-Hormons. Arch Gynakol 68: 438–545

Fraenkel L, Cohn F (1901) Experimentelle Untersuchungen über den Einfluß des Corpus luteum auf die Insertion des Eies (Theorie von Born). Anat Anz 20: 294–99

Frager MS, Pieper DR, Tonetta SA, Ducan JA, Marshall JC (1981). Pituitary gonadotropin-releasing hormone (GnRH) receptors: effects of castration, steroid replacement and the role of GnRH in modulating receptors in the rat. J Clin Invest 67: 615–623

Francis GL, Brown TJ, Bercu BB (1981) Control of Sertoli cell response to FSH: regulation by homologous hormone exposure. Biol Reprod 24: 955–961

Francis RL (1977) 22-kHz calls by isolated rats. Nature 265: 236–237

Fraser HM, Baker TG (1978) Changes in the ovaries of rats after immunization against luteinizing hormone-releasing hormone. J Endocrinol 77: 85–94

Frisch RE (1980) Comments. Recent Prog Horm Res 30: 46

Fritz IB, Griswold MD, Louis BG, Dorrington JH (1976) Similarity of responses of cultured Sertoli cells to cholera toxin and FSH. Mol Cell Endocrinol 5: 289–294

Fuchs JL, Moore RY (1980) Development of circadian rhythmicity and light responsiveness in the rat suprachiasmatic nucleus. A study using 2-deoxy-(1-^{14}C) glucose method. Proc Natl Acad Sci USA 77: 1204–1205

Gabriel SM, Simpkins JW, Kalra SP (1985) Chronic morphine treatment induces hypersensitivity to testosterone negative feedback in castrated male rats. Neuroendocrinology 40: 39–44

Gabriel SM, Berglund LA, Kalra SP, Kalra PS, Simpkins JW (1986) The influence of chronic morphine treatment on the negative feedback regulation of gonadotrophin secretion by gonadal steroids. Endocrinology 119: 2762–2767

Gallo RV (1980) Neuroendocrine regulation of pulsatile luteinizing hormone release in the rat. Neuroendocrinology 30: 122–131

Gallo RV (1981) Pulsatile LH release during periods of low level LH secretion in the rat estrous cycle. Biol Reprod 24: 771–776

Gallo RV, Bona-Gallo A (1985) Lack of ovarian steroid negative feedback on pulsatile luteinizing hormone release between estrus and diestrous day 1 in the rat estrous cycle. Endocrinology 116: 1525–1528

Garcia A, Herbon L, Barkan A, Papavasiliou S, Marshall JC (1985) Hyperprolactinemia inhibits gonadotrophin-releasing hormone (GnRH) stimulation of the number of pituitary GnRH receptors. Endocrinology 117: 954–959

Garcia M, Rochefort H (1977) Androgens on the estrogen receptor. II. Correlation between nuclear translocation and uterine protein synthesis. Steroids 29: 111–118

Gardner JH (1949) Effects of inunction of estradiol on testes and thyroids of albino rats. Proc Soc Exp Biol Med 72: 306–310

Garzo G, Dorrington JH (1984) Aromatase activity in human granulosa cells during follicular development and the modulation by follicle-stimulating hormone and insulin. Am J Obstet Gynec 148: 657–662

Gaunt R, Parkins WH (1933) The alleged interrelationship of the adrenal cortical hormone and the gonads. Am J Physiol 103: 511–519

Gautron JP, Pattou E, Kordon C (1981) Occurence of higher molecular forms of LHRH in fractionated extracts from rat hypothalamus, cortex and placenta. Moll Cell Endocrinol 24: 1–15

Gay VL, Sheth NA (1972) Evidence for a periodic release of LH in castrated male and female rats. Endocrinology 90: 158–162

George IW, Noble JF (1984) Androgen receptors are similar in fetal and adult rabbits. Endocrinology 115: 1451–1458

Gerendai I, Shaha C, Thau R, Bardin CW (1984) Do testicular opiates regulate Leydig cell function? Endocrinology 115: 1645–1647

Ghosh DK, Dunham WR, Sands RH, Menon KMJ (1987) Regulation of cholesterol side-chain cleavage enzyme activity by gonadotropin in rat corpus luteum. Endocrinology 121: 21–27

Gibson MJ, Krieger DT, Charlton HM, Zimmerman EA, Silverman A-J, Perlow MJ (1984) Mating and pregnancy can occur in genetically hypogonadal mice and preoptic area brain grafts. Science 225: 949–951

Giguere V, Lefevre FA, Labrie F (1981) Androgens decrease LHRH binding sites in rat anterior pituitary cells in culture. Endocrinology 108: 350–352

Gillette S, Gillette RW (1979) Changes in thymic estrogen receptor expression following orchidectomy. Cell Immunol 42: 194–199

Glascock RF, Hoekstra WG (1959) Selective accumulation of tritium-labelled hexoestrol by the reproductive organs of immature female goats and sheep. Biochem J 72: 673–682

Glass JD, Ferreira S, Deaver DR (1988) Photoperiodic adjustments in hypothalamic amines, gonadotropin-releasing hormone, and -endorphin in the white-footed mouse. Endocrinology 123: 1119–1127

Gnodde HP, Schuiling GA (1976) Involvement of catecholaminergic and cholinergic mechanisms in the pulsatile release of LH in the long term ovariectomized rat. Neuroendocrinology 20: 212–223

Goddard RF (1948) Anatomic and physiologic studies in young rats with propylthiouracil-induced dwarfism. Anat Rec 101: 539–547

Goff AK, Leung PCK, Armstrong DT (1979) Stimulatory effect of follicle-stimulating hormone and androgens on the responsiveness of rat granulosa cells to gonodotrophins in vitro. Endocrinology 104: 1124–1129

Goldring NB, Durica JM, Lifka J, Hedin L, Ratoosh SL, Miller WL, Orly J, Richards JA (1987) Cholesterol side-chain cleavage P450 messenger ribonucleic acid: evidence for hormonal regulation in rat ovarian follicles and constitutive expression in corpora lutea. Endocrinology 120: 1942–1950

Goldstein AL, Guha A, Zatz MM, White A (1972) Purification and biological activity of thymosin: a hormone of the thymus gland. Proc Nat Acad Sci USA 69: 1800–1803

Goncharov NP, Taranov AG, Antonichev AV, Gorlushkin VM, Aso T, Cekan SK, Diczfalusy E (1979) Effect of stress on the profile of plasma steroids in baboons (Papio hamadryas). Acta Endocrinol (Copenh) 90: 372–384

Goodman AL, Hodgen GD (1983) The ovarian triad of the primate menstrual cycle. Recent Prog Horm Res 39: 1–74

Goodman RL (1978) Quantitative analysis of the physiological role of estradiol and progesterone in the control of tonic and surge secretion of luteinizing hormone in the rat. Endocrinology 102: 142–147

Gordon GG, Southren AL, Tochimoto S, Rand JJ, Olivo J (1969) Effects of hyperthyroidism and hypothyroidism on the metabolism of testosterone and androstenedione in man. J Clin Endocrinol Metab 29: 164–170

Gordon JH, Reichlin S (1974) Changes in pituitary responsiveness to luteinizing hormone-releasing factor during the rat estrous cycle. Endocrinology 94: 974–978

Gore BZ, Caldwell BV, Speroff L (1973) Estrogen-induced human luteolysis. J Clin Endocrinol Metab 36: 615–617

Gore-Langton RE, LaCroix M, Dorrington JH (1981) Differential effects of luteinizing hormone-releasing hormone on follicle stimulating hormone-dependent responses in rat granulosa cells and Sertoli cells in vitro. Endocrinology 108: 812–819

Grady RR, Charlesworth MC, Schwartz NB (1982) Characterization of FSH-suppressing activity in follicular fluid. Recent Prog Horm Res 38: 409-456

Grandison L, Hudson C, Chen HJ, Advis J, Simpkins J, Meites J (1977) Inhibition by prolactin of post-castration rise in LH. Neuroendocrinology 23: 312-322

Gray GD, Smith ER, Damassa DA, Ehrenkranz JRL, Davidson JM (1978) Neuroendocrine mechanisms mediating the suppression of circulating testosterone levels associated with chronic stress in male rats. Neuroendocrinology 25: 247-256

Greenwald GS (1965) Anti-ovulatory potency of various steroids determined by single injection into female hamsters. J Endocrinol 33: 25-32

Grodin JM, Siiteri PK, MacDonald PC (1973) Source of estrogen production in the postmenopausal woman. J Clin Endocrinol Metab 36: 207-214

Grossman CJ (1985) Interactions between the gonadal steroids and the immune system. Science 227: 257-261

Gupta C, Bullock LP, Bardin CW (1978) Further studies on the androgenic, anti-androgenic, and synandrogenic actions of progestins. Endocrinology 120: 736-744

Gustafson AW, Damassa DA (1986) Annual variations in plasma sex binding protein and testosterone concentrations in the adult male little brown bat *Myotis lucifugus lucifugus:* relation to the asynchronous recrudescence of the testis and accessory reproductive organs. Biol Reprod 33: 1126-1137

Gwynne JT, Strauss JF III (1982) The role of lipoproteins in steroidogenesis and cholesterol metabolism in steroidogenic glands. Endocr Rev 3: 299-244

Hagino N, Watanabe M, Goldzeiher JW (1969) Inhibition by adrenocorticotrophin of gonadotrophin-induced ovulation in immature female rats. Endocrinology 84: 308-314

Hahnel R, Twaddle E, Ratajczak T (1973) The specificity of the estrogen receptor of human uterus. J Steroid Biochem 4: 21-31

Halász B, Gorski RA (1967) Gonadotrophic hormone secretion in the female rat after partial or total interruption of neural afferents to the medial basal hypothalamus. Endocrinology 80: 608-622

Halász B, Pupp L, Uhlarik S, Tima L (1965) Further studies on the hormone secretion of the anterior pituitary transplanted into the hypophysiotrophic area of the rat hypothalamus. Endocrinology 77: 343-355

Hall NR, McGillis JP, Spangelo BL, Palaszynski E, Moody T, Goldstein AL (1982) Evidence for a neuroendocrine-thymus axis mediated by thymosin polypeptides. In: Serrou B, Rosenfeld C, Daniels JC, Saunders JP (eds) Current concepts in human immunology and cancer immunomodulation, Elsevier, Amsterdam, p 653

Hall PF (1985) Trophic stimulation of steriodogenesis: in search of the elusive trigger. Recent Prog Horm Res 41: 1-45

Hamada H, Neumann F, Junkmann K (1963) Intrauterine antimaskuline Beeinflussung von Rattenfeten durch ein stark gestagen wirksames Steroid. Acta Endocrinol (Copenh) 44: 380-388

Hamada Y, Bronson RA, Wright KH, Wallach EE (1977) Ovulation in the perfused rabbit ovary: the influences of prostaglandins and prostaglandin inhibitors. Biol Reprod 17: 58-63

Hamernik DL, Nett TM (1988) Gonadotropin-releasing hormone increases the amount of messenger ribonucleic acid for gonadotropins in ovariectomized ewes after hypothalamic-pituitary disconnection. Endocrinology 122: 959-966

Hammar M, Petersson F (1986) Testosterone production *in vitro* in human testicular tissue. Andrologia 18: 196-200

Hammond JM, Hersey RM, Walega MA, Weisz J (1986) Catecholestrogen production by porcine ovarian cells. Endocrinology 118: 2292-2299

Hanley AF, Schomberg DW (1978) Steroidal modulation of progesterone secretion by granulosa cells from large porcine follicles: a role for androgens and estrogens in controlling steroidogenesis. Biol Reprod 19: 242-248

Harris GW (1948) Neural control of the pituitary gland. Physiol Rev 28: 139-179

Harris GW (1964) Sex hormones, brain development and brain function. Endocrinology 75: 627-648

Hart GH, Mead SW, Reagan WM (1946) Stimulating the sex drive of bovine males in artificial insemination. Endocrinology 39: 221-228

Hart IC (1975) Seasonal factors affecting the release of prolactin in goats in response to milking. J Endocrinol 64: 313–322

Harwood JP, Clayton RN, Catt KJ (1980) Ovarian gonadotrophin releasing hormone receptors. I. Properties and inhibition of luteal cell function. Endocrinology 107: 407–413

Hauger RL, Karsch IJ, Foster DL (1977) A new concept for control of the estrous cycle of the ewe based on the temporal relationships between luteinizing hormone, estradiol and progesterone in peripherol serum and evidence that progesterone inhibits tonic LH secretion. Endocrinology 101: 807–817

Haynes NB, Hafs HD, Manns JG (1977) Effect of chronic administration of gonadotrophin releasing hormone and thyrotrophin releasing hormone to pubertal bulls on plasma luteinizing hormone, prolatin and testosterone concentrations, the number of epididymal sperm and body weight. J Endocrinol 73: 227–234

Healy DL, Hodgen GD, Schulte HM, Chrousos GP, Loriaux DL, Hall NR, Goldstein AL (1983) The thymus-adrenal connection: thymosin has corticotropin-releasing activity in primates. Science 222: 1353–1354

Heaulme M, Dray F (1984) Noradrenaline and prostaglandin E2 stimulate LH-RF release from rat median eminence through distinct 1-alpha-adrenergic and PGE_2 receptors. Neuroendocrinology 39: 403–407

Hedin L, Gaddy-Kurten D, Kurten R, De Witt DL, Smith WL, Richards JS (1987) Prostaglandin endoperoxide synthase in rat ovarian follicles: content, celluar distribution and evidence for hormonal induction preceding ovulation. Endocrinology 121: 722–731

Heiman ML, Ben-Jonathan N (1982) Dopaminergic receptors in the rat anterior pituitary change during the estrous cycle. Endocrinology 111: 37–41

Hellman L, Bradlow HL, Zumoff B, Fukushima DK, Gallagher TF (1959) Thyroid-androgen interrelations and the hypocholesteremic effect of androsterone. J Clin Endocrinol 19: 936–942

Henderson SB, Ciaccio LA, Kincl FA (1979) Dynamics of estradiol and testosterone uptake in the brain of adult male rats. J Steroid Biochem 11: 1601–1607

Henderson SR, Bonnar J, Moore A, Mackinnon PCB (1976) Luteinizing hormone-releasing hormone for induction of follicular maturation and ovulation in women with infertility and amenorrhea. Fertil Steril 27: 621–627

Henzl MR, Spaur CL, Magoun RE, Kincl FA (1970) A note on endocrine functions of neonatally pinealectomized rats. Endocrinol Exp (Bratisl) 4: 77–82

Herbert DC (1975) Localization of antisera to LH and FSH in the rat pituitary gland. Am J Anat 144: 379–385

Herbert DC (1976) Immunocytochemical evidence that luteinizing hormone (LH) and follicle stimulating hormone (FSH) are present in the same cell type in the Rhesus monkey pituitary gland. Endocrinology 98: 1554–1557

Higuchi T, Kawakami M (1982) Changes in the characteristics of pulsatile luteinizing hormone secretion during the oestrous cycle and after ovariectomy and oestrogen treatment in female rats. J Endocrinol 94: 177–182

Hillard J, Pang C-N, Sawyer CH (1976) Effects of luteinizing hormone-releasing hormone on fetal survival in pregnant rabbits. Fertil Steril 27: 421–425

Hillier SG (1987) Paracrine control in the ovaries. Res Reprod 19(4):1–2

Hillier SG, DeZwart FA (1981) Evidence that granulosa cell aromatase induction/activation by follicle-stimulating hormone is an androgen receptor-regulated process in vitro. Endocrinology 109: 1303–1305

Hirono M, Igarashi M, Matsumoto S (1970) Short and auto feedback control of pituitary FSH secretion. Neuroendocrinology 6: 274–281

Hiroshige T, Honma K, Watanabe K (1982) Prenatal onset and material modification of the circadian rhythm of plasma corticosterone in blinded infantile rats. J Physiol 325: 521–532

Hirschfield AN, Midgley AR (1978) The role of FSH in the selection of large ovarian follicles in the rat. Biol Reprod 19: 606–611

Hisaw FL (1926) Experimental relaxation of the pubic ligament of the guinea pig. Proc Soc Exp Biol Med 23: 661–669

Hoar RM, Goy RW, Young WC (1957) Loci of action of thyroid hormone on reproduction in the female guinea pig. Endocrinology 60: 337–343

Hoffmann F (1960) Untersuchungen über die hormonale Beeinflussung der Lebensdauer des Corpus luteum im Zyklus der Frau. Geburtshilfe Frauenheilkd 20: 1153-1157

Hohlweg W (1934) Veränderungen des Hypophysenvorderlappens und des Ovariums nach Behandlung mit großen Dosen von Follikelhormon. Klin Wochenschr 13: 92-95

Hohlweg W, Junkmann K (1932) Die Hormonale-nervose Regulierung der Funktion des Hypophysenvorderlappens. Klin Wochenschr 11: 321-329

Honna K, Wuttke W (1980) Norepinephrine and dopamine turnover rates in the medial preoptic area and the mediobasal hypothalmus in the rat brain after various endocrinological manipulations. Endocrinology 106: 1848-1853

Horton TH (1984) Growth and reproductive development in Microtus montanus is affected by the prenatal photoperiod. Biol Reprod 31: 499-504

Hsueh AJW, Erickson GF (1979) Extrapituitary action of gonadotropin releasing hormone: direct inhibition of ovarian steroidogenesis. Science 204: 854

Hsueh AJW, Jones PBC (1981) Extrapituitary actions of gonadotrophin-releasing hormone. Endocr Rev 2: 437-461

Hsueh AJW, Dufau ML, Catt KJ (1976) Regulation of luteinizing hormone receptors in testicular interstitial cells by gonadotropin. Biochem Biophys Res Commun 72: 1145-1152

Hsueh AJW, Dufau ML, Catt KJ (1978) Direct inhibitory effect of estrogen on Leydig cell function of hypophysectomized rats. Endocrinology 103: 1096-1102

Hsueh AJW, Erickson GF, Lu KH (1979) Changes in uterine estrogen receptor and morphology in aging female rats. Biol Reprod 21: 793-798

Hsueh AJW, Joneu PBC, Adashi EY, Wang C, Zhuang L, Weish TH Jr (1983) Introvarian mechanisms in the hormonal control of granulosa cell differentiation. J Reprod Fertil 69: 325-329

Hsueh AJW, Adashi EY, Jones PBC, Welsh TH (1984) Hormonal regulation of the differentiation of cultured ovarian granulosa cells. Endocr Rev 5: 76-103

Hsueh AJW, Dahl KD, Vaughan J, Tucker E, Rivier J, Bardin CW, Vale W (1987) Heterodimers and homodimers of inhibin subunits have different paracrine action in the modulation of luteinizing hormone-stimulated androgen biosynthesis. Proc Natl Acad Sci USA 84: 5082-5085

Hsueh AJW, Liu Y-X, Cajander S, Peng X-R, Dahl K, Kristensen P, Tor NYT (1988) Gonadotropin-releasing hormone induces ovulation in hypophysectomized rats: studies on ovarian tissue-type plasminogen activator activity, messenger ribonucleic acid content, and cellular localization. Endocrinology 122: 1486-1495

Huang ESR, Miller WL (1980) Effects of estradiol-17 on basal and luteinizing hormone and follicle stimulating hormone by ovine pituitary cell culture. Biol Reprod 23: 124-128

Hudson KE, Hillier SG (1985) Catechol oestradiol control of FSH-stimulated granulosa cell steroidogenesis. J Endocrinol 106: R1

Huggins C (1956) Augmentation and depression of estriol-induced growth of the uterus by progesterone. Proc Soc Exp Biol Med 92: 304-305

Hughes Wiley M, Howton MM, Siperstein MD (1979) Sex differences in the sterol and nonsterol metabolism of the rat. J Biol Chem 254: 837-842

Humphrey RR, Windsor BL, Jones DC, Reel JR, Edgren RA (1977 a) The progesterone-sensitive period of rat pregnancy: some effects of LHRH and ovariectomy. Proc Exp Biol Med 156: 345-348

Humphrey RR, Windsor BL, Reel JR, Edgren RA (1977 b) The effects of luteinizing hormone releasing hormone (LH-RH) in pregnant rats. I. Postnidatory effects. Biol Reprod 16: 614-621

Humphrey RR, Windsor BL, Jones DC, Reel JR, Edgren RA (1978) The effects of luteinizing hormone releasing hormone (LHRH) in pregnant rats. 2. Prenidatory effects of delayed parturition. Biol Reprod 19: 84-91

Hutchison JS, Kubik CJ, Nelson PB, Zeleznik AJ (1987 a) Estrogen induces premature luteal regression in rhesus monkeys during spontaneous menstrual cycles, but not in cycles driven by exogenous gonadotrophin-releasing hormone. Endocrinology 121: 466-474

Hutchinson LA, Findlay Jr, de Vos FL, Robertson DM (1987 b) Effects of bovine inhibin, transforming growth factor and bovine activin-A on granulosa cell differentiation. Biochem Biophys Res Commun 146: 1405-1410

Hutson JM, Donahoe PK (1986) The hormonal control of testicular descent. Endocr Rev 7: 270-283

Hwan JC, Freeman ME (1987) Partial purification of a hypothalamic factor that inhibits gonadotrophin-releasing hormone-stimulated luteinizing hormone release. Endocrinology 120:
483-490

Hyde SF, Ben-Jonathan N (1988) Characterization of prolactin-releasing factor in the posterior
pituitary. Endocrinology 122: 2533-2539

Ilenchuk TT, Walters MR (1987) Rat uterine progesterone receptor analyzed by [^{3}H]R5020 photo-
affinity labeling: evidence that the A and B subunits are not equimolar. Endocrinology 120:
1449-1456

Imperato-McGinley J, Peterson RE (1976) Male pseudohermaphroditism, the complexities of
male phenotypic development. Am J Med 61: 251-272

Jackson GL (1975) Blockage of progesterone-induced release of LH by intrabrain implants of actinomycin D. Neuroendocrinology 17: 236-244

Jaffee RB (1969) Testosterone metabolism in target tissue hypothalamic and pituitary tissues of
the adult rat and human fetus, and the immature rat epiphysis. Steroids 14: 483-498

Jaffee RB, Keye Jr WR (1974) Modulation of pituitary gonadotropin response to gonadotropin-
releasing hormone by estradiol. J Clin Endocrinol Metab 39: 850-855

Janes RG (1944) Occurrence of follicular cysts in thyroidectomized rats treated with diethylstilbestrol. Anat Rec 90: 93-101

Jänne O, Kontula K, Vihko R (1976) Progestin receptors in the human tissue: concentrations and
binding kinetics. J Steroid Biochem 7: 1061-1068

Jarjour LT, Handelsman DJ, Swerdloff RS (1986) Effects of aging on the *in vitro* release of gonadotropin-releasing hormone. Endocrinology 119: 1113-1117

Jensen EV, Desombre ER (1972) Mechanism of action of the female sex hormones. Ann Rev Biochem 41: 203

Jensen EV, Jacobson HI (1962) Basic guide to the mechanism of oestrogen action. Recent Prog
Horm Res 18: 387-408

Jensen EV, Green GL, Closs LE, DeSombre ER, Nadji M (1982) Receptors reconsidered: a
20-year perspective. Recent Prog Horm Res 38: 1-40

Jia X-C, Kessel B, Welsh THV Jr, Hsueh AJW (1985) Androgen inhibition of follicle-stimulating
hormone-stimulated luteinizing hormone receptor formation in cultured rat granulosa cells. Endocrinology 117: 13-22

Johnson DC (1974) Temporal augmentation of LH by prolactin in stimulation of androgen production by the testis of hypophysectomized male rats. Proc Soc Exp Biol Med 145: 610-614

Jones PBC, Hsueh AJW (1981) Direct stimulation of ovarian progesterone metabolizing enzyme by gonadotropin-releasing hormone in cultured granulosa cells. J Biol Chem 256: 1248-
1253

Jones PBC, Hsueh AJW (1982 a) Regulation of ovarian 3 β-hydroxysteroid dehydrogenase activity by gonadotropin-releasing hormone and follicle-stimulating hormone in cultured rat granulosa cells. Endocrinology 110: 1663-1671

Jones PBC, Hsueh AJW (1982 b) Pregnenolone biosynthesis by cultured rat granulosa cells: modulation by follicle-stimulating hormone and gonadotropin-releasing hormone. Endocrinology
110: 713-721

Jones RC (1979 a) The fate of fertilized ova in rats receiving luteinizing hormone-releasing hormone (LRH). Endocr Res Commun 6: 159-167

Jones RC (1979 b) Reversal of the anti-implantational effect of luteinizing hormone-releasing hormone (LRH) by estradiol-17 and progesterone, alone and in sequence. Contraception 19:
239-245

Jones RC (1979 c) Local antifertility effect of luteinizing hormone-releasing hormone (LRH).
Contraception 20: 569-578

Jones RC (1980) Local antifertility effect of luteinizing hormone-releasing hormone (LRH) in the
rat: mechanism of action. Endocr Res Commun 7: 107-112

Jones RC, Cole G, Koch K (1979) Anti-implantational effect of luteinizing hormone releasing
hormone (LRH) in the rat. Int J Fertil 24: 202-205

Junkmann K (1952) Über protrahiert wirksame Androgene. Arch Exp Pathol Pharmakol 215:
85-92

Junkmann K (1953) Über protrahiert wirksame Östrogene. Arch Exp Pathol Pharmakol 220:
358-364

Junkmann K (1954 a) Über protrahiert wirksame Gestagene. Arch Exp Pathol Pharmakol 223: 224-253

Junkmann K (1954 b) Über protrahiert wirksame Corticoide. Arch Exp Pathol Pharmakol 223: 280-284

Junkmann K (1957) Long-acting steroids in reproduction. Recent Prog Horm Res 13: 389-428

Kaji H, Chihara K, Abe H, Minamitani N, Kodama H, Kita T, Fujita T, Tatemoto K (1984) Stimulatory effect of peptide histidine isoleucine amide 1-27 on prolactin release in the rat. Life Sci 35: 641-647

Kalland GA, Vera A, Peterson M, Swerdloff RS (1978) Reproductive hormonal axis of the male rat in experimental hypothyroidism. Endocrinology 102: 476-484

Kalra SP, Kalra PS (1983) Neural regulation of luteinizing hormone secretion in the rat. Endocr Rev 4: 311-351

Kalra SP, McCann SM (1974) Effects of drugs modifying catecholamine synthesis on plasma LH and ovulation in the rat. Neuroendocrinology 15: 79-91

Kalra SP, Simpkins JW (1981) Evidence for noradrenergic mediation of opioid effects on luteinizing hormone secretion. Endocrinology 109: 776-782

Kalra SP, Ajika K, Krulich L, Fawcett CP, Quijada M, McCann SM (1971) Effects of hypothalamic and preoptic electrochemical stimulation on gonadotropin and prolactin release in proestrous rats. Endocrinology 88: 115-1158

Kalra PS, Kalra SP, Krulich L, Fawcett CP, McCann SM (1972) Involvement of norepinephrine in transmission of the stimulatory influence of progesterone on gonadotropin release. Endocrinology 90: 1168-1176

Kalra PS, Sahu A, Kalra SP (1988) Opiate-induced hypersensitivity to testosterone feedback: pituitary involvement. Endocrinology 122: 997-1003

Kamel F, Krey LC (1982) Gonadal steroid modulation of LHRH stimulated LH secretion by pituitary cell cultures. Mol Cell Endocrinol 26: 151-156

Kamel F, Kubajak CL (1987) Modulation of gonadotropin secretion by corticosterone: interaction with gonadal steroids and mechanism of action. Endocrinology 121: 561-568

Kamel F, Balz JA, Kubajak CL, Schneider VA (1987 a) Effects of luteinizing hormone (LH)-releasing hormone pulse amplitude and frequency on LH secretion by perifused rat anterior pituitary cells. Endocrinology 120: 1644-1650

Kamel F, Balz JA, Kubajak CL, Schneider VA (1987 b) Gonadal steroids modulate pulsatile luteinizing hormone secretion by perifused rat anterior pituitary cells. Endocrinology 120: 1651-1657

Kaneyuki T, Kohsaka M, Shohmori T (1979) Sex hormone metabolism in the brain: influence of central acting drugs on 5a-reduction in rat diencephalon. Endocrinol Jpn 26: 345-351

Kanungo MS, Patnaik SK, Koul O (1975) Decrease in 17β-oestradiol receptor in brain of ageing rats. Nature 253: 366-369

Karg H, Schams D (1974) Prolactin release in cattle. J Reprod Fertil 39: 463-472

Karsch FJ, Krey LC, Weick RF, Dierschke DJ, Knobil E (1973) Functional luteolysis in the rhesus monkey: the role of estrogen. Endocrinology 92: 1148-1152

Karsch FJ, Bittman EL, Foster DL, Goodman RL, Legan SJ, Robinson JE (1984) Neuroendocrine basis of seasonal reproduction. Recent Prog Horm Res 40: 185-232

Karsch FJ, Foster DL, Bittman EL, Goodman RL (1985) A role for estradiol in enhancing luteinizing hormone pulse frequency during the follicular phase of the estrous cycle of sheep. Endocrinology 113: 1333-1339

Kasson BG, Adashi EY, Hsueh AJW (1986) Arginine vasopressin in the testis: an intragonadal peptide control system. Endocr Rev 7: 156-168

Kato J (1976) Cytosol and nuclear receptors for 5α-dihydrotestosterone and testosterone in the hypothalamus and hypophysis, and testosterone receptors isolated from neonatal female rat hypothalamus. J Steroid Biochem 7: 1179-1187

Kato J, Onouchi T (1977) Specific progesterone receptors in the hypothalamus and anterior hypophysis of the rat. Endocrinology 101: 920-928

Kato J, Onouchi T (1981) Progesterone receptors in the cerebral cortex of neonatal female rats. Dev Neurosci 4: 427-432

Kato J, Kobayashi T, Villee CA (1968) Effect of clomiphene on the uptake of oestradiol by the anterior hypothalamus and hypophysis. Endocrinology 82: 1049-1052

Kato Y, Iwasaki Y, Iwasaki J, Abe H, Yanaihara N, Imura H (1978) Prolactin release by vasoactive intestinal polypeptide in rats. Endocrinology 103: 554–558

Kawakami M, Ando S (1981) Lateral hypothalamic mediation of midbrain catecholaminergic influences on preovulatory surges of serum gonadotropin and prolactin in female rats. Endocrinology 108: 66–71

Kawakami M, Higuchi T (1979) Effects of active and passive immunization with LH-RH on gonadotropin secretion and reproductive function in female rats. Acta Endocrinol (Copenh) 91: 616–628

Kawakami M, Sawyer CH (1959) Neuroendocrine correlates of changes in brain activity thresholds by sex steroids and pituitary hormones. Endocrinology 65: 652–668

Kawakami M, Arita J, Kimura F, Hayashi R (1979) The stimulatory roles of catecholamines and acetylcholine in the regulation of gonadotropin release in ovariectomized estrogen-primed rats. Endocrinol Jpn 26: 275–281

Kawashima S, Takahashi S, Ueda K (1981) Effects of pinealectomy on the entrainability of locomotor activity rhythm to changing photoperiod. Jikeikai Med J 28 (Suppl 1): 23–28

Kelly BH (1937) Studies in fertility of sheep. CSIRO Bull (Australia) 112

Kennaway DJ, Obst JM, Dunstan EA, Friesen HG (1981) Ultradian and seasonal rhythms in plasma gonadotropins, prolactin, cortisol, and testosterone pinealectomized rams. Endocrinology 108: 639–646

Kennedy TG (1980) Estrogen and uterine sensitization for the decidual cell reaction: role of prostaglandins. Biol Reprod 23: 955–962

Kesner JS, Convey EM, Anderson CR (1981) Evidence that estradiol induces the preovulatory LH surge in cattle by increasing pituitary sensitivity to LHRH and then increasing LHRH release. Endocrinology 108: 1386–1391

Kesner JS, Riebold TW, Convey EM (1982) Effect of dosage and frequency of injection of luteinizing hormone releasing hormone on release of luteinizing hormone and follicle stimulating hormone in estradiol-treated steers. J Anim Sci 54: 1023–1027

Kesner JS, Kraeling RR, Rampacek GB, Johnson B (1987) Absence of an estradiol-induced surge of luteinizing hormone in pigs receiving unvarying pulsatile gonadotropin-releasing hormone stimulation. Endocrinology 121: 1862–1869

Ketchel MM, Pincus G (1964) In vitro exposure of rabbit ova to estrogens. Proc Soc Exp Biol Med 115: 419–421

Ketelslegers J-M, Hetzel WD, Sherins RJ, Catt KJ (1978) Developmental changes in testicular gonadotropin receptors: plasma gonadotropins and plasma testosterone in the rat. Endocrinology 103: 212–222

Khodr GS, Siler-Khodr TM (1980) Placental luteinizing hormone-releasing factor and its synthesis. Science 207: 315–317

Khorram O, McCann SM (1986) Interaction of α-melanocyte-stimulating hormone with β-endorphin to influence anterior pituitary hormone secretion in the female rat. Endocrinology 119: 1071–1075

Khorram O, Bedran DeCastro JC, McCann SM (1985) Stress-induced secretion of α-melanocyte-stimulating hormone and its physiological role in modulating the secretion of prolactin and luteinizing hormone in the female rat. Endocrinology 117: 2483–2489

Kicovic PM, Luisi M, Cortes-Prieto J, Franchi F (1980) Fluctuations in plasma estradiol levels in normal men. Reproduccion 4: 49–53

Kincl FA (1963) Notiz über den Mechanismus der Anti-Ovulation mit 6-chloro-Δ^6-17α-Acetoxyprogesteron in Kaninchen. Endokrinologie 44: 66–71

Kincl FA (1964) Copulatory reflex response to steroids. In: Dorfman RI (ed) Methods in hormone research, vol III. Academic Press, New York, pp 477–484

Kincl FA (1965) Anabolic steroids. In: Dorfman RI (ed) Methods in hormone research, vol IV. Academic Press, New York, pp 21–76

Kincl FA (1966) Failure of various steroids to block ovulation at the ovarian level. In: Cassano C (ed) Research on steroids. Il Pensero Scientifico, Rome, pp 353–356

Kincl FA (1971 a) Progesterone as testosterone antagonist. In: Tausk M (ed) Pharmacology of the endocrine system and related drugs: progesterone, progestational drugs and antifertility agents. Pergamon, Oxford. International Encyclopedia of Pharmacology and Therapeutics, Sect 48 vol 1, pp 325–329

Kincl FA (1971 b) Synergism with and antagonism to estrogens. In: Tausk M (ed) Pharmacology of the endocrine system and related drugs: progesterone, progestational drugs and antifertility agents. Pergamon, Oxford. International Encyclopedia of Pharmacology and Therapeutics, Sect 48, vol 1, pp 275–290

Kincl FA (1972) Estrogens as antifertility agents. In: Tausk M (ed) Pharmacology of the endocrine system and related drugs: progesterone, progestational drugs and antifertility agents. Pergamon Press, Oxford. International Encyclopedia of Pharmacology and Therapeutics, Sect 48, vol 2, pp 347–384

Kincl FA, Benagiano G (1967) The failure of the pineal gland removal in neonatal animals to influence reproduction. Acta Endocrinol (Copenh) 54: 189–192

Kincl FA, Ciaccio LA (1980) Suppression of immune response by progesterone. Endocr Exp 14: 27–33

Kincl FA, Dorfman RI (1963) Anti-ovulatory activity of steroids in the adult oestrous rabbit. Acta Endocrinol (Copenh) 42 (Suppl 73): 3–30

Kincl FA, Rudel HW (1971) Sustained release hormonal preparations. Acta Endocrinol (Copenh) (Supl) 151: 6–45

Kincl FA, Chang CC, Zbuzkova V (1970) Observation of the influence of changing photoperiods on spontaneous wheel-running activity of neonatally pinealectomized rats. Endocrinology 87: 38–42

Kirton KT, Kimball FA, Porteus SE (1976) Reproductive physiology prostaglandin associated events. In: Samuelson B, Paoletti R (eds) Advances in prostaglandin and thromboxane research. vol 2. Raven Press, New York, pp 621–675

Klotz IM (1982) Numbers of receptor sites from scatchard graphs: facts and fantasies. Science 217: 1247–1249

Knauer E (1900) Die Ovarientransplantation. Arch Gynaekol 60: 322–376

Knecht M, Amsterdam A, Catt KJ (1982) Inhibition of granulosa cell differentiation by gonadotropin-releasing hormone. Endocrinology 110: 865–872

Kniewald Z, Milkovic S (1973) Testosterone: a regulator of 5α reductase activity in the pituitary of male and female rats. Endocrinology 92: 1772–1775

Knobil E (1980) The neuroendocrine control of the menstrual cycle. Recent Prog Horm Res 36: 53–88

Kochakian CD, Murlin JR (1935) Effect of male hormone on protein metabolism of castrate dogs. Proc Soc Exp Biol Med 32: 1064–1065

Kochakian CD, Murlin JR (1936) The relationship of the synthetic male hormone, androstenedione, to the protein and energy metabolism of castrate dogs and the protein metabolism of a normal dog. Am J Physiol 117: 642–654

Kontula KO, Janne O, Janne J, Vihko R (1972) Partial purification and characterization of progesterone-binding protein from pregnant guinea-pig uterus. Biochem Biophys Res Commun 47: 596–603

Koos RD, Feirtag MA, Brodie AMH, LeMaire WJ (1984) Inhibition of estrogen synthesis does not inhibit luteinizing hormone-induced ovulation. Am J Obstet Gynecol 148: 939–945

Koprowski JA, Tucker HA (1973) Serum prolactin during various physiological states and its relationship to milk production in the bovine. Endocrinology 92: 1480–1487

Korach KS, Muldoon TG (1973) Comparison of specific 17β-estradiol-receptor interactions in the anterior pituitary of male and female rats. Endocrinology 92: 322–326

Korach KS, Muldoon TG (1974a) Studies on the nature of the hypothalamic estradiol-concentrating mechanism in the male and female rat. Endocrinology 94: 785–793

Korach KS, Muldoon TG (1974b) Characterization of the interaction between 17β-estradiol and its cytoplasmic receptor in the rat anterior pituitary gland. Biochemistry 13: 1932–1936

Krähenbühl C, Desaulles P (1964) The action of sex hormones on gonadotrophin induced ovulation in hypophysectomized prepubertal rats. Acta Endocrinol (Copenh) 47: 457–465

Kreuz LE, Rose RM, Jennings JR (1972) Suppression of plasma testosterone levels and psychological stress. Arch Gen Psychiatry 26: 479–485

Krey LC, Silverman AJ (1978) The luteinizing hormone releasing hormone neuronal networks of the guinea-pig brain. Part 2. The regulation on gonadotropin secretion and the origin of terminals in the median eminence. Brain Res 157: 247–256

Krey LC, Silverman AJ (1981) LHRH neuronal networks of the guinea-pig brain regulation of cyclic gonadotropin secretion. Brain Res 229: 429–444

Krulich L (1979) Central neurotransmitters and the secretion of prolactin, GH, LH and TSH. Ann Rev Physiol 41: 603-615

Krulich L, Hefco E, Illner P, Read CB (1974) The effects of acute stress on the secretion of LH, FSH, prolactin and GH in the normal male rat, with comments on their statistical evaluation. Neuroendocrinology 16: 293-311

Laborde NP, Wolfsen AR, Odell WD (1981) Short loop feedback system for the control of follicle-stimulating hormone in the rabbit. Endocrinology 108: 72-75

Labrie F, Ferland L, Denizeau F, Beaulieu M (1980) Sex steroids interact with dopamine at the hypothalamic and pituitary levels to modulate prolactin secretion. J Steroid Biochem 12: 323-330

Lang U, Aubert ML, Rivest RW, Vinas-Bradtke JC, Sizonenko PC (1984) Daily afternoon administration of melatonin does not irreversibly inhibit sexual maturation in the male rat. Endocrinology 115: 2303-2310

Lasley BL, Wang CF, Yen SSC (1975) The effects of estrogen and progesterone on the functional capacity of the gonadotrophs. J Clin Endocrinol Metab 41: 820-826

Leathem JH (1961) Nutritional effects on endocrine secretions. In: Yound WC (ed) Sex and internal secretions. Williams and Wilkins, Baltimore, pp 666-706

Lee VWK, Keogh EJ, Burger HG, Hudson B, de Kretser DM (1976) Studies on the relationship between FSH and germ cells: evidence for selective suppression of FSH by testicular extracts. J Reprod Fertil [Suppl] 24: 1-16

Lee VWK, McMaster J, Quigg H, Findlay J, Leversha L (1981) Ovarian and peripheral blood inhibin concentrations increase with gonadotrophin treatment in immature rats. Endocrinology 108: 2403-2405

Lee VWK, McMaster J, Quigg H, Leversha L (1982) Ovarian and circulating inhibin levels in immature female rats treated with gonadotrophin and after castration. Endocrinology 111: 1849-1854

Legan S (1980) Comment. Recent Prog Horm Res 36: 51

Leining RA, Bourne RA, Tucker AH (1979) Prolactin response to duration of wavelength of light in prepubertal bulls. Endocrinology 104: 289-294

Leipheimer RE, Bona-Gallo A, Gallo RV (1986) Ovarian steroid regulation of basal pulsatile luteinizing hormone release between the mornings of proestrus and estrus in the rat. Endocrinology 118: 2083-2090

LeMaire WJ, Clark MR, Marsh JM (1978) Biochemical mechanism of ovulation. In: Hafez ESE (ed) Human ovulation: mechanism, detection and regulation. Elsevier/North Holland, Amsterdam, pp 159-175

Lenahan SE, Seibel HR, Johnson JH (1987) Opiate-serotonin synergism stimulating luteinizing hormone release in estrogen-progesterone-primed ovariectomized rats: mediation by serotonin$_2$ receptors. Endocrinology 120: 1498-1502

Lenzi E, Marino C (1947) Modificazioni istologiche prodotte nel testicolo e nell'ovaio dalla tireotossicosi sperimentale subacuta e eronica (con particolare riguardo alla biometrica delle cellule interstiziali). Arch De Vecchi Anat Pathol 9: 711-720

Leung PCK, Goff AK, Armstrong DT (1979) Stimulatory effect of androgen administration *in vivo* on ovarian responsiveness to gonadotropins. Endocrinology 104: 1119-1123

Levine JE, Ramirez VD (1982) Luteinizing hormone-releasing hormone release during the rat estrous cycle and after ovariectomy, as estimated with push-pull cannulae. Endocrinology 111: 1349-1448

Lewy AJ, Sack RL, Miller SL, Hoban TM (1987) Antidepressant and circadian phase-shifting effects of light. Science 235: 352-354

Lieberburg I, Maclusky NJ, McEwen BS (1977) 5-dihydrotestosterone (DHT) receptors in rat brain and pituitary cell nuclei. Endocrinology 100: 598-607

Light KE (1984) Analyzing nonlinear scatchard plots. Science 223: 76-77

Lin WW, Ramirez VD (1988) Effect of pulsatile infusion of progesterone on the in vivo activity of the luteinizing hormone-releasing hormone neural apparatus of awake unrestrained female and male rabbits. Endocrinology 122: 868-876

Lincoln DW, Fraser HM, Lincoln GA, Martin GB, McNeilly AS (1985) Hypothalamic pulse generators. Recent Prog Horm Res 41: 369-420

Lincoln GA, Short RV (1980) Seasonal breeding: nature's contraceptive. Recent Prog Horm Res 36: 1-43

Lincoln GA, McNeilly AS, Camerson CL (1978) The effects of a sudden decrease or increase in daylength on prolactin secretion in the ram. J Reprod Fertil 52: 305–311

Lindsay DR (1965) The importance of olfactory stimuli in the mating behavior of the ram. Animal Behav 13: 75–78

Ling N, Ying S-Y, Ueno N, Esch F, Denoroy L, Guillemin R (1985) Isolation and partial characterization of a Mr 32 000 protein with inhibin activity from porcine follicular fluid. Proc Natl Acad Sci USA 82: 7217–7219

Ling N, Ying S-Y, Ueno N, Shimasaki S, Esch F, Hotta M, Guillemin R (1986 a) Pituitary FSH is released by a heterodimer of the β-subunits from the two forms of inhibin. Nature 321: 779–781

Ling N, Ying S-Y, Ueno N, Shimasaki S, Esch F, Hotta M, Guillemin R (1986 b) A homodimer of the β-subunits of inhibin A stimulates the secretion of pituitary follicle stimulating hormone. Biochem Biophys Res Commun 138: 1129–1134

Lipner H, Wendelken L (1971) Inhibition of ovulation by inhibition of steroidogenesis in immature rats. Proc Soc Exp Biol Med 136: 1141–1145

Liu FTY, Lin HS, Johnson DC (1972) Serum FSH, LH and the ovarian response to exogenous gonadotropins in alloxan diabetic immature female rats. Endocrinology 91: 1172–1179

Lloyd JM, Childs GV (1988) Differential storage and release of luteinizing hormone and follicle-releasing hormone from individual gonadotropes separated by centrifugal elutriation. Endocrinology 122: 1282–1290

Loeken MR, Channing CP, D'Eletto R, Weiss G (1983) Stimulatory effect of luteinizing hormone upon relaxin secretion by cultured porcine preovulatory granulosa cells. Endocrinology 112: 769–773

Logeat F, Sartor P, Hai MTV (1980) Local effect of the blastocyst on estrogen and progesterone receptors in the rat endometrium. Science 207: 1083–1085

Lostroh AJ, Johnson RE (1966) Amounts of interstitial cell-stimulating hormone and follicle-stimulating hormone required for follicular development, uterine growth, and ovulation in the hypophysectomized rat. Endocrinology 79: 991–996

Lott DF, Hopwood JH (1972) Olfactory pregnancy-block in mice *(Mus musculus):* an unusual response acquisition paradigm. Anim Behav 20: 263–267

Lu KH, Chen HT, Gardison L, Huang HH, Meites J (1976) Reduced luteinizing hormone release by synthetic luteinizing hormone releasing hormone (LHRH) in postpartum lactating rats. Endocrinology 98: 1235–1240

Lubahn DB, Joseph DR, Sulivan PM, Willard HF, French FS, Wilson EM (1988) Cloning of human androgen receptor complementary DNA and localization to the X chromosome. Science 240: 327–330

Lucky AW, Schreiber JR, Hillier SG, Schulman JD, Ross GT (1977) Progesterone production by cultured preantral rat granulosa cells: stimulation by androgens. Endocrinology 100: 128–133

Lumpkin MD, Negro-Vilar A, Franchimont P, McCann S (1981) Evidence for a hypothalamic site of action of inhibin to suppress FSH release. Endocrinology 108: 1101–1104

Lumpkin MD, DePaolo LV, Negro-Vilar A (1984) Pulsatile release of follicle-stimulating hormone in ovariectomized rats is inhibited by porcine follicular fluid (inhibin). Endocrinology 114: 201–206

Lynch KM Jr (1952) Recovery of the rat testis following estrogen therapy. Ann NY Acad Sci 55: 734–742

MacLaughlin DT, Richardson GS (1976) Progesterone binding by normal and abnormal human endometrium. J Clin Endocrinol Metab 42: 667–678

MacLeod RM, Lehmeyer JE (1974) Studies on the mechanism of the dopamine-mediated inhibition of prolactin secretion. Endocrinology 94: 1077–1085

MacLeod RM, Login IS (1977) Regulation of prolactin secretion through dopamine, serotonin and the cerebrospinal fluid. Adv Biochem Psychopharmacol 16: 147–162

MacLusky NJ, McEwen BS (1978) Oestrogen modulates progestin receptors concentration in some rat brain regions but not in others. Nature 274: 276

MacLusky NJ, McEwen BS (1980) Progestin receptors in rat brain: distribution and properties of cytoplasmic progestion-binding sites. Endocrinology 106: 192–202

Magoffin DA, Erickson GF (1981) Mechanism by which 17β-estradiol inhibits ovarian androgen production in the rat. Endocrinology 108: 962–969

Mainwaring WIP, Mangan FR (1973) A study of the androgen receptors in a variety of androgen-sensitive tissues. J Endocrinol 59: 121–127

Majewska MD, Harrison NL, Schwartz RD, Barker JL, Paul SM (1986) Steroid hormone metabolites are barbiturate-like modulators of the GABA receptor. Science 232: 1004–1007

Makepeace AW, Weinstein GL, Friedman MH (1937) Effect of progestin and progesterone on ovulation in the rabbit. Am J Physiol 119: 512–516

Mandl AM (1954) The sensitivity of adrenalectomized rats to gonadotrophins. J Endocrinol 11: 359–365

Mann DR, Evans D, Edoimioya F, Kamel F, Butterstein GM (1985) A detailed examination of the in vivo and in vitro effects of ACTH on gonadotropin secretion in the adult rat. Neuroendocrinology 40: 297–302

Mann DR, Free C, Nelson C, Scott C, Collins DC (1987) Mutually independent effects of adrenocorticotropin on luteinizing hormone and testosterone secretion. Endocrinology 120: 1542–1550

Maqsood M (1955) Influence of thyroid status on sexual development and fertility in the male rabbit and sheep. Zootechnia 5: 258–288

Maqsood M, Reineke EP (1950) Influence of environmental temperatures and thyroid status on sexual development in male mouse. Am J Physiol 162: 24–31

Marchetti B, Labrie F (1982) Prolactin inhibits pituitary luteinizing hormone-releasing hormone receptors in the rat. Endocrinology 111: 1209–1216

Marian J, Cooper RL, Conn PM (1981) Regulation of the rat pituitary gonadotropin-releasing hormone receptor. Mol Pharmacol 19: 399–405

Martin GB, Wallace JM, Taylor PL, Fraser HM, Tsonis CG, McNeilly AS (1986) The roles of inhibin and gonadotrophin-releasing hormone in the control of gonadotrophin secretion in the ewe. J Endocrinol 111: 287–296

Martin JB, Durand D, Gurd W, Faille G, Audet J, Brazeau P (1978) Neuropharmacological regulation of episodic growth hormone and prolactin secretion in the rat. Endocrinology 102: 106–113

Martinovic JV, McCann SM (1977) Effect of lesions in the ventral noradrenergic tract produced by microinjection of 6-hydroxydopamine on gonadotropin release in the rat. Endocrinology 100: 1206–1213

Martins T, Rocha A (1931) The regulation of the hypophysis by the testicle and some problems of sexual dynamics. Endocrinology 15: 421–429

Marut EL, Williams RF, Cowan BD, Lynch A, Lerner SP, Hodgen GD (1981) Pulsatile pituitary gonadotropin secretion during maturation of the dominant follicle in monkeys: estrogen positive feedback enhances the biological activity of luteinizing hormone. Endocrinology 109: 2270–2276

Masotto C, Negro-Vilar A (1988) Gonadectomy influences the inhibitory effect of the endogenous opiate system on pulsatile gonadotropin secretion. Endocrinology 123: 747–752

Massiocotte J, Vailleux R, Lavoie M, Labrie F (1980) An LHRH agonist inhibits FSH-induced cyclic AMP accumulation and steroidogenesis in porcine granulosa cells in culture. Biochem Biophys Res Commun 94: 1362–1366

Masson G, Romanchuck M (1945) Observations sur certain changements morphologiques produits par la thyroxine et la testostérone. Rev Can Biol 4: 206–212

Maurel D, Boissin J (1982) Peripheral metabolism of testosterone in relation to the annual cycle of plasma testosterone and 5α-dihydrotestosterone in the badger Meles meles and the fox Vulpes vulpes. Can J Zool 60: 406–416

May JV, Schomberg DW (1981) Granulosa cell differentiation in vitro: effect of insulin on growth and functional integrity. Biol Reprod 25: 421–431

May RM (1988) How many species are there on earth? Science 241: 1441–1449

McCormack CE, Meyer RK (1963) Ovulating hormone release in gonadotropin-treated immature rats. Proc Soc Exp Biol Med 110: 343–346

McCraken JA (1974) Discussion. Recent Prog Horm Res 30: 39–40

McCullagh DR (1932) Dual endocrine activity of testis. Science 76: 19–29

McEwen BS, Lieberburg I, Chaptal C, Krey LC (1977) Aromatization: important for sexual differentiation of the neonatal rat brain. Horm Behav 9: 249–263

McEwen BS, Biegon A, Davis PG, Krey LC, Luine VN, McGinnis MY, Paden CM, Parsons B,

Rainbow TC (1982) Steroid hormones: humoral signals which alter brain cell properties and functions. Recent Prog Horm Res 38: 41-92

McGillis JP, Hall NR, Vahouny GV, Goldstein AL (1985) Thymosin fraction 5 causes increased serum corticosterone in rodents *in vivo*. J Immunol 134: 3952-3955

McGuire JL, Bariso CD (1972) Isolation and preliminary characterization of a progestogen specific binding macromolecule from the 273 000 × g supernatant of rat and rabbin uteri. Endocrinology 90: 496-506

McIntosh EN, Uzgiris VI, Alonso C, Salhanick HA (1971) Spectral properties, respiratory activity, and enzyme systems of bovine corpus luteum mitochondria. Biochemistry 10: 2909-2913

McNatty KP, Sawers RS, McNeilly AS (1974) A possible role for prolactin in control of steroid secretion by the human Graafian follicle. Nature 250: 653-655

McNatty KP, Neal P, Baker TG (1976) Effect of prolactin on the production of progesterone by mouse ovaries *in vitro*. Reprod Fertil 47: 155-161

McNeill TH, Sladek JR Jr (1978) Fluorescence-immunocytochemistry: simultaneous localization of catecholamines and gonadotropin-releasing hormone. Science 200: 72-74

McNeilly AS, Sharpe RM, Davidson DW, Fraser HM (1978) Inhibition of gonadotrophin secretion by induced hyperprolactinemia in the male rat. J Endocrinol 79: 59-68

McNeilly AS, Sharpe RM, Fraser HM (1983) Increased sensitivity to the negative feedback effects of testosterone induced by hyperprolactinemia in the adult male rat. Endocrinology 112: 22-27

Means AR, Tindal DJ (1975) FSH-induction of androgen binding protein in testes of Sertoli cell-only rats. In: French FS, Hansson V, Ritzen EM, Nayfeh SN (eds) Current topics in molecular endocrinology, vol 2. Plenum Press, NY, p 383

Melrose P, Gross L (1987) Steroid effects on the secretory modalities of gonadotropin-releasing hormone release. Endocrinology 121: 190-199

Melrose P, Gross L, Cruse I, Rush M (1987) Isolated gonadotropin-releasing hormone neurons harvested from adult male rats secret biologically active neuropeptide in a regular repetitive manner. Endocrinology 121: 182-189

Mendelson CR, Cleland WH, Smith ME, Simpson ER (1982) Regulation of aromatase activity of stromal cells derived from human adipose tissue. Endocrinology 111: 1077-1085

Mendelson JH, Meyer RE, Pakeh VD (1975) Plasma testosterone levels in heroin addiction and during methadone maintenance. J Pharmacol Exp Ther 192: 211-215

Metcalf D (1966) The thymus. In: Rentchnick P (ed) Recent results in cancer research, vol 5. Springer, Berlin Heidelberg New York

Meyerson BJ, Sawyer CH (1968) Monoamines and ovulation in the rat. Endocrinology 83: 170-176

Michael RP, Keverne EB (1968) Pheromones in the communication of sexual status in primates. Nature 218: 746-749

Michael RP, Keverne EB (1970) Primate sex pheromones of vaginal origin. Nature 225: 84-85

Michael RP, Keverne EB, Bonsall RW (1971) Pheromones: isolation of male sex attractants from a female primate. Science 172: 964-966

Michael RP, Bonsal RW, Warner P (1974) Human vaginal secretions: volatile fatty acid content. Science 186: 1217-1219

Millar RP, Wormald PJ, De L, Milton RC (1986) Stimulation of gonadotropin release by a non-GnRH peptide sequence of the GnRH precursor. Science 232: 68-70

Miller C, Ulloa-Aguirre A, Hyland L, Chappel SC (1983) Pituitary FSH heterogeneity: assessment of biological activities of each FSH form. Fertil Steril 40: 242-247

Miller WL (1988) Molecular biology of steroid hormone synthesis. Endocr Rev 9: 295-318

Miller WL, Huang ES-R (1985) Secretion of ovine luteinizing hormone *in vitro*: differential positive control by 17β-estradiol and a preparation of porcine ovarian inhibin. Endocrinology 117: 907-911

Minamitani N, Minamitani T, Lechan RM, Bollinger-Gruber J, Reichlin S (1987) Paraventricular nucleus mediates prolactin secretory responses to restraint stress, ether stress and 5-hydroxy-l-tryptophan injection in the rat. Endocrinology 120: 860-867

Mirarchi RE, Howland BE, Scanlon PF, Kirkpatrick RL, Sanford SM (1978) Seasonal variation in plasma LH, FSH, prolactin, and testosterone concentrations in adult male white-tailed deer. Can J Zool 56: 121-127

Miyake A, Tasaka K, Kawamura Y, Sakumoto T, Aono T (1972) Progesterone facilitates the LHRH releasing action of oestrogens. Acta Endocrinol (Copenh) 101: 321–324

Miyake A, Aono T, Tanizawa O, Kinugasa T, Kurachi K (1977) Influence of human chorionic gonadotrophin on response of luteinizing hormone to luteinizing hormone-releasing hormone in gonadectomized women. J Endocrinol 74: 499–505

Miyake A, Kawamura Y, Aono T, Kurachi K (1980) Changes in plasma LRH during the normal menstrual cycle in a woman. Acta Endocrinol (Copenh) 93: 257–263

Miyake T, Pincus G (1958) Anti-progestational activity of estrogens in rabbit endometrium. Proc Soc Exp Biol Med 99: 478–483

Mock EJ, Frankel AI (1978 a) A seasonal influence on testes weight and serum gonadotropin levels of the mature male laboratory rat. Biol Reprod 18: 772–778

Mock EJ, Frankel AI (1978 b) A shifting circannual rhythm in serum testosterone concentration in male laboratory rats. Biol Reprod 19: 927–930

Mock EJ, Frankel AI (1980) Influence of month of birth on the serum hormone concentrations and weights of the accessory sex organs and testes during maturation of the male laboratory rats. Biol Reprod 22: 119–133

Molitch M, Edmonds M, Jones E, Odell WD (1976) Shortloop feedback control of luteinizing hormone in the rabbit. Am J Physiol 230: 907–910

Mondain-Monval M, Bonnin M, Scholler R, Canivenc R (1979) Androgens in peripheral blood of the red fox *Vulpes vulpes* during the reproductive season and the anestrus. J Steroid Biochem 11: 1315–1322

Mooradian AD, Morley JE, Korenman SG (1987) Biological actions of androgens. Endocr Rev 8: 1–28

Moore CR, Price D (1932) Gonad hormone functions and the reciprocal influence between gonads and hypophysis with its bearing on the problem of sex hormone antagonism. Am J Anat 50: 13–17

Mori H, Tsutsui K, Kawashima S (1985) Developmental changes in testicular luteinizing hormone receptors and Leydig cell number in mice with reference to changes in plasma gonadotropins and testosterone. J Sci Hiroshima Univ Ser B 32: 143–156

Morville R, Pesquies PC, Guezennec CY, Serrurier BD, Guignard M (1979) Plasma variations in testicular and adrenal androgens during prolonged physical exercise in man. Ann Endocrinol (Paris) 40: 501–505

Moulton BC, Koenig BB (1984) Uterine deoxyribonucleic acid synthesis during preimplantation in precursors of stromal cell differentiation during decidualization. Endocrinology 115: 1302–1307

Mowszowicz I, Bieber DE, Chung KW, Bullock LP, Bardin CW (1974) Synandrogenic and antiandrogenic effect of progestins: comparison with nonprogestational antiandrogens. Endocrinology 95: 1589–1599

Mulder E, Peters MJ, De Vries J, Van Der Molen HJ (1975) Characterization of a nuclear receptor for testosterone in seminiferous tubules of mature rat testes. Mol Cell Endocrinol 2: 171–176

Muller-Schwarze D (1971) Pheromones in black-tailed deer *(Odocoileus hemionus columbianus)*. Anim Behav 19: 141–152

Murphy MR, Schneider GE (1970) Olfactory bulb removal eliminates sexual behavior in the male hamster. Science 167: 302–303

Musto N, Hafiez AA, Bartke A (1972) Prolactin increases 17-hydroxysteroid dehydrogenase activity in the testis. Endocrinology 91: 1106–1108

Naftolin F, Ryan KJ, Petro Z (1971 a) Aromatization of androstenedione by the diencephalon. J Clin Endocrinol Metab 33: 368–370

Naftolin F, Ryan KH, Petro Z (1971 b) Aromatization of androstenedione by limbic system tissue from human foetuses. J Endocrinol 51: 795–796

Naftolin F, Judd HL, Yen SSC (1973) Pulsatile patterns of gonadotropins and testosterone in man: the effects of clomiphene with and without testosterone. J Clin Endocrinol Metab 36: 285–290

Naftolin F, Ryan KJ, David IJ, Reddy VV, Flores F, Petro Z, Kuhn M, White IJ, Takoa Y, Wolin L (1975 a) The formation of estrogens by central neuroendocrine tissue. Recent Prog Horm Res 31: 295–319

Naftolin F, Morishita H, Davies IJ, Todd R, Ryan KJ, Fishman J (1975 b) 2-Hydroxyestrone induced rise in serum LH in the immature male rat. Biochem Biophys Res Commun 64: 905-910

Naik SI, Young LS, Charlton HM, Clayton RN (1984 a) Pituitary gonadotropin-releasing hormone receptor regulation in mice. I: males. Endocrinology 115: 106-113

Naik SI, Young LS, Charlton HM, Clayton RN (1984 b) Pituitary gonadotropin-releasing hormone receptor regulation in mice. II: females. Endocrinology 115: 114-120

Nakashima A, Kashiyama K, Uozumi T, Monden Y, Hamanaka Y, Kuracki K, Aono T, Mizutani S, Matsumoto K (1975) Effects of general anesthesia and severity of surgical stress on serum LH and testosterone in males. Acta Endocrinol (Copenh) 78: 258-269

Nalbandov AV (1974) Control of luteal function in mammals. In: Greep RO, Astwood B (eds) Handbook of physiology, Endocrinology II, part 1. The Physiological Society, Washington, DC, pp 153-167

Nass TE, Terasawa E, Dierschke DJ, Goy RW (1984 a) Developmental changes in luteinizing hormone secretion in the female guinea pig. I. Effects of ovariectomy, estrogen, and luteinizing hormone-releasing hormone. Endocrinology 115: 220-226

Nass TE, Terasawa E, Dierschke DJ, Goy RW (1984 b) Developmental changes in luteinizing hormone secretion in the guinea pig. II. Positive feedback effects of ovarian steroids. Endocrinology 115: 227-232

Negro-Vilar A, Advis JP, Ojeda SR, McCann SM (1982) Pulsatile luteinizing hormone (LH) patterns in ovariectomized rats: involvement of norepinephrine and dopamine in the release of LH-releasing hormone and LH. Endocrinology 111: 932-938

Neill JD, Patton JM, Dailey RA, Tsou RC, Tindall GT (1977) Luteinizing hormone releasing hormone (LH-RH) in pituitary stalk blood of rhesus monkeys: relationship to level of LH release. Endocrinology 101: 430-435

Neill JD, Frawley LS, Plotsky PM, Tindall GT (1981) Dopamine in hypophysial stalk blood of the rhesus monkey and its role in regulating prolactin secretion. Endocrinology 108: 489-494

Neumann F, Steinbeck H (1974) Antiandrogens. In: Eichler O, Farah A, Herken H, Welch AD (eds) Handbook of experimental pharmacology, vol 35/2. Springer, Berlin Heidelberg New York pp 235-484

Nguyen BL, Giambiagi N, Mayrand C, Lecerf F, Pasqualine JR (1986) Estrogen and progesterone receptors in the fetal and newborn vagina of guinea pig: biological, morphological, and ultrastructural responses to tamoxifen and estradiol. Endocrinology 119: 978-988

Nikolics K, Mason AJ, Szonyi E, Ramachandran J, Seeburg PH (1985) A prolactin-inhibiting factor within the precursor for human gonadotropin-releasing hormone. Nature 316: 511-517

Nillius SJ, Wide L (1972) Variation in LH and FSH response to LH-releasing hormone during the menstrual cycle. J Obstet Gynaecol Br Commonw 79: 865-868

Nimrod A (1977) Studies on the synergistic effect of androgen on the stimulation of progestin secretion by FSH in cultured rat granulosa cells: a search for the mechanism of action. Mol Cell Endocrinol 8: 201-211

Nimrod A, Lindner HR (1976) A synergistic effect of androgen on the stimulation of progesterone secretion by FSH in cultured rat granulosa cells. Mol Cell Endocrinol 5: 315-320

Nishizuka Y, Sakakura T (1971) Ovarian dysgenesis induced by neonatal thymectomy in the mouse. Endocrinology 89: 886-893

Nishizuka Y, Sakakura T, Tsujimura T, Matsumoto K (1973) Steroid biosynthesis *in vitro* by dysgenetic ovaries induced by neonatal thymectomy in mice. Endocrinology 93: 786-792

Niswender GD, Schwall RH, Fitz TA, Farin CE, Sawyer HR (1985) Regulation of luteal function in domestic ruminants: new concepts. Recent Prog Horm Res 41: 101-151

Noirot E (1972) Ultrasounds and maternal behavior in small rodents. Dev Psychobiol 5: 371-387

Norman RL, Spies HC (1986) Cyclic ovarian function in a male macaque: additional evidence for a lack of sexual differentiation in the physiological mechanisms that regulate the cyclic release of gonadotropins in primates. Endocrinology 118: 2606-2610

Norman RL, Lindstrom SA, Bangsberg D, Ellinwood WE, Gliessman P, Spies HG (1984) Pulsatile secretion of luteinizing hormone during the menstrual cycle of rhesus macaques. Endocrinology 115: 261-266

Norton BI, Miyairi S, Fishman J (1988) 19-Hydroxylation of androgens by rat granulosa cells. Endocrinology 122: 1047-1052

Notides AC, Hamilton DE, Muechler EK (1976) A molecular analysis of the human estrogen receptor. J Steroid Biochem 7: 1025–1030

Novotny M, Jemiolo B, Harvey S, Wiesler D, Marchlewska-Koj A (1986) Adrenal-mediated endogenous metabolites inhibit puberty in female mice. Science 231: 722–725

Odell WD, Swerdloff RS (1976) Etiologies of sexual maturation: a model system based on the sexually maturing rat. Recent Prog Horm Res 32: 245–288

Ogle TF (1977) Modification of serum luteinizing hormone and prolactin concentrations by corticotropin and adrenalectomy in ovariectomized rats. Endocrinology 101: 494–497

Ogren L, Vertes M, Woolley D (1976) In vivo nuclear 3H-estradiol binding in brain areas of the rat: reduction by endogenous and exogenous androgens. Neuroendocrinology 21: 350–365

Ojeda SR, Ramirez VD (1969) Automatic control of LH and FSH secretion by short feedback circuits in immature rats. Endocrinology 84: 786–797

Ojeda SR, Castro-Vazquez A, Jameson HE (1977 a) Prolactin release in response to blockade of dopaminergic receptors and to TRH injection in developing and adult rats: role of estrogen in determining sex differences. Endocrinology 100: 427–437

Ojeda SR, Jameson HE, McCann SM (1977 b) Developmental changes in pituitary responsiveness to luteinizing hormone-releasing hormone (LH-RH) in the female rat: ovarian-adrenal influence during the infantile period. Endocrinology 100: 440–452

Okatani Y, Fishman J (1984 a) Suppression of the preovulatory luteinizing hormone surge in the rat by 2-hydroxyestrone: relationship to endogenous estradiol levels. Endocrinology 115: 1082–1089

Okatani J, Fishman J (1984 b) Effects of 2-hydroxyestradiol-17β, 2-hydroxyestradiol-17α, and 4-hydroxyestrone on the preovulatory luteinizing hormone surge in the rat: agonist and antagonist actions. Endocrinology 115: 1090–1094

Okatani Y, Fishman J (1986) Inhibition of the preovulatory prolactin surge in the rat by catechol estrogens: functional and temporal specificity. Endocrinology 119: 261–267

O'Malley BW, Means AR (1976) The mechanism of steroid-hormone regulation of transcription of specific eukaryotic cells. In: Cohn WE, Volkin E (eds) Progress in nucleic acid research and molecular biology, vol 19. Academic Press, New York, p 403

Ono N, Lumpkin MD, Samson WK, McDonald JK, McCann SM (1984) Intrahypothalamic action of corticotropin-releasing factor (CRF) to inhibit growth hormone and LH release in the rat. Life Sci 35: 1117–1121

Ono N, Samson WK, McDonald JK, Lumpkin MD, Bedran de Castro JC, McCann SM (1985) Effects of intravenous and intraventricular injection of antisera directed against corticotropin-releasing factor on the secretion of anterior pituitary hormones. Proc Natl Acad Sci USA 82: 7787–7789

Orgebin-Crist MC, Danzo BJ, Davies J (1975) Endocrine control of the development and maintenance of sperm fertilizing ability in the epididymis. Handbook of physiology, Sect 7, vol V. The Physiological society, Washington DC, pp 319–338

Orth JM, Gunsalus GL, Lamperti AA (1988) Evidence from sertoli cell-depleted rats indicates that spermatid number in adults depends on numbers of sertoli cells produced during perinatal development. Endocrinology 122: 787–794

O'Shaughnessy PJ, Brown PS (1978) Reduction in FSH receptors in the rat testis by injection of homologous hormone. Mol Cell Endocrinol 12: 9–15

Oshima H, Nankin HR, Fan D-F, Troen P, Yanaihara T, Niizato N, Yoshida KI, Ochiai KI (1975) Delay in sexual maturation of rats caused by synthetic LH-releasing hormone: enhancement of steroid Δ^4-5α-hydrogenase in testes. Biol Reprod 12: 491–497

Padmanabhan V, Kesner JS, Convey EM (1978) Effects of estradiol on basal and luteinizing hormone releasing hormone (LHRH) induced release of luteinizing hormone (LH) from bovine pituitary cells in culture. Biol Reprod 18: 608–611

Padmanabhan V, Keech C, Convey EM (1983) Cortisol inhibits and adrenocorticotropin has no effect on luteinizing hormone-releasing hormone-induced release of luteinizing hormone from bovine pituitary cells. Endocrinology 112: 1782–1787

Pallos KV (1941) Über den Einfluß längere Zeit hindurch verabreichter großer Mengen natürlicher und synthetischer oestrogener Stoff auf die Hoden erwachsener männlicher Ratten. Arch Gynak 171: 471–478

P'an SY, Laubach GD (1964) Steroid central depressants. In: Dorfman RI (ed) Methods in hormone research, vol III. Academic Press, New York, pp 415–475

Pang CN, Zimmermann E, Sawyer CH (1977) Morphine inhibition of the preovulatory surges of plasma luteinizing hormone and follicle stimulating hormone in the rat. Endocrinology 101: 1726-1732

Parisi MN, Vitale ML, Villar MJ, Estivariz FE, Chiocchio SR, Tramezzani JH (1987) Serotonergic terminals in the anterior hypothalamic nucleus involved in the prolactin release during suckling. Endocrinology 120: 2404-2412

Parkes AS, Bruce HM (1962) Pregnancy block in female mice placed in boxes soiled by males. J Reprod Fertil 4: 303-308

Pasqualini JR, Cosquer-Clavreul C (1978) Purification of the cytosol oestradiol-receptor complex from foetal guinea-pig uterus using electrofocusing on polyacrylamide plates. Experientia 34: 268-269

Pasqualini JR, Kincl FA (1985) Hormones and the fetus, vol 1. Pergamon Press, Oxford

Pasqualini JR, Nguyen BL (1980) Progesterone receptors in the fetal uterus and ovary of the guinea pig: evolution during fetal development and induction and stimulation in estradiol-primed animals. Endocrinology 106: 1160-1165

Pasqualini JR, Sumida C, Gelly C (1976) Cytosol and nuclear [^{3}H] oestradiol binding in the foetal tissues of guinea pig. Acta Endocrinol (Copenh) 83: 811-828

Patritti-Laborde N, Odell WD (1978) Shortloop feedback of luteinizing hormone: dose-response relationships and specificity. Fertil Steril 30: 456-460

Patritti-Laborde N, Wolfsen A, Heber D, Odell WD (1979) Pituitary gland. Site of shortloop feedback for luteinizing hormone in the rabbit. J Clin Invest 64: 1066-1069

Patritti-Laborde N, Ash RH, Pauerstein CJ, Odell WD (1982) Prevention of the postcoital luteinizing hormone surge by ultrashortloop feedback control. Fertil Steril 38: 349-353

Payne AH, Downing JR, Wong K-L (1980) Luteinizing hormone receptors and testosterone synthesis in two distinct populations of Leydig cells. Endocrinology 106: 1424-1429

Payne DW, Katzenellenbogen JA (1980) Differential effects of estrogens in tissues: a comparison of estrogen receptor in rabbit uterus and vagina. Endocrinology 106: 1345-1350

Peckham WD, Knobil E (1975) Qualitative changes in the pituitary gonadotropins of the male rhesus monkey following castration. Endocrinology 98: 1061-1064

Peckham WD, Knobil E (1976) The effects of ovariectomy, estrogen replacement and neuraminidase treatment on the properties of the adenohypophysial glycoprotein hormones of the rhesus monkey. Endocrinology 98: 1054-1060

Peckham WD, Yamaji T, Dierschke DJ, Knobil E (1973) Gonadal function and the biological physicochemical properties of follicle-stimulating hormone. Endocrinology 92: 1660-1666

Pelletier J (1973) Evidence for photoperiodic control of prolactin release in rams. J Reprod Fertil 35: 143-146

Peters RR, Tucker AH (1978) Prolactin and growth hormone responses to photoperiod in heifers. Endocrinology 103: 229-234

Petraglia F, Vale W, Rivier C (1986) Opioids act centrally to modulate stress-induced decrease in luteinizing hormone in the rat. Endocrinology 119: 2445-2450

Petraglia F, Sutton S, Vale W, Plotsky P (1987) Corticotropin-releasing factor decreases plasma luteinizing hormone levels in female rats by inhibiting gonadotropin-releasing hormone release into hypophysial-portal circulation. Endocrinology 120: 1083-1088

Pfaff DW (1968) Autoradiographic localization of heterologous sex hormones in the rat brain. Experientia 24: 958-960

Pfaff DW (1983) Impact of estrogens on hypothalamic nerve cells: ultrastructural, chemical, and electrical effects. Recent Prog Horm Res 39: 127-180

Pickering AJ-MC, Fink G (1979) Variation in size of the 'readily releasable pool' of luteinizing hormone during the oestrous cycle of the rat. J Endocrinol 83: 53-59

Pieper DR, Gala RR, Regiani SR, Marshall JC (1982) Dependence of pituitary gonadotropin-releasing hormone (GnRH) receptors on GnRH secretion from the hypothalamus. Endocrinology 110: 749-754

Pincus G, Chang MC (1953) The effects of progesterone and related compounds on ovulation and early development in the rabbit. Acta Physiol Lat Am 3: 177-183

Plant TM (1982) Effects of orchidectomy and testosterone replacement treatment on pulsatile luteinizing hormone secretion in the adult rhesus monkey *(Macaca mulatta)*. Endocrinology 110: 1905-1911

Plant TM, Dubey AK (1984) Evidence from the rhesus monkey *(Macaca mulatta)* for the view that negative feedback control of luteinizing hormone secretion by the testis is mediated by a deceleration of hypothalamic gonadotropin-releasing hormone pulse frequency. Endocrinology 115: 2145-2153

Plant TM, Zorub DS (1982) The role of nongonadal restraint of gonadotropin secretion in the delay of the onset of puberty in the rhesus monkey *(Macaca mulatta).* J Anim Sci [Suppl 12] 55: 43-56

Plapinger L, McEwen BS (1973) Ontogeny of estradiol-binding sites in rat brain. I. Appearance of presumptive adult receptors in cytosol and nuclei. Endocrinology 93: 1119-1128

Plotsky P, Vale W (1984) Hemorrhage-induced secretion of corticotropin-releasing factor-like immunoreactivity into the rat portal circulation and its inhibition by glucocorticoids. Endocrinology 114: 164-169

Poretsky L, Smith D, Seibel M, Pazianes A, Moses AC, Flier JS (1984) Specific insulin binding sites in human ovary. J Clin Endocrinol Metab 59: 809-814

Poretsky L, Grigorescu F, Seibel M, Moses AC, Flier JS (1985) Distribution and characterization of insulin and insulin-like growth factor I receptors in normal human ovary. J Clin Endocrinol Metab 61: 728-734

Porter RW, Cavanaugh EB, Critchlow BV, Sawyer CH (1957) Localized changes in electrical activity of the hypothalamus in estrous cats following vaginal stimulation. Am J Physiol 189: 145-151

Preslock JP (1980) A review in *in vitro* testicular steroidogenesis in rodents, monkeys and humans. J Steroid Biochem 13: 965-976

Preston SL, Parmer TG, Behrman HR (1987) Adenosine reverses calcium-dependent inhibition of follicle-stimulating hormone action and induction of maturation in cumulus-enclosed rat oocytes. Endocrinology 120: 1346-1353

Prins GS (1987) Prolactin influence on cytosol and nuclear androgen receptors in the ventral, dorsal, and lateral lobes of the rat prostate. Endocrinology 120: 1457-1464

Psychoyos A (1973) Hormonal control of ovoimplantation. Vitam Horm 31: 201-256

Psychoyos A (1974) Endocrine control of egg implantation. Handbook of Physiology, Endocrinology II, part 2. The Physiological Society, Washington DC, pp 187-215

Pucca GA, Bresciani F (1968) Receptor molecule for oestrogens from rat uterus. Nature 218: 967-969

Pucca GA, Medici N, Molinari AM, Moncharmont B, Nola E, Sica V (1980) Estrogen receptor of calf uterus: an easy and fast purification procedure. J Steroid Biochem 12: 105-111

Purvis K, Cusan L, Hansson V (1981) Regulation of steroidogenesis and steroid action in Leydig cells. J Steroid Biochem 15: 77-86

Quay WB (1972) Pineal homeostatic regulation of shifts in the circadian activity rhythm during maturation and aging. Trans NY Acad Sci 34: 239-254

Quirk SM, Hilbert JL, Fortune JE (1986) Progesterone secretion by granulosa cells from rats with four- or five-day estrous cycles: the development of responses to follicle-stimulating hormone, luteinizing hormone, and testosterone. Endocrinology 118: 2402-2410

Radford HM, Wallace ALC (1974) Central nervous blockade of oestradiol-stimulated release of luteinizing hormone in the ewe. J Endocrinol 60: 247-252

Ramirez VD, Kim K, Dluzen D (1985) Progesterone action on the LHRH and the nigrostriatal dopamine neuronal systems: *in vitro* and *in vivo* studies. Recent Prog Horm Res 41: 421-474

Rani CSS, Salhanick AR, Armstrong DT (1981) Follicle-stimulating hormone induction of luteinizing hormone receptor in cultured rat granulosa cells: an examination of the need for steroids in the induction process. Endocrinology 108: 1379-1385

Raynaud JP, Ojasoo T, Bouton MM, Bignon E, Pons M, de Paulet AC (1985) Structure-activity relationships of steroid estrogens. In: McLachlan JA (ed) Estrogens in the environment. Elsevier Science, Amsterdam, pp 24-42

Reame N, Sauder SE, Kelch RP, Marshall JC (1984) Pulsatile gonadotropin secretion during the human menstrual cycle: evidence for altered frequency of gonadotropin-releasing hormone secretion. J Clin Endocrinol Metab 59: 328-337

Rebar RW, Miyake A, Low TLK, Goldstein AL (1981) Thymosin stimulates secretion of luteinizing hormone-releasing factor. Science 214: 669-671

Reddy PV, Azhar S, Menon KMJ (1980) Multiple inhibitory actions of luteinizing hormone-releasing hormone agonist on luteinizing hormone/human chorionic gonadotropin receptor-mediated ovarian responses. Endocrinology 107: 930–936

Reeves JJ, Arimura A, Schally AV (1971) Changes in pituitary responsiveness to luteinizing hormone-releasing hormone (LH-RH) in anestrous ewes pretreated with estradiol benzoate. Biol Reprod 4: 88–92

Reich R, Miskin R, Tsafriri A (1985) Follicular plasminogen activator involvement in ovulation. Endocrinology 116: 516–521

Reichman ME, Villee CA (1978) Estradiol binding by rat thymus cytosol. J Steroid Biochem 9: 637–642

Reiter RJ (1968) The pineal gland and gonadal development in male rats and hamsters. Fertil Steril 19: 1009–1014

Reiter RJ, Fraschini F (1969) Endocrine aspects of the mammalian pineal gland. Neuroendocrinology 5: 219–226

Reppert SM, Schwartz WJ (1984) Functional activity of the suprachiasmatic nuclei in the fetal primate. Neurosci Lett 46: 145–149

Reppert SM, Schwartz WJ (1986) Maternal endocrine extirpations do not abolish maternal coordination of the fetal circadian clock. Endocrinology 119: 1763–1767

Reppert SM, Weaver DR, Rivkees SA, Stopa EG (1988) Putative melatonin receptors in a human biological clock. Science 242: 78–81

Rezek DL (1975) The localization and retention of testosterone and its metabolites in the male rat brain after the intravenous administration of tritiated testosterone: possible implications for male behavior. Thesis, U California, Irvine, pp 1–100

Richards JS, Bogovich K (1982) Effects of human chorionic gonadotropin and progesterone on follicular development in the immature rat. Endocrinology 111: 1429–1438

Richards JS, Williams JJ (1976) Luteal cell receptor content for prolactin (PRL) and luteinizing hormone (LH): regulation by LH and PRL. Endocrinology 99: 1571–1577

Richter KM (1944) Some new observations bearing on the effect of hyperthyroidism on genital structure and function. J Morphol 74: 375–386

Richter KM, Winter CA (1947) A quantitative study of the effect of hyperthyroidism on genital structure and function. Am J Physiol 150: 95–101

Ringstrom SJ, Schwartz NB (1984) Examination of prolactin and pituitary-adrenal axis components as intervening variables in the adrenalectomy-induced inhibition of gonadotropin response to castration. Endocrinology 114: 880–887

Ringstrom SJ, Schwartz NB (1985) Cortisol suppresses the luteinizing hormone, but not the follicle-stimulating hormone, response to gonadotropin-releasing hormone after orchidectomy. Endocrinology 116: 472–474

Rippel RH, Johnson ES (1976) Inhibition of HCG-induced ovarian and uterine weight augmentation in the immature rat by analogs of GnRH. Proc Soc Exp Biol Med 152: 432–436

Rivier C, Vale W (1984) Influence of corticotropin-releasing factor on reproductive functions in the rat. Endocrinology 114: 914–921

Rivier C, Vale W (1985) Involvement of corticotropin-releasing factor and somatostatin in stress-induced inhibition of growth hormone secretion in the rat. Endocrinology 117: 2478–2482

Rivier C, Brown M, Vale W (1977) Effect of neurotensin, substance P and morphine sulfate on the secretion of prolactin and growth hormone in the rat. Endocrinology 100: 751–754

Rivier C, Rivier J, Vale W (1979) Chronic effects of [D-Trp^6,Pro^9-NEt]luteinizing hormone-releasing factor on reproductive processes in the male rat. Endocrinology 105: 1191–1201

Rivier J, Spiess J, McClintock R, Vaughan J, Vale W (1985) Purification and partial characterization of inhibin from porcine follicular fluid. Biochem Biophys Res Commun 133: 120–125

Robertson DM, Van Damme M-P, Diczfalusy E (1977) Biological and immunological characterization of human luteinizing hormone: I. Biological profile in pituitary and plasma samples after electrofocusing. Mol Cell Endocrinol 9: 45–56

Robertson DM, Foulds LM, Ellis S (1982) Heterogeneity of rat pituitary gonadotropins on electrofocusing; differences between sexes and after castration. Endocrinology 111: 385–391

Rochefort H, Garcia M (1976) Androgen on the estrogen receptor. I. Binding and in vivo nuclear translocation. Steroids 28: 549–556

Rochefort H, Lignon F, Capony F (1972) Formation of estrogen nuclear receptor in uterus: effect of androgens, estrone and nafoxidine. Biochem Biophys Res Commun 47: 662–670

Rolland R, Lequin RM, Schellekens LA, de Jong FH (1975) The role of prolactin in restoration of ovarian function during the early post-partum period in the human female. I. A study during physiological lactation. Clin Endocrinol (Oxf) 4: 15–25

Rosenfield RL, Helke JC (1974) Small diurnal and episodic fluctuations of the plasma free testosterone level in normal women. Am J Obstet Gynecol 120: 461–465

Roy SK Jr, Laumas KR (1969) 1,2-³H-Testosterone: distribution and uptake in neural and genital tissues of intact male, castrate male and female rats. Acta Endocrinol (Copenh) 61: 629–640

Roy SN, Kar AB, Datta SN (1955) The influence of testosterone propionate on the response of the testis of young rats to thyroxine. Arch Int Pharmacodyn 102: 450–457

Roy S, Mahesh VB, Greenblatt RB (1964) Effects of clomiphene on the physiology of reproduction in the rat. III. Inhibition of uptake of radioactive oestradiol by the uterus and the pituitary gland of the immature rat. Acta Endocrinol (Copenh) 47: 669–675

Ruberg M, Rotsztejn WH, Arancibia S, Besson J, Enjalbert A (1978) Stimulation of prolactin release by vasoactive intestinal peptide (VIP). Eur J Pharmacol 51: 319–324

Rubinstein L, Sawyer CH (1970) Role of catecholamines in stimulating the release of pituitary ovulatory hormones in rats. Endocrinology 86: 988–995

Rudel HW, Kincl FA (1966) The biology of anti-fertility steroids. Acta Endocrinol (Copenh) 31 (Suppl 105): 7–45

Rudel HW, Kincl FA (1971) Long acting steroid formulations. In: Diczfalusy, Borell U (eds) Control of human fertility, Nobel Symposium 15. Alqvist and Wiksell, Stockholm, pp 39–50

Ruh TS, Ruh MF (1975) Androgen induction of a specific uterine protein. Endocrinology 97: 1144–1149

Ruh TS, Wassilak SG, Ruh MF (1974) Androgen-induced nuclear accumulation of the estrogen receptor. Steroids 24: 209–214

Ryan KJ, Naftolin F, Reddy V, Flores F, Petro Z (1972) Estrogen formation in the brain. Am J Obstet Gynecol 114: 454–460

Saez JM, Morera AM, Haour F, Evain D (1977) Effects in in vivo administration of dexamethasone, corticotropin and human chorionic gonadotropin on steroidogenesis and protein and DNA synthesis of testicular interstitial cells in prepuberal rats. Endocrinology 101: 1256–1263

Sales GD (1972) Ultrasound and mating behaviour in rodents with some observations on other behavioral situations. J Zool 168: 149–164

Sales GD, Pye D (1974) Ultrasonic communication by animals. Chapman and Hall, London

Salhanick AI, Wiebe JP (1980) FSH receptors in isolated Sertoli cells: changes in concentration of binding sites at the onset of sexual maturation. Life Sci 26: 2281–2288

Samperez S, Thieulant ML, Mercier L, Jouan P (1974) A specific testosterone receptor in the cytosol of rat anterior hypophysis. J Steroid Biochem 5: 911–915

Samson WK, McCann SM, Chud L, Dudley CA, Moss RL (1980) Intra- and extrahypothalamic luteinizing hormone-releasing hormone (LHRH) distribution in the rat with special reference to mesencephalic sites which contain both LHRH and single neurons responsive to LHRH. Neuroendocrinology 31: 66–72

Samson WK, Aguila MC, Bianchi R (1988) Atrial natriuretic factor inhibits luteinizing hormone secretion in the rat: evidence for a hypothalamic site of action. Endocrinology 122: 1573–1582

Sanborn BM, Elkington JSH, Steinberger A, Steinberger E (1975) Androgen binding in the testis: in vitro production of androgen binding protein (ABP) by Sertoli cell cultures and measurement of nuclear bound androgen by a nuclear exchange assay. In: French FS, Hansson V, Ritzen EM, Nayfeh SN (eds) Current topics in molecular endocrinology, vol 2. Plenum Press, New York, pp 293–315

Sander HJ, Meijs-Roelof HMA, van Leeuwen ECM, Kramer P, van Cappellen WA (1986) Inhibin increases in the ovaries of female rats approaching first ovulation: relationships with follicle growth and serum FSH concentrations. J Endocrinol 111: 159–166

Sandow J, Hahn M (1978) Chronic treatment with LH-RH in golden hamsters. Acta Endocrinol (Copenh) 88: 601–610

Sanford LM, Beaton DB, Howland BE, Palmer WM (1978) Photoperiod-induced changes in LH, FSH, prolactin and testosterone secretion in the ram. Can J Anim Sci 58: 123–131

Sar M, Stumpf WE (1973) Pituitary gonadotrophs: nuclear concentration of radioactivity after injection of [³H] testosterone. Science 179: 389–391

Sarff M, Gorski J (1971) Control of estrogen binding protein concentration under basal conditions and after estrogen administration. Biochemistry 10: 2557–2560

Sarkar DK, Chiappa SA, Fink G, Sherwood NM (1976) Gonadotropin-releasing hormone surge in pro-oestrous rats. Nature 264: 461–463

Sarkar DK, Chiappa SA, Fink G, Sherwood NM (1981) Gonadotrophin-releasing hormone surge in pro-oestrous rats. Nature 264: 461–464

Savion N, Liu GM, Laherty R, Gospodarowicz D (1981) Factors controlling proliferation and progesterone production by bovine granulosa cells in serum-free medium. Endocrinology 109: 409–420

Sawyer CH (1952) Stimulation of ovulation in the rabbit by the intraventricular injection of epinephrine or norepinephrine. Anat Rec 112: 385 (Abs)

Sawyer CH (1959) Seasonal variations in the incidence of spontaneous ovulation in rabbits following estrogen treatment. Endocrinology 65: 523–525

Sawyer CH, Kawakami M (1959) Characteristics of behavioral and electroencephalographic after-reactions to copulation and vaginal stimulation in the female rabbit. Endocrinology 65: 622–630

Sawyer CH, Everett JW, Markee JE (1949) A neural factor in the mechanism by which estrogen induces the release of luteinizing hormone in the rat. Endocrinology 44: 218–233

Schally AV, Bowers CY, Carter WH, Arimura A, Redding TW, Saito M (1969) Effect of actinomycin D on the inhibitory response of estrogen on LH release. Endocrinology 85: 290–299

Schally AV, Arimura A, Kastin A, Matsuo H, Baba Y, Redding Y, Nair R, Debeljuk L, White W (1971) Gonadotropin releasing hormone: one polypeptide regulates secretion of luteinizing and follicle-stimulating hormones. Science 173: 1036–1039

Schams D, Reinhardt V (1974) Influence of the season on plasma prolactin levels in cattle from birth to maturity. Horm Res 5: 217–226

Schatz F, Markiewicz L, Gurpide E (1987) Differential effects of estradiol, arachidonic acid, and A23187 on prostaglandin F_2 output by epithelial and stromal cells of human endometrium. Endocrinology 120: 1465–1471

Schipper H, Brawer JR, Nelson JF, Felicio LS, Finch CE (1981) The role of the gonads in histologic aging of the hypothalamic arcuate nucleus. Biol Reprod 25: 413–419

Schreiber JR, Reid R, Ross GT (1976) A receptor-like testosterone-binding protein in ovaries from estrogen-stimulated hypophysectomized immature female rats. Endocrinology 98: 1206–1213

Schulte BA, Seal US, Plotka ED, Letellier MA, Verme LJ, Ozoga JJ, Parson JA (1981) The effect of pinealectomy on seasonal changes in prolactin secretion in the white-tailed deer *(Odocoileus virginianus borealis)*. Endocrinology 108: 173–177

Schulze H, Barrach CK (1987) Immunocytochemical localization of estrogen receptors in the normal male and female canine urinary tract and prostate. Endocrinology 121: 1773–1783

Schwartz NB (1974) The role of FSH and LH and of their antibodies on follicle growth and on ovulation. Biol Reprod 10: 236–272

Schwarz W, Merker HR (1965) Die Hodenzwischenzellen der Ratte nach Hypophysectomie und nach Behandlung mit Choriongonadotropin und Amphenon. B Zellforsch Mikrosk Anat 65: 272–279

Scow RO, Marx W (1945) Response to pituitary growth hormone of rats thyroidectomized on the day of birth. Anat Rec 91: 227–232

Scow RO, Simpson ME (1945) Thyroidectomy in the newborn rat. Anat Rec 91: 209–216

Screpanti I, Gulino A, Pasqualini JR (1982) The fetal thymus of guinea pig as an estrogen target organ. Endocrinology 111: 1552–1561

Selye H (1939) The effect of adaptation to various damaging agents on the female sex organs in the rat. Endocrinology 25: 615–623

Sempere AJ, Lacroix A (1982) Temporal and seasonal relationships between luteinizing hormone, testosterone and antlers in fawn and adult male roe deer *(Capreolus capreolus)*. A longitudinal study from birth to 4 years of age. Acta Endocrinol (Copenh) 99: 295–301

Shaar CJ, Clemens JA (1974) The role of catecholamines in the release of anterior pituitary prolactin *in vitro*. Endocrinology 95: 1202–1212

Shani (Mishkinsky) J, Givant Y, Sulman FG, Eylath U, Eckstein B (1971) Competition of phenothiazines with oestradiol for oestradiol receptors in rat brain. Neuroendocrinology 8: 307–316

Sharpe RM, Fraser HM (1980) HCG stimulation of testicular LHRHlike activity. Nature 287: 642-643

Sharpe RM, McNeilly AS (1978) Gonadotrophin-induced reduction in LH-receptors and steroido-genic responsiveness of the immature rat testis. Int J Androl (Suppl) 2: 264-275

Sharpe RM, Fraser HM, Cooper I, Rommerts FFG (1981) Sertoli-Leydig cell communication via an LHRH-like factor. Nature 290: 785-787

Shaykh M, LeMaire WJ, Papkoff H, Curry TE Jr, Sogn JH, Koos RD (1985) Ovulation in rat ova-ries perfused in vitro with follicle stimulating hormone. Biol Reprod 33: 629-634

Shelton M (1960) Influence of the presence of a male goat on the initiation of oestrus cycling and ovulation of angora does. Animal Sci 19: 368-375

Sherwood NM, Fink G (1980) Effect of ovariectomy and adrenalectomy on luteinizing hormone-releasing hormone in pituitary stalk blood from female rats. Endocrinology 106: 363-367

Shorey HH (1976) Animal communication by pheromones. Academic Press, New York

Shupnik MA, Gharib SD, Chin WW (1988) Estrogen suppresses rat gonadotropin gene transcrip-tion in vivo. Endocrinology 122: 1842-1846

Shyamala G, Gorski J (1969) Estrogen receptors in the rat uterus. Studies on the interaction of cytosol and nuclear binding sites. J Biol Chem 244: 1097-1101

Siegel HI, Bast JD, Greenwald GS (1976) The effects of phenobarbital and gonadal steroids on periovulatory serum levels of LH and FSH in the hamster. Endocrinology 98: 48-55

Siiteri PK, Febres F, Clemens LE, Chang JR, Gondos B, Stites D (1977) Progesterone and main-tenance of pregnancy: is progesterone nature's immunosuppressant? Ann NY Acad Sci 286: 384-379

Sikorski R, Tuszkiewicz M, Zbroja W, Pleszczynska E (1975) Incidence of antiovarian antibodies in women with primary and secondary amenorrhoea. Cesk Gynekol 40: 591-594

Silverman AJ, Antunes JL, Abrams GM, Nilaver G, Thau R, Robinson JA, Fern M, Krey LC (1982) The luteinizing hormone-releasing hormone pathways in rhesus (Macaca mulatta) and pigtailed (Macaca nemestrina) monkeys. New observation on thick, unembedded sections. J Comp Neurol 35: 309-317

Simonnet H, Thieblot L, Melik T (1951) Influence de l'epiphyse sur l'ovaire de la jeune rate. Ann Endocr (Paris) 12: 202-205

Sirinathsinghji DJS, Rees LH, Rivier J, Vale W (1983) Corticotropin-releasing factor is a potent inhibitor of sexual receptivity in the female rat. Nature 305: 232-234

Sisk CL, Desjardins C (1986) Pulsatile release of luteinizing hormone and testosterone in male ferrets. Endocrinology 119: 1195-1203

Slaunwhite WR Jr., Samuels LT (1956) Progesterone as a precursor of testicular androgens. J Biol Chem 220: 341-352

Sluss M, Schneyer AL, Franke MA, Reichert LE Jr (1987) Porcine follicular fluid contains both follicle-stimulating hormone agonist and antagonist activities. Endocrinology 120: 1477-1481

Sluss PM, Reichert LE (1984) Porcine follicular fluid contains several low molecular weight inhib-itors of follicle stimulating binding receptors. Biol Reprod 30: 1091-1097

Smelser GK (1939) The effect of thyroidectomy on the reproductive system of hypophysis of the adult male rat. Anat Rec 74: 7-34

Smith MS (1978) A comparison of pituitary responsiveness to luteinizing hormone-releasing hor-mone during lactation and the estrous cycle of the rat. Endocrinology 102: 114-120

Smith MS (1982) Effect of pulsatile gonadotropin-releasing hormone on the release of luteinizing hormone and follicle stimulating hormone in vitro by anterior pituitaries from lactating and cycling rats. Endocrinology 110: 882-891

Smith MS, Freeman ME, Neill JD (1975) The control of progesterone secretion during the estrous cycle and early pseudopregnancy in the rat: prolactin, gonadotropin and steroid levels associated with the rescue of the corpus luteum of pseudopregnancy. Endocrinology 96: 219-226

Snell GD, Fekete E, Hummel KP, Law IW (1940) The relation of mating, ovulation and the estrus smear in the house mouse to time of day. Anat Rec 76: 39-54

Soper BD, Weick RF (1980) Hypothalamic and extrahypothalamic mediation of pulsatile dis-charge of luteinizing hormone in the ovariectomized rat. Endocrinology 106: 348-355

Soto AM, Sonnenschein C (1987) Cell proliferation of estrogen sensitive cells: the case for nega-tive control. Endocr Rev 8: 44-52

Spicer LJ, Hammond JM (1987) 2-Hydroxyestradiol modulates a facilitative action of catechol-amines on porcine granulosa cells. Endocrinology 120: 2375–2382

Srere PA, Chaikoff IL, Treitman SS, Burstein LS (1950) The extrahepatic synthesis of cholesterol. J Biol Chem 182: 629–635

Starzec A, Counis R, Jutisz M (1986) Gonadotropin-releasing hormone stimulates the synthesis of the polypeptide chains of luteinizing hormone. Endocrinology 119: 561–565

Stearns EL, Winter JSD, Faiman C (1973) Positive feedback effect of progestin upon serum go-nadotropins in estrogen-primed castrated men. J Clin Endocrinol Metab 37: 635–638

Steger RW, Peluso JJ (1978) Gonadotropin regulation in the lactating rat. Acta Endocrinol (Copenh) 88: 668–675

Steinberger A, Steinberger E (1976) Secretion of an FSH inhibiting factor by cultured Sertoli cells. Endocrinology 99: 918–921

Steinberger A, Thanki KH, Siegal B (1974) FSH binding in rat testes during maturation and following hypophysectomy. Cellular localization of FSH receptors. In: Dufau ML, Means AR (eds) Current topics in molecular endocrinology, vol 1. Plenum Press, New York, pp 177–192

Steiner RA, Clifton DK, Spies HG, Resko JA (1976) Sexual differentiation and feedback control of luteinizing hormone secretion in the rhesus monkey. Biol Reprod 15: 206–212

Steiner RA, Peterson AP, Yu JYL, Conner H, Gilbert M, Penning B, Bremner WJ (1980) Ultra-dian luteinizing hormone and testosterone rhythms in the adult male monkey Macaca fascicula-ris. Endocrinology 107: 1489–1493

Steiner RA, Bremner WJ, Clifton DK (1982) Regulation of luteinizing hormone pulse frequency and amplitude by testosterone in the adult male rat. Endocrinology 111: 2055–2061

Stetson MH, Elliott JA, Goldman BD (1986) Maternal transfer of photoperiodic information influences the photoperiodic response of prepubertal Djungarian hamsters (Phodopus sungorus sungorus). Biol Reprod 34: 664–669

Stimson WH, Hunter IC (1976) An investigation into the immunosuppressive properties of oes-trogen. J Endocrinol 69: 42–47

Stockard CR, Papanicolaou GN (1917) The existence of a typical oestrous cycle in the guinea pig, with a study of its histological and physiological changes. Am J Anat 22: 225–83

Strickland S, Beers WH (1976) Studies on the role of plasminogen activator in ovulation. J Biol Chem 281: 5694–5697

Strobl JS, Thompson EB (1985) Mechanism of steroid hormone action. In: Auricchio F (ed) Sex steroid receptors. Field Educational, Halia Acta Medica, Rome, pp 9–36

Strollo F, Harlin J, Hernandez-Montes H, Robertson DM, Zaidi AA, Diczfalusy E (1981) Qualita-tive and quantitative differences in the isoelectrofocusing profile of biologically active lutropin in the blood of normally menstruating and post-menopausal women. Acta Endocrinol (Copenh) 97: 166–175

Su T-P, London ED, Jaffe JH (1988) Steroid binding at o receptors suggests a link between endo-crine, nervous, and immune systems. Science 240: 219–221

Šulcová J, Jirásek JE, Neuwirth J, Raboch J, Stárka L (1978) Peripheral conversion and uptake of androgens in a XXY-man with Klinefelter's syndrome. Endokrinologie 72: 304–310

Sulimovici S, Boyd GS (1969) The effect of ascorbic acid in vitro on the rat ovarian cholesterol side chain cleavage enzyme system. Steroids 12: 127–133

Sulman FG (1956) Experiments on the mechanism of the push and pull principle. J Endocrinol 14 (Proc):27–28

Suter DE, Schwartz NB (1985) Effects of glucocorticoids on secretion of luteinizing hormone and follicle-stimulating hormone by female rat pituitary cells in vitro. Endocrinology 117: 849–854

Swerdloff RS, Odell WD (1977) Modulating influences of FSH, GH and prolactin on LH-stimu-lated testosterone secretion. In: Troen P, Nankin HR (eds) The testis in normal and infertile men. Raven Press, New York, pp 395–407

Szenthagothai J, Flerko B, Mess B, Halasz B (1968) Hypothalamic control of the anterior pitu-itary, 3 ed. Akademiai Kiado, Budapest

Tache Y, Ducharme JR, Charpenet G, Haour F, Saez J, Collu R (1980) Effect of chronic intermit-tent immobilization stress on hypophysogonadal function of rats. Acta Endocrinol (Copenh) 93: 168–174

Takagi T (1986) Longitudinal study on circadian rhythms of plasma hormone levels during Japanese antarctic research expedition. Hokkaido J Med Sci 61: 121–133 (Biosis No 82002797)

Takahashi K, Hayafuji C, Murakami N (1982) Foster mother rat entrains circadian adrenocortical rhythm in blinded pups. Am J Physiol 243:E443–449

Takeyama M, Nagareda T, Takatsuka D, Namiki M, Koizumi K, Aono T, Matsumoto K (1986) Stimulatory effect of prolactin on luteinizing hormone-induced testicular 5α-reductase activity in hypophysectomized adult rats. Endocrinology 118: 2268–2275

Tamada H, Ichikawa S (1980) The effect of estrogen on fetal survival in progesterone-treated ovariectomized rats. Endocrinol Jpn 27: 163–167

Tang LKL (1980) Effect of serum sex steroids on pituitary LH response to LHRH and LH synthesis. Am J Physiol 238:E458–461

Tannenbaum GS (1984) Growth hormone-releasing factor: direct effects on growth hormone, glucose, and behavior via the brain. Science 226: 464–466

Taranov AG, Goncharov NP (1985) Annual blood level of androgens in male hamadryas baboon. Bull Eksp Biol Med 100: 647–649 (Biosis No 82022513)

Tardy J, Pasqualini JR (1983) Localization of ^{3}H-estradiol and gonadotropin-releasing hormone (GnRH) in the hypothalamus of the fetal guinea pig. Exp Brain Res 49: 77–83

Tashjian AH Jr, Barowsky NV, Jensen DK (1971) Thyrotropin releasing hormone: direct evidence for stimulation of prolactin production by pituitary cells in culture. Biochem Biophys Res Commun 43: 516–520

Tausk M (1972) Pharmocology of the endocrine system and related drugs: progesterone, progestational drugs and antifertility agents. In: Encyc Pharm Therap, vol 1, sect 2. Pergamon Press, Oxford

Terasawa E, Bridson WE, Nass TE, Noonan JJ, Dierschke DJ (1984) Developmental changes in the luteinizing hormone secretory pattern in peripubertal female rhesus monkeys: comparisons between gonadally intact and ovariectomized animals. Endocrinology 115: 2233–2240

Terasawa E, Krook C, Eman S, Watanabe G, Bridson WE, Sholl SA, Hei DL (1987) Pulsatile luteinizing hormone (LH) release during the progesterone-induced LH surge in the female rhesus monkey. Endocrinology 120: 2265–2271

Terry LC, Martin JB (1981) Evidence for adrenergic regulation of episodic growth hormone and prolactin secretion in the undisturbed male rat. Endocrinology 108: 1869–1874

Testart J, Thebault A, Lefevre B (1983) *In vitro* ovulation of rabbit ovarian follicles isolated after the endogenous gonadotrophin surge. J Reprod Fertil 68: 413–418

Thomas DA, Barfield RJ (1985) Ultrasonic vocalization of the female rat (*Rattus norvegicus*) during mating. Anim Behav 33: 720–725

Thorner MO, McNeilly AS, Hagen C, Besser GM (1974) Long-term treatment of galactorrhoea and hypogonadism with bromocriptine. Br Med J 4: 419–423

Thorner MO, Perryman RL, Rogol AD, Conway BP, Macleod RM, Login IS, Morris JL (1981) Rapid changes of prolactinoma volume after withdrawal and reinstitution of bromocriptine. J Clin Endocrinol Metab 53: 480–483

Tibolt RE, Childs GV (1985) Cytochemical and cytophysiological studies of gonadotropin-releasing hormone (GnRH) target cells in the male rat pituitary: differential effects of androgens and corticosterone on GnRH binding and gonadotropin release. Endocrinology 117: 396–404

Tindall DJ, Miller DA, Means AR (1977) Characterization of androgen receptor in Sertoli cell-enriched testis. Endocrinology 101: 13–23

Toft D, O'Malley BW (1972) Target tissue receptors for progesterone: the influence of estrogen treatment. Endocrinology 90: 1041–1045

Tomatis ME, Taleisnik S (1969) Pituitary melanocytes-stimulating hormone and its hypothalamic factors during the estrous cysle. Acta Physiol Lat Am 18: 96–101

Tsafriri A, Koch Y, Lindner HR (1973) Ovulation rate and serum LH levels in rats treated with indomethacin or prostaglandin E_2. Prostaglandins 3: 461–467

Tsou RC, Dailey RA, McLanahan CS, Parent AD, Tindall CT, Neill JD (1977) Luteinizing hormone releasing hormone (LHRH) levels in pituitary stalk plasma during the preovulatory gonadotropin surge of rabbits. Endocrinology 101: 534–539

Tsutsui K, Shimizu A, Kawamoto K, Kawashima S (1985) Developmental changes in the binding of follicle-stimulating hormone (FSH) to testicular preparations of mice and the effects of hypophysectomy and administration of FSH on the binding. Endocrinology 117: 2534–2543

Tuomisto J, Mannisto P (1985) Neurotransmitter regulation of anterior pituitary hormones. Pharmacol Rev 37: 249–269

Turek FW, Swann J, Earnest DJ (1984) Role of circadian system in reproductive phenomena. Recent Prog Horm Res 40: 143–183

Turgeon JS (1979) Estradiol-luteinizing hormone relationship during the proestrous gonadotropin surge. Endocrinology 105: 731–742

Tyndale-Biscoe CH, Hinds LA, McConnel SJ (1986) Seasonal breeding in a marsupial: opportunities of a new species for an old problem. Recent Prog Horm Res 42: 471–512

Ueno N, Ling N, Ying S-Y, Esch F, Shimasaki S, Guillemin R (1987) Isolation and partial characterization of follistatin, a novel Mr 35,000 monomeric protein which inhibits the release of follicle stimulating hormone. Proc Natl Acad Sci USA 84: 8282–8286

Uilenbroek JThJ, Richards JS (1979) Ovarian follicular development during the rat estrous cycle: gonadotropin receptors and follicular responsiveness. Biol Reprod 20: 1159–1165

Uilenbroek JThJ, van der Linden R (1983) Changes in gonadotropin binding to rat ovaries during sexual maturation. Acta Endocrinol (Copenh) 103: 413–419

Ulloa-Aguirre A, Chappel SC (1982) Multiple species of follicle stimulating hormone exist within the anterior pituitary gland of male golden hamsters. J Endocrinol 95: 257–266

Urbanski HF, Ojeda SR (1985a) The juvenile-peripubertal transition period in the female rat: establishment of a diurnal pattern of pulsatile luteinizing hormone secretion. Endocrinology 117: 644–649

Urbanski HF, Ojeda SR (1985b) In vitro simulation of prepubertal changes in pulsatile luteinizing hormone release enhances progesterone and 17 -estradiol secretion from immature rat ovaries. Endocrinology 117: 638–643

van Beurden WMO, Roodnat B, van der Molen HJ (1978) Effects of oestrogens and FSH on LH stimulation of steroid production by testis Leydig cells from immature rats. Int J Androl (Suppl) 2: 374–383

Vancauter EW, Virasoro E, Leclercq R, Copinschi G (1981) Seasonal circadian and episodic variations of human immuno reactive beta melanocyte stimulating hormone, ACTH and cortisol. Int J Pept Protein Res 17: 3–13

van Dijk S, Steenbergen J, Gielen JTH, de Jong FH (1986) Sexual dimorphism in immunoneutralization of bioactivity of rat and ovine inhibin. J Endocrinol 111: 225–261

van Sickle M, Oberwetter JM, Birnbaumer L, Means AR (1981) Developmental changes in the hormonal regulation of rat testis Sertoli cell adenylyl cyclase. Endocrinology 109: 1270–1286

Vazquez AM, Kenny FM (1973) Ovarian failure and antiovarian antibodies in association with hypoparathyroidism, moniliasis and Addison's diseases. Obstet Gynecol 41: 414–419

Velardo JT, Hishaw FL (1951) Quantitative inhibition of progesterone by estrogens in development of deciduomata. Endocrinology 49: 530–537

Veldhuis JD, Klase PA, Strauss JF III, Hammond JM (1982a) The role of estradiol as a biological amplifier of the actions of follicle-stimulating hormone: in vitro studies in swine granulosa cells. Endocrinology 111: 144–151

Veldhuis JD, Klase PA, Strauss JF III, Hammond JM (1982b) Facilitative interactions between estradiol and luteinizing hormone in the regulation of progesterone production by cultured swine granulosa cells: relation to cellular cholesterol metabolism. Endocrinology 111: 441–447

Verhoeven G, Cailleau J (1988) Follicle-stimulating hormone and androgens increase the concentration of the androgen receptor in sertoli cells. Endocrinology 122: 1541–1550

Vidgoff B, Vehrs H (1941) Studies on the inhibitory hormone of the testis. Endocrinology 26: 656–661

Vidgoff B, Hill R, Vehrs H, Kubin R (1939) Studies on the inhibitory hormone of the testis II. Preparation and weighed changes in the sex organs of the adult male, white rat. Endocrinology 25: 391–399

Vijayan E, Samson WK, Said SI, McCann SM (1979) Vasoactive intestinal peptide: evidence for a hypothalamic site of action to release growth hormone, lutenizing hormone and prolactin in conscious, ovariectomized rats. Endocrinology 104: 53–57

Villee CA (1961) Some problems of the metabolism and mechanism of action of steroid sex hormones. In: Young WC (ed) Sex and internal secretions, 3rd edn. Williams and Wilkins, Baltimore, pp 643–665

Viswanathan N, Chandrashekaran MK (1984) Mother mouse sets the circadian clock of pups. Proc Indian Acad Sci (Anim Sci) 93: 235–241

Vitale ML, Parisi MN, Chiocchio SR, Tramezzani JH (1986) Serotonin induces gonadotrophin release through stimulation of LH-releasing hormone release from the median eminence. J Endocrinol 111: 309–315

Vito CC, Fox TO (1979) Embryonic rodent brain contains estrogen receptors. Science 204: 517–519

Vreeburg JTM, Schretlen PJM, Baum MJ (1975) Specific, high affinity binding of 17 -estradiol in cytosols from several brain regions and pituitary of intact and castrated adult male rats. Endocrinology 97: 969–975

Vreeburg JTM, de Greef WJ, Ooms MP, van Wouw P, Weber RFA (1984) Effects of adrenocorticotropin and corticosterone on the negative feedback action of testosterone in the adult male rat. Endocrinology 115: 977–983

Waever DR, Reppert SM (1986) Maternal melatonin communicates daylength to the fetus in djungarian hamsters. Endocrinology 119: 2861–2863

Wagner TOF, Adams TE, Nett TM (1979) GnRH interaction with anterior pituitary. I. Determination of the affinity and number of receptors for GnRH in ovine anterior pituitary. Biol Reprod 20: 140–149

Wallace JM, McNeilly AS (1986) Changes in FSH and the pulsatile secretion of LH during treatment of ewes with bovine follicular fluid throughout the luteal phase of the oestrous cycle. J Endocrinol 111: 317–327

Walters MR (1985) Steroid hormone receptors and the nucleus. Endocr Rev 6: 512–543

Walton C, Kelly WF, Laing I, Bullock DE (1983) Endocrine abnormalities in idiopathic haemochromatosis. Q J Med 205: 99–105

Wang QF, Farnworth PG, Findlay JK, Burger HG (1988) Effect of purified 31K bovine inhibin on the specific binding of gonadotropin-releasing hormone to rat anterior pituitary cells in culture. Endocrinology 123: 2161–2166

Watanobe H, Takebe K (1987 a) Involvement of postnatal gonads in the maturation of dopaminergic regulation of prolactin secretion in male rats. Endocrinology 120: 2205–2211

Watanobe H, Takebe K (1987 b) Involvement of postnatal gonads in the maturation of dopaminergic regulation of prolactin secretion in female rats. Endocrinology 120: 2212–2219

Wehrenberg WB, Wardlaw SL, Frantz AG, Ferin M (1982) β-endorphin in hypophyseal portal blood: variations throughout the menstrual cycle. Endocrinology 111: 879–881

Weick RF (1978) Acute effects of adrenergic receptor blocking drugs and neuroleptic agents on pulsatile discharge of luteinizing hormone in the ovariectomized rat. Neuroendocrinology 26: 108–117

Weick RF, Pitelka V, Thompson DL (1983) Separate negative feedback effects of estrogen on the pituitary and the central nervous system in the ovariectomized rhesus monkey. Endocrinology 112: 1862–1864

Weiland NG, Wise PM (1987) Estrogen alters the diurnal rhythm of α-adrenergic receptor densities in selected brain regions. Endocrinology 121: 1751–1758

Weiner RI, Ganong WF (1978) Role of brain monoamines and histamine in regulation of anterior pituitary secretion. Physiol Rev 58: 905–923

Welschen R (1973) Amounts of gonadotropins required for normal follicular growth in hypophysectomized adult rats. Acta Endocrinol (Copenh) 72: 137–143

Welsh MJ, Wiebe JP (1978) Sertoli cell capacity to metabolize C_{19} steroids: variation with age and the effect of follicle stimulating hormone. Endocrinology 103: 838–844

Welsh TH Jr, Johnson BH (1981) Stress induced alteratious in secretion of corticosteroids, progesterone, luteinizing hormone, and testosterone in bulls. Endocrinology 109: 185–190

Welsh TH Jr, Jones PBC, Ruiz de Galarreta CM, Fanjul LF, Hsueh AJW (1982) Androgen regulation of progestin biosynthetic enzymes in FSH-treated rat granulosa cells in vitro. Steroids 40: 691–700

Werner S, Hulting AL, Hokfelt T, Eneroth P, Tatemoto K, Mutt V, Moroder L, Wunsch E (1983) Effect of the peptide PHI-2 on prolactin release in vitro. Neuroendocrinology 37: 476–478

Westley BR, Thomas PJ, Salaman DF, Knight A, Barley J (1976) Properties and partial purification of an oestrogen receptor from neonatal rat brain. Brain Res 113: 441–446

Westphal U (1971) Steroid-protein interactions. Springer, Berlin Heidelberg New York

Wheaton JE, Krulich L, McCann SM (1975) Localization of luteinizing hormone-releasing hormone in the preoptic area and hypothalamus of the rat using radioimmunoassay. Endocrinology 97: 30–38

Wiegerinck MAHM, Poortman J, Agema AR, Thijssen JHH (1980) Estrogen receptors in human vaginal tissue. Maturitas 2: 59–66

Wilcox JN, Feder HH (1983) Long-term priming with a low dosage of estradiol benzoate or an antiestrogen (clomiphene) increases nuclear progestin receptor levels in brain. Brain Res 266: 243–252

Williams AT, Lipner H (1982) The contribution of gonadostatin (Inhibin-F) to the control of gonadotropin secretion in a simulated estrous cycle in steroid-treated ovariectomized rats. Endocrinology 111: 231–237

Wilson JD (1988) Androgen abuse by athletes. Endocr Rev 9: 181–199

Wilson ME, Gordon TP, Collins DC (1986) Ontogeny of luteinizing hormone secretion and first ovulation in seasonal breeding rhesus monkeys. Endocrinology 118: 293–301

Wira CR, Sandoe CP (1980) Hormonal regulation of immunoglobulins: influence of estradiol on immunoglobulins A and G in the rat uterus. Endocrinology 106: 1020–1026

Wise PM, Rance N, Barraclough CA (1981) Effects of estradiol and progesterone on catecholamine turnover rates in discrete hypothalamic regions in ovariectomized rats. Endocrinology 108: 2186–2193

Wisner JR, Stalvey JRD (1980) Alteration in the normal pattern of serum testosterone and 5α-androstane-3β,17β-diol in the immature male rat following chronic treatment with luteinizing hormone releasing hormone. Steroids 36: 337–348

Witkin JW, Paden CM, Silverman A-J (1982) The luteinizing hormone-releasing hormone (LHRH) systems in the rat brain. Neuroendocrinology 35: 429–438

Woodbury DM, Vernadakis A (1966) Effects of steroids on the central nervous system. In: Dorfman RI (ed) Methods in hormone research, vol V. Academic Press, New York, pp 1–57

Wynne-Edwards KE, Terranova PF, Lisk RD (1987) Cyclic djungarian hamsters, *Phodopus campbelli*, lack the progesterone surge normally associated with ovulation and behavioral receptivity. Endocrinology 120: 1308–1316

Yamazaki K, Boyse EA, Mike V, Thaler HT, Mathieson BJ, Abbott J, Boyse J, Zayas ZA, Thomas L (1976) Control of mating preferences in mice by genes in the major histocompatibility complex. J Exp Med 144: 1324–1335

Yamazaki K, Beauchamp GK, Kupniewski D, Bard J, Thomas L, Boyse EA (1988) Familial imprinting determines H-2 selective mating preferences. Science 240: 1331–1332

Yanaihara T, Troen P (1972) Studies of the human testis. I. Biosynthetic pathways for androgen formation in human testicular tissue *in vitro*. J Clin Endocrinol Metab 34: 783–800

Yen SSC, Tsai CC, Naftolin F, Vandenberg G, Ajabor L (1972) Pulsatile patterns of gonadotropin release in subjects with and without ovarian function. J Clin Endocrinol Metab 34: 671–675

Ying S-Y (1988) Inhibins, activins, and follistatins: gonadal proteins modulating the secretion of follicle-stimulating hormone. Endocr Rev 9: 267–293

Ying S-Y, Ling N, Bohlen P, Guillemin R (1981) Gonadocrinins: peptides in ovarian follicular fluid stimulating the secretion of pituitary gonadotropins. Endocrinology 108: 1206–1215

Ying S-Y, Becker A, Ling N, Ueno N, Guillemin R (1986) Inhibin and beta type transforming growth factor (TGFα) have opposite modulating effects on the follicle stimulating hormone (FSH)induced aromatase activity of cultured rat granulosa cells. Biochem Biophys Res Commun 136: 969–972

Ying S-Y, Becker A, Swanson G, Tan P, Ling N, Esch F, Ueno N, Shimasaki S, Guillemin R (1987) Follistatin specifically inhibits pituitary follicle stimulating hormone release *in vitro*. Biochem Biophys Res Commun 149: 133–137

Yoshimura Y, Hosoi Y, Bongiovanni AM, Santulli R, Atlas SJ, Wallach EE (1987) Are ovarian steroids required for ovum maturation and fertilization? Effects of cyanoketone in the *in vitro* perfused rabbit ovary. Endocrinology 120: 2555–2561

Yoshizawa I, Fishman J (1971) Radioimmunoassay of 2-hydroxyestrone in human plasma. J Clin Endocrinol Metab 32: 3–6

Zanisi M, Messi E, Martini L (1983) Influence of adrenalectomy and ovariectomy on gonadotropin secretion. J Steroid Biochem 19: 455–460

Zeleznik AJ, Midgley AR Jr, Reichert LE Jr (1974) Granulosa cell maturation in the rat: increased binding of human chorionic gonadotropin following treatment with follicle-stimulating hormone *in vivo*. Endocrinology 95: 818-824

Zeleznik AJ, Hutchison JS, Schuler HM (1985) Interference with the gonadotropin-suppressing actions of estradiol in macaques overrides the selection of a single preovulatory follicle. Endocrinology 117: 991-999

Zhiwen Z, Carson RS, Herington AC, Lee VWK, Burgen HG (1987) Follicle-stimulating hormone and somatomedin-C stimulate inhibin production by rat granulosa cells *in vitro*. Endocrinology 120: 1633-1638

Ziecik AJ, Britt JH, Esbenshade KL (1988) Short loop feedback control of the estrogen-induced luteinizing hormone surge in pigs. Endocrinology 122: 1658-1662

Zolman JC (1983) Peptide-receptor protein relationships: steroid feedback in GnRH stimulation of the anterior pituitary. J Steroid Biochem 18: 297-301

3 Teratogenicity of Hormones In Utero

3.1 Introduction

Many drugs taken by pregnant and nursing women might affect the baby (see Hutchings, 1989). Drug teratogenicity was not a major concern until the thalidomide tragedy in the early 1960's. The tranquilizer, produced in Germany (by Chemie Grünenthal) and improperly tested, was widely prescribed in Europe to combat "morning sickness" in pregnancy. The birth of hundreds of infants with phocomelia, poorly developed limbs (the "seal flipper" deformity), was the result. The drug sold even in the face of mounting reports of the malformation, hitherto one of the rarest. Only when the US Food and Drug Administration did not clear the drug for the US market was thalidomide banned in Europe.

The prenatal period is the most vulnerable time. It was once thought that the placenta protected the fetus, but this it not so. Improper prenatal care accounts for many birth defects. In the USA of about 3 million annual live births, 200,000 (7%) are believed to have some birth defects. The incidence of malformations is about 2% in England (Chaplin and Smith, 1987).

In humans the effects of drugs on the ovum and zygote before it attaches to the uterine wall remain unexplored. Data obtained in rabbits show that eggs are sensitive to estrogens (Ketchel and Pincus, 1964).

Once the zygote becomes implanted (about 5-7 days after fertilization) drugs taken by the mother can reach the embryo by way of the placental circulation. The next seven weeks are critical. During the early days the embryo may be either resistant to a drug's effect, or the drug may be too toxic and cause fetal death and abortion. Organs subjected to damage during the early part of the first trimester are the skeletal structures, eye, ear and heart. Damages to genital organs, teeth and central nervous system may develop somewhat later. Anticoagulants, oral hypoglycemic, anti-viral drugs, thyroid suppressing agents, antibiotics, immunosuppressive medications, iodine, tranquilizers, drugs (reserpine, antihypertensives, magnesium sulfate) given to manage hypertension (toxemia) and eclamptic convulsions and narcotics given during delivery may cause toxic effects during the second and third trimesters. Viral infections may cause mental retardation, and smoking may lead to decreased infant weight (Table 3.1).

It is the physician's responsibility to weigh the risk-benefit factor, but that is not always the case. Physicians prescribe antibiotics for viral infections. The use of thalidomide to prevent "morning sickness" and diethylstilbestrol to prevent miscarriages in the 1950's despite questionable effectiveness was negligent.

Table 3.1 Effects of Teratogens on the Developing Human Fetus.

Drug	Defect after exposure	
	First trimester	Second/third trimester
Alcohol	mental retardation still birth perinatal mortality	
Androgens	virilization of female fetus	
Antibiotics		
streptomycin	deafness	
tetracyclines		impaired bone growth stained teeth
Anticonvulsants	mental retardation facial defects	
β-blockers		hypoglycemia, bradycardia
Cytotoxic	multiple, growth retardation still birth, abortion	
Estrogens	cardiovascular, termination of male fetus	
DES	carcinogenesis, uterine lesions	
Iodine	goitre	
Ionizing radiation	skeletal and others	
Lithium	cardiovascular	hypotonia, hyporeflexia
Narcotics		respiratory depression
Nicotine	growth retardation, fetal death	
Retinoids	cardiac, CNS, craniofacial	
Salicylates		hemorrhage
Sulphonamides		hyperbilirubinaenira permicterus
Thalidomide	limb reduction and others	
Tranquilizers		respiratory depression, impaired thermoregulation
Warfarin	skeletal	fetal hemorrhage, CNS abnormalities

Modified from Chaplin and Smith (1987).

I use the term "teratogenesis" to denote any adverse effects on the fetus whether structural deformation, biochemical or physiological impairment, behavioral changes or developmental retardation in the neonate.

For a teratogen to become toxic to the fetus several conditions must converge:

(i)   *Placental transfer:* Teratogen must cross the placental barrier. Virtually all low molecular weight compounds in the range of 250–400 (most drugs) diffuse across the placenta; the rate is enhanced for lipid soluble drugs and those which exhibit low protein binding.

(ii) *Maternal toxicity:* teratogenic agents harmless to the mothers possess greater risks for the fetus than drugs which are teratogenic at doeses toxic to the mother.

(iii) *Dose response:* there is usually a threshold dose below which a teratogen has no observable effects. The dose response curve may be steep; small increments in dose above the threshold may result in large increases in effect.

(iv) *Genetic determination:* a predisposition to a teratogen may be inherited. Some teratogens induce chromosomal abnormalities (Ingalls *et al.*, 1963).

3.2. Steroid Hormones: Malformation of the External Genitalia and Interference with the Reproductive Function

3.2.1 Gonadal Hormones

Exposure of fish and amphibian larvae to androgens changes all offspring into functional males; estradiol will produce females. In birds treatment of fertile eggs with gonadal hormones induce akin changes but only temporarily. After some time, inverted animals will revert to their genetic sex. In mammals genetic sex determines gonadal sex, and exposure of embryos to sex hormones will change only the somatic and behavioral sex. During the undifferentiated stage two anlage for the development of male (Wolffian) and female (Müllerian) ducts are present in the fetus. The process of differentiation consists of two components. In males Müllerian ducts regress and Wolffian ducts proliferate as a result of testosterone secretion by the embryonic testis. Dihydrotestosterone is not the active intracellular hormone (Wilson, 1973). Wolffian ducts give rise to epididymis, vas deferens and seminal vesicles. The urogenital sinus develops into the prostate and the urogenital tubercle into the urogenital apparatus.

Androgens cause proliferation of the duct in both sexes. Estrogens may cause a partial involution, but in general the anlage is insensitive to estrogens (Raynaud, 1942). In castrated males Müllerian ducts persist. If castrated fetuses receive testosterone Wolffian ducts will develop into a normal male reproductive tract; in addition the animal will have a set of female reproductive organs. The findings shows that the embryonic testes must produce a Müllerian inhibiting factor (MIF) which actively inhibits the Müllerian anlage. Studies of castrated male embryos, treated in addition with anti-androgens, demonstrated that testosterone is not the MIF. Exposure of intact embryos to anti-androgens results in underdeveloped internal genitalia while the external genitalia become feminized, and a vaginal opening forms (Neumann et al., 1966).

In females the absence of testosterone causes involution of the Wolffian portion and proliferation of the Müllerian part. The oviduct, uterus, cervix and a small anterior part of the vagina arise from the Müllerian ducts; the rest of the vagina develops from the urogenital sinus. The sinus is more sensitive to estrogens during perinatal development (Zuckerman, 1940). This may explain the preponderance of lesions in organs derived from this tissue after DES exposure. Androgens cause suppression of proliferation of regional parts (vaginal segments in mice) or failure to form a vagina (Raynaud, 1942). The prostate buds are stimulated, and the genitalia develop into a male-like form. Estrogens precipitate precocious development in females. In males estrogens cause differentiation to varying degrees depending upon the status of the duct, the dose, and the species. A vaginal cord may develop, the growth of the prostate is suppressed, and the external genitalia may become female-like.

The influence of sex hormones on the formation of genital organs was studied in the late 1930's. Mme. Dantchakoff working in Bratislava, and later in Paris, published between 1936 and 1938 a series of observations pertaining to the consequence of exposing guinea pig fetuses to steroid hormones. A French physiologist

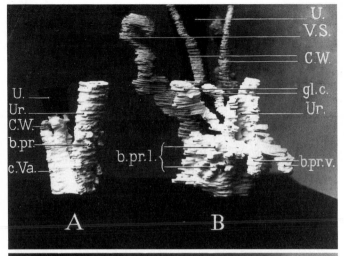

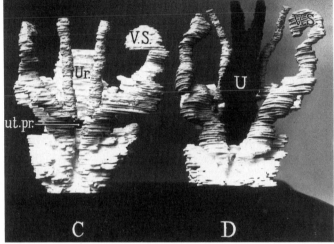

Figure 3.1.

Raynaud (1939 a,b; 1942) studied the influence of androgens and estrogens on the organization of sex organs in mice (Figure 3.1). Deanesly (1939) evaluated the reaction of rabbit embryos to both androgens and estrogens.

Test substances can be injected into the mother, or directly into the fetus. Either method has disadvantages. Estrogens dispensed to the mother reduce intrauterine pressure (Ichikawa and Tamada, 1989), induce adrenal hyperplasia in both the mother and the fetus (Seeliger *et al.*, 1974) and cause considerable fetal wastage (Chapter 2). Direct injection into the fetus leads to high mortality. Even spontaneous malformations often induce fetal death and resorption (Kalter, 1978) further limiting the number of offspring.

During pregnancy estrogens are biosynthesized in ever increasing amounts. The presence of high concentrations of α-fetoprotein may "protect" the fetus from the deleterious consequences of estrogens circulating in the mother (Vannier and

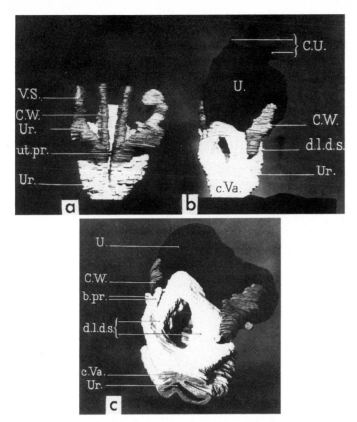

Figure 3.1. Reconstruction of the Genital Tract of Newborn Mouse. **A** normal female; **B** normal male; **C** and **D** genital tract mouse virilized by testosterone propionate; **a** normal male; **b**, **c** mouse feminized by injections of estradiol; cVa, vagina; CW, Wolffian ducts; VS, seminal vesicles; U, uterus; Ur, urethra. Courtesy A Raynaud.

Raynaud, 1975; Payne and Katzenellenbogen, 1979). α-Fetoprotein most likely does not play a major role in other species. The protein is present mainly in rodents (rats) while in many species estradiol does reach the fetus but only as a conjugate, either a glucuronide or a sulfate (see Chapter 5).

In the genetic male rat Wolffian ducts display a transitory increase in lumen size and diameter when the embryo is 16½ days old. In females the ducts begin to involute 24 h later. In females androgens will maintain the ducts if injected before this time and will produce inconsistent changes if injected after involution has begun. The portion still present on day 18 is insensitive to androgens. The presence of androgens on day 15½ has no effect on the development of the ducts (Stinnakre, 1975). The sensitivity of Wolffian ducts to hormonal influence varies from species to species.

A rapid and reliable method for assessing the effect of steroid hormones in rodents is the measurement of the distance between the anus and the urinary papilla (anogenital distance); the interval is shorter in females and longer in males.

Androgenic hormones cause an increase in the distance in females (virilization) whereas estrogens have the opposite effect (feminization). Microscopic examination of the internal genital apparatus is necessary to reveal more significant malformations.

3.2.1.1 Effects on Genetic Males

Androgens
Androgens produce permanent hyperplasia of accessory sex organs. Dantchakoff (1938a) was the first to study the influence of sex hormones in fetal guinea pigs. She began injecting gravid animals (day 18) with 0.5-1 mg of testosterone propionate (TP), increased the dose to 2 mg 5 days later and implanted in the newborn 2 mg of the crystalline substance. She found in adolescent animals (31 days old) active spermatogenesis and hypertrophy of the epididymis, seminal vesicles and penis. Slob *et al.* (1973) found marginally increased body growth in male guinea pigs born to mothers injected daily (from day 24 to day 41) with 5 mg of TP. The weight increase was transitory and disappeared during postnatal growth.

Kincl injected gravid rats daily (day 17 to 21) with 0.5 mg of TP. At the age of 80 days the males were lighter, and sex organs were smaller, but there was no evidence of inhibition of spermatogenesis. Swanson and van der Werff ten Bosch (1964; 1965) also reported active spermatogenesis in adults exposed *in utero* to androgens. Brown-Grant and Sherwood (1971) found no aftermath in guinea pigs born to mothers treated with 12.5 mg TP on days 33, 35 and 37 of gestation. In contrast, females became virilized (Section 3.2.1.2).

Goldman *et al.* (1972) generated in rabbits an antibody to testosterone conjugated at C-3 to bovine serum albumin via 3-carboxymethyloxime. Injected into pregnant rats (days 13 to 20) the antibody blocked in part the effect of testosterone. In genetic males the anogenital distance was shorter, and testes were atrophied.

Antiandrogens
In males compounds which compete with the action of androgens block the development of external genitalia and allow a partial development of the Müllerian ducts but do not prevent the development of Wolffian ducts or the differentiation of the gonads. A German group led by Neumann (Neumann *et al.*, 1966; Forsberg *et al.*, 1968; see also Neumann and Steinbeck, 1974) explored in rats the action of an anti-androgen (cyproterone acetate, CA). The steroid, injected daily during days 17-20 of gestation, caused in genetic male offspring a shortening of the anogenital distance (the length was similar to that in females) and prevented the development of seminal vesicles and of the penis. The urethra was feminine and the urogenital sinus formed as a female vagina. Wolffian ducts regressed only partially. The development of the vagina in rats is incomplete at birth. To induce full feminine formation, the group treated newborn male offspring, in addition to *in utero* exposure, during the first 10-14 days after birth. Cyproterone acetate also induces in males female-like differentiation of mammary glands.

In feminized mice (6 mg CA from day 14 to day 20 of gestation) the posterior two-thirds of the vaginal epithelium became cornified (Suzuki and Arai, 1977).

Estrogens

The concentration of estrogens in the amniotic fluid of female embryos (house mouse) is higher than in males. As a result male fetuses, which develop between two female embryos, become feminized. Adult males are less aggressive, are sexually more active, and have smaller seminal vesicles (vom Saal *et al.*, 1983).

Exogenous estrogens produce severe pathological changes. Greene *et al.* (1939 c,d; 1940) injected estradiol dipropionate into pregnant rats and found in the male offspring at puberty cryptorchidism, atrophied testes and retained Wolffian ducts. Stinnakre (1972) confirmed the paradoxical retention of Wolffian ducts in animals exposed to high estrogen dose. Falconi and Rossi (1965) reported inhibition of sex organ formation by estrogens and noted that doses which induced the changes were close to embryotoxic levels.

In mice Raynaud (1942) described feminization of the genital apparatus (see Figure 3.1) and atrophy of gubernacula (Raynaud, 1957; 1958). Injection on day 13 of gestation with 62 or 94 µg of estradiol dipropionate was enough to cause atrophy (Figure 3.2). The gubernaculum testis is a fibrous cord extending from the fetal testis to scrotal swelling; it occupies the potential inguinal canal and guides the testis in its descent.

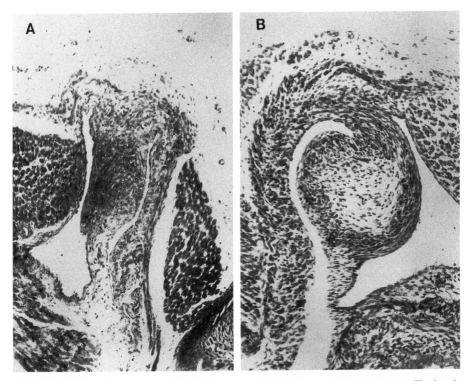

Figure 3.2. Effect of Estrogens on the Development of the Gubernacula Apparatus. **A** The inguinal canal of a 18-day-old untreated male mouse fetus; **B** atrophied inguinal canal of a 18-day-old mouse fetus whose mother was injected with 94 µg of estradiol dipropionate on day 13 of gestation. Courtesy A Raynaud.

Hadziselimovic and Guggenheim (1980) prevented atrophy of the epididymis in male mice, induced by estrogens during gestation, by simultaneous administration of human chorionic gonadotropin.

Rats

Vannier and Raynaud (1975; 1980) used graded doses of estradiol (50, 250 and 1250 μg daily, days 16-20) in rats and reported that fetal wastage was dose related. Sixty six percent of embryos died in animals injected with 1250 μg; the loss was 41% in the low dose group. A dose of 50 μg was sufficient to induce a decrease in the anogenital distance. Higher amounts led to greater teratogenic effects. Microscopic examination of fetuses (50 and 250 μg dose) revealed underdevelopment of seminal vesicles and coagulating glands, inhibition of ventral and dorsal prostate buds and epithelial proliferation of urethral walls. Males exposed to the high dose were sterile. In all males prostates were atrophic, penile development depressed (hypospadia), and testes were cryptorchid.

Kincl (see Table 3.8) confirmed a decreased embryo survival in rats injected once or five times, beginning on day 11 of gestation. The anogenital distance was shorter in embryos of both sexes (delivered by Cesarean before expected birth).

Boylan (1978) injected a total dose of 1.2 to 12,000 μg of estradiol during different times of gestation. Injections of 4 μg (or more) between days 10 and 13 were toxic to most embryos. The same daily dose was less toxic (50% fetal survival) if given on days 15, 17, and 19. When the dose was decreased to 0.4 μg 97% of the offspring survived, but the dose was apparently too low to have any adverse effects.

Other Estrogens

Vannier and Raynaud (1975) studied the activity of a synthetic estrogen, 11β-methoxy-17α-ethynyl-1,3,5(10)-estratriene-3,17β-diol (R-2858). The amounts used were 0.4, 2, 10, or 50 μg/day from day 16 to day 22 of gestation. This estrogen was more toxic than estradiol. A dose of 2 μg resulted in a decrease of anogenital distance, atrophy of the seminal vesicles and ejaculatory ducts and inhibition of the growth of prostatic buds. A dose of 10 μg induced death in 60% of the fetuses; 250 μg of estradiol was needed to produce a comparable result.

3.2.1.2 Effects on Genetic Females

In genetic females (rodents, ungulates and primates) androgens prevent the regression of Wolffian ducts (virilization). The animals become sterile if the hormone is injected directly into the fetus to bypass metabolite formation in the mother. Mortality is very high in these cases needing a large number of gravid females to obtain meaningful data. Estrogens also imperil fetal survival.

Greene et al. (1939b) found progesterone, a precursor in androgen biosynthesis, androgenic. Later studies did not confirm the results: Revesz and Chappel (1966) injected subcutaneously 100 mg to gravid rats and found no evidence of virilization. Courrier and Jost (1942) found virilization with pregnenolone.

Androgens

In cattle (Keller and Tandler, 1916; Lillie, 1916; 1917), swine (Hughes, 1929; Andersson, 1956), sheep, and goats, female fetuses become exposed to a male hormonal milieu if they share a common placental circulation with a male twin (freemartin). The intersexual female sibling may develop external female genitalia and typically male germ cords. Germ cells and some elements of vasa deferentia are absent. Females develop normally if blood circulation does not anastomose. The noted British biologist Aldous Huxley narrated, in an anecdotal manner (*Brave New World*, published in 1932), the impact of male hormone on female embryos. A supervisor, Mr. Foster, conducts a tour of a factory where babies are artificially incubated and explains to students ..."other female (embryos) get a dose of male sex-hormone...result: they are decanted as free martins–structurally quite normal...(except he had to admit that they have the slightest tendency to grow beards)...but sterile; guaranteed sterile..."

In mice a female fetus developing between two male embryos may become influenced by the proximity. In adults the average cycle length increased by 1½ days (vom Saal and Bronson, 1980), and their behavior resembled in some respects male responses (Hauser and Gandelman, 1983).

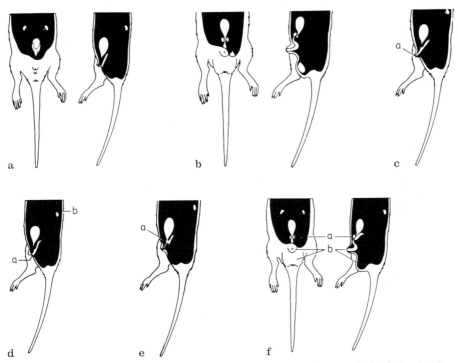

Figure 3.3. Schematic Representation of Virilization Changes in the Female Rat. The drawings show the arrangement of internal and external sex organs in normal female **A**, normal male **B**, and virilized females: **C** undifferentiated urogenital sinus, common orifice in urinary papilla; **D** (a) undifferentiated urogenital sinus, common orifice at the base of urinary papilla; (b) oviducts anterior to ovary; **E** blind uterus, fistula connecting uterus to urethra; **F** (a) enlarged female prostate; (b) typical male penis and scrotal sac, no testes present.

A typical finding in virilized females is an increase in the anogenital distance, the presence of a prostate, and more severe morphological changes (Figure 3.3). Female fetuses of many mammalian species, including primates, retain small cords of cells which represent a homologue of the male prostate buds. In some species (insectivores, bats, rodents and lagomorphs) the glands develop after birth into a female prostate resembling a male organ. The lobes are present but do not grow as large. Rauther (1909) was the first to describe the female prostate in a species of an African mouse. Marx (1931; 1932) noted the sporadic spontaneous occurrence in other rodents. The gland responds to androgen treatment (Korenchevsky and Dennison, 1936; Korenchevsky, 1937). In the inbred laboratory white rat (Wistar strain) spontaneous occurrence of the atrophic gland may vary between 28 and 99 percent (see Price and Williams-Ashman, 1961).

Guinea Pigs
Dantchakoff (1936; 1937 a,b,c) injected directly (0.05 mg) TP into 20–23 day old fetuses and gave a second injection (3 mg) 6-10 days later. The results were severe. Female embryos (45 days old) had, in addition to ovaries, two sets of secretory ducts (oviducts and sperm duct) located in a mid-point between the kidneys and the bladder (as in males), prostate, seminal vesicles and a rudimentary penis. The ovaries of adult (6 weeks old) females were cystic and contained some maturing follicles with degenerating ova; corpora lutea (CL) were absent (Dantchakoff, 1938 b).

Phoenix et al. (1959) confirmed the virilizing influence of testosterone given to mothers. The does were injected before day 24 of gestation, in some cases up to delivery (65 days). The authors stated that the formation of external genitalia and behavior were "abnormal" (see also Chapter 6). In two later reports (Goy et al., 1964; Young et al., 1964) reconfirmed Dantchakoff's result. Both reports state that sex organs were masculinized and CL did not form in the ovaries, but again no details were given. Brown-Grant and Sherwood (1971) used 12.5 mg of TP injected on days 29, 31 and 33 of embryonic life. The treatment induced gross malformations of the lower genitalia in female offspring; the urethra opening was at the end of an enlarged phallus, and the vaginal opening was absent. Fertility was not influenced. The ovaries appeared normal with old and new CL present.

Slob et al. (1973) found no changes in postnatal body growth of females born to does injected with TP (5 mg daily) between days 24 and 41 of gravidity.

Rats
The ovary of perinatal rats responds to gonadotropin stimulation with the production of androgens. Androgens stimulate in infants the growth of the clitoris which enlarges into a penis-like structure and may develop the cartilage anlage of the os penis (Greene et al., 1939 a).

Cagnoni et al. (1964) injected TP (10 mg) on day 16 and evaluated the effects in 120 day old offspring. All females (only 4 survived) became virilized. Kincl (Table 3.2) injected graded doses of TP daily between days 15–19 and measured the anogenital distance on day 1, again on day 24 of life, and autopsied the animals when 60–80 days old. The degree of virilization was dose related. A daily dose of 250 µg was sufficient to increase the anogenital distance in the genetic

Table 3.2 Influence of Testosterone Propionate Injected from Day 15 to Day 19 of Gestation on the Development of Female Offspring.

Treatment	Daily dose μg	Final body wt. g ± S. E.	Anogenital distance mm ± S. E.		Total no. of pups	No. of abnormal pubs (virilization grade)	Ventral prostate mg ± S. E.
			Day 1	Day 24			
0	0	63.4 ± 1.4	1.3 ± 0.1	9.75 ± 0.52	16	0	0.0 ± 0.0
Testosterone	250	52.3 ± 2.8	2.7 ± 0.1	10.9 ± 0.4	18	18(1); 13(2); 1(3); 2(4)	9.4 ± 1.4
propionate	750	46.6 ± 5.6	2.5 ± 0.0	14.6 ± 1.0	5	5(2); 5(3)	12.4 ± 1.2
	2250	52.5 ± 6.5	3.0 ± 0.2	14.3 ± 0.9	4	4(3); 4(4)	21.8 ± 4.1

The numbers in parentheses indicate virilization grade found on visual inspection. An individual animal could exhibit one or several malformations at the same time. Virilization grades: (1) vaginal opening at base of urinary papilla; (2) prostate present; (3) common vaginal-urinary opening; (4) blind vagina. Kincl (unpublished data).

Table 3.3 Influence of Testosterone Propionate (TP) Injected Subcutaneously to Female Rats from Day 15 to Day 19 of Pregnancy on the Development of Female Offspring (Autopsy at 50 Days of Age).

Treatment	Daily dose μg	Ovarian response*	Virilized	Virilization grade (number)	Average no. of CL (Range)	Organ weights, mg ± S. E. (range)		
						Ovaries	Prostate	Anogenital distance mm ± S. E.
0	0	22/23	0	0	12 (0–24)	45.9 ± 2.7	0.0	15 ± 0.3
TP	250	19/21	8	1, 2 (7)	10 (0–20)	43.9 ± 2.7	7.8 (0.8–15.6)	15 ± 0.3
	1000	24/24	24	1 (2); 2 (24) 3 (22); 4 (3)	11 (5–22)	47.5 ± 2.3	17.8 (9.1–38.5)	19 ± 0.4

Virilization grade: see Table 3.2 for description; *ovarian response: number of rats with CL/total number of rats. CL, corpus luteum. Kincl (unpublished data).

females and to produce more severe malformations. The function of the ovaries was not influenced. In adult females the number of corpora lutea was similar to the number found in the control group (Table 3.3). Swanson and van der Werff ten Bosch (1964) also observed ovarian cyclicity in 21–26 week old females born to mothers exposed to 2.5 mg of TP between days 19–22.

Smith (1970) observed a "delayed androgenization" syndrome in rats born to mothers injected with 1.5 mg on day 17 of gestation. Few animals (5) were born with a patent vaginal orifice. The five mated when they were 3, 6, 9 and 12 months old. All females littered after mating when 3 months old. Only 1 became pregnant when mated at 6 months, and none were fertile thereafter. During the fertile period the intervals of estrus were longer than normal. At autopsy only old CL were present. The ovaries of animals with absent vaginal orifice showed similar abnormalities.

Table 3.4 Survival of Fetuses Injected *In Utero* with Testosterone Propionate or Diluent Alone.

Treatment	Number used		Number surviving
	litters	fetuses	
Oil	12	122	none
Saline	8	90	none
DMSO[a]	11	120	none
TP in oil[b]	16	195	1
TP in saline	8	90	3
TP in DMSO	11	120	none

[a] DMSO, dimethylsulfoxide; [b] TP, testosterone propionate. Kincl (unpublished data).

Direct Injection into the Embryo

Hormones injected into the mother reach the fetus after exposure to the metabolizing influence of the feto-placental unit. Maternal metabolism can be circumvented by injecting androgens directly into fetuses, but this results in high mortality (Table 3.4). Mortality was high regardless of diluent (50 μl) which was oil, saline, or dimethyl sulfoxide (DMSO), a solvent in use to preserve live cells. Most fetuses died either *in utero* or, if born, were destroyed by the mother. Few died within the first 24 hours of life. Only 3 females survived to 50 days. One female (injected with 250 μg TP in oil) was a hermaphrodite (Kincl and Maqueo, 1965). Two females (50 and 250 μg TP in saline, respectively) had ovaries which contained fresh CL.

Others exposed the fetus to testosterone later during fetal life with greater success. Swanson and van der Werff ten Bosch (1965) used 20 or 100 μg of TP between days 19 and 22 and reported the development of anovulatory cycles. Gorski (1968) stated that 10 μg of TP injected one or a few days before delivery induced anovulatory cycles in 9/12 animals but provided no other data. Flerkó *et al.* (1967) injected TP (500 μg) on day 19 and found high mortality. In a group of 19 mothers, only 6 delivered and raised a total of 15 female pups. All were born masculinized (vaginal opening absent), and all had ovaries containing only growing and cystic follicles but no CL.

Fels and Bosch (1971) tested fertility after injecting either TP or testosterone hemisuccinate (a water-soluble derivative) into the mother or into the fetus. The hemisuccinate was not effective even when injected into the intra-amniotic fluid. Fetal wastage was high (81%), but all females which reached puberty (32) were fertile. Injection of testosterone propionate into mothers produced virilization but no apparent effect on the function of the ovaries. Intra-amniotic injection (0.5–1 mg) induced high fetal mortality. Only 11/61 mothers delivered (in the oil injected group the mortality was 66%), and most offspring were sterile. Twenty-five females reached puberty, but only 3 (12%) became pregnant.

Josso (1970) studied the virilizing properties of testosterone in organ cultures.

Mice

TP (25 and 250 μg) injected into 17-day-old fetuses caused in adults (examined when 5–10 months old) virilization of the genitalia, absence of vaginal orifice and irreversible proliferation of the vaginal epithelium. There was no effect on the reproductive capacity. Injection of TP at birth did not masculinize the genitalia but blocked the reproductive capacity. The ovaries of 13/18 animals contained only cystic follicles and no CL (Taguchi *et al.*, 1977).

Sheep
Clarke *et al.* (1976) demonstrated that in sheep genitalia differentiate between day 40 and 50 of fetal life. Implantation of 1 g bolus of testosterone between days 30 and 80 (releasing about 7 mg daily) caused masculinization of the external genitalia. The ewes were born with a penis and an empty scrotal sac. A vaginal opening absent at birth developed slowly. In 10-month-old animals a small opening became visible in addition to the one present at the tip of the fused labia. Small seminal vesicles and bulbo-urethral glands were also present. Exposure between days 50 and 100 caused only hypertrophy of the clitoris and restricted vaginal opening. Virilization was less severe in fetuses subjected to the same testosterone dose between days 70 and 120.

Exogenous testosterone between days 30 and 80 had only a marginal effect on the ovarian function; 5/6 adults ovulated the first year, and the average ovulation rate (2.4) was similar to that seen in controls (1.4). A later exposure (days 50–100) induced sterility in 70% of females; only 2/7 had luteinized ovaries. All females exposed to testosterone between days 90 and 140 ovulated (Clarke *et al.*, 1977). The nature of the lesion was not clear in the sterilized animals. The concentration of LH in the anovulatory ewes was higher but still below castration levels. Mean ovarian venous concentrations (pg/ml) of estradiol (217 ± 51), androstenedione (454 ± 990), and testosterone (359 ± 31) were not different from values found in control animals (116 ± 25, 434 ± 95, and 376 ± 41, respectively; Clark and Scaramuzzi, 1977).

Hamsters
In female hamsters the development of the urogenital apparatus is completed after birth. The endodermal sinovaginal cord regresses, a vaginal introitus and a pair of vaginal pouches form during the first few days of postnatal life. Injection of 1 mg or less of testosterone propionate on day 9 of gestation speeds up the development of female sex organs. Injection of TP (5 mg or more) between days 10 and 12 inhibits the downgrowth of the vagina which remains short and opens into the urogenital sinus "...almost complete male type differentiation of the organs derived from urogenital sinus..." The females remain sterile, and no corpora lutea form (Bruner and Witschi, 1946).

Pigs
Elsaesser *et al.* (1978) observed a dual response depending on the timing. A direct injection of testosterone esters (TP and T enanthate corresponding to 20 mg of testosterone) into fetuses before day 40 of gestation virilized external genitalia of females. Exposure to androgens after day 40 and before day 90 inhibited ovulation, possibly by impairing the functioning of estrogen feedback mechanism on LH release.

Other Species

Androgenization of female fetuses was reported in hamsters (see Chapter 6), dogs (Beach and Kuehn, 1970; Beach 1975), and monkeys (Goy and Phoenix, 1971). In most studies changes in behavior were the end point used to assess virilization changes (Chapter 6). In a later study Young *et al.* (1964) described the development of a hermaphrodite monkey (a female with well developed phallus). The daily TP dose was 25 mg from postcoital day 40, 20 mg from day 51 to 70 and 10 mg from day 71 to 90.

Other Androgens

Dihydrotestosterone, the "active" androgenic hormone in the male (Chapter 2), causes virilization. The hormone forms from testosterone in the urogenital sinus and tubercle before sex differentiation (rabbits and rats) and in the Wolffian and Müllerian ducts after the sex has become established. Possibly the former tissues have an inherent enzymatic capacity for the reduction while the sex cords need be exposed to testosterone before they become capable of a response (Wilson and Lasnitzki, 1971).

DHT and 5α-androstane-$3\alpha,17\beta$-diol (5 mg) injected into pregnant guinea pigs (day 28-36), or 1 mg (days 37-57), are equally potent to virilize female fetuses as the same dose of TP (van der Werff ten Bosch and Goldfoot, 1975). Goldman and Baker (1971) used two doses, 10 mg/kg and 50 mg/kg maternal body weight (or about 2 and 10 mg per pregnant rat), delivered the fetuses by cesarean section on day 22 and measured the anogenital distance. Androstenedione was an active substance; in genetic females the average distance increased from 1.26 ± 0.10 mm (controls) to 1.70 ± 0.19 mm (low dose group) and 2.71 ± 0.12 mm (fetuses exposed to the high dose). Schultz and Wilson (1974) could not confirm the virilizing properties of androstenedione. The group used daily injections between days 14 and 21 and judged the effect by macroscopic examination of the stained (toluidine blue) urogenital tract of 1 day old pups. DHT was androgenic at a dose of 4 mg; a daily dose of 1 mg produced no effect. Three other steroids (5α-androstane-$3\alpha,17\beta$-diol, 17α-methyl testosterone and 17α-methyl DHT) were active (Table 3.5), and five were not inactive (Table 3.6). The authors stressed that lack of virilization could be the result of other factors (absorption from injection site, transport and binding in the blood, metabolic rate and diffusion across the placenta) and that it is not possible to draw quantitative conclusions from studies employing only a single dose. Elger *et al.* (1970) confirmed the virilizing property of methyl testosterone. Goldman (1970) found that 100 mg/kg of DHA induced slight clitoral enlargement and increases in the anogenital distance. The sulfate was not active.

Kawashima *et al.* (1978a) relied on the measurement of the length of the urovaginal septum sensitive to androgenic stimulation between days 18 and 19 of gestation. Testosterone (50 µg dissolved in 1 µl olive oil) injected directly into the fetus induced a significant shortening in females (from 0.25 ± 0.02 mm/g BW $\pm$ SE in controls to 0.12 ± 0.03 mm in the treated group). The same dose injected into fetuses on day 17, or 20, was not active. DHT and $5\alpha(?)$-androstane-$3\beta,17\beta$-diol (50 µg) were active. The 3α-isomer ($5(?)$-androstane-$3\alpha,17\beta$-diol) was active at a dose of 100 µg. Androstenedione and epiandrosterone were inactive at a dose of 200 µg. Kawashima *et al.* (1978b) described the reduction of the urovaginal sep-

Table 3.5 Virilizing Potency of Various Androgens (Female Fetuses) When Injected into the Gravid Rat.

Steroid studied	Daily dose, mg	Injection period[a]	Estimated potency[b]	Reference
DHT	4	14–21		1
	1, 10/kg	13–20	~1.2	2
5α-A-3α 17β-diol	4	14–21		1
	1, 5, 10/kg	13–20	1	2
A-dione	10, 50/kg	13–20	0.4	2
5α-A-3β 17β-diol	1, 10/kg	13–20	~0.2	2
Androsterone	10/kg	13–20	0.1	2
17-MeT	4	14–21		1
	1, 3, 10	15–21		3
17-MeDHT	4	14–21		1

[a] Injection period during gestation (days); [b] testosterone = 1; ref. 1, Schultz and Wilson (1974) end point based on gross examination of sex organs; ref. 2, Goldman and Baker (1971) end point based on microscopic measurement of anogenital distance; ref. 3, Elger *et al.* (1970) end point based on microscopic evaluation of sex organs. DHT = dihydrotestosterone; 5α-A-3α,17β-diol = 5α-androstane-3α,17β-diol; A-dione = 4-androstene-3,17-dione; 17-MeT = 17α-methyl testosterone; 17-MeDHT = 17α-methyldihydrotestosterone.

Table 3.6 Androgens Found Inactive (Female Fetuses) When Injected into Gravid Rats.

Steroid studied	Daily dose, mg	Injection period[a]	Reference
5β-DHT	4	14–21	Schultz and Wilson (1974)
3β,17β-5α-A-diol	4	14–21	
5α-A-dione	4	14–21	
A-dione	4	14–21	
DHA	30, 100/kg	13–20	Goldman and Baker (1971)
Androsterone	4	14–21	

5β-DHT = 17β-hydroxy-5β-androstan-3-one; 3β,17β-5α-A-diol = 5α-androstane-3β,17β-diol; 5α-A-dione = 5α-androstane-3,17-dione; DHA = dehydroepiandrosterone. [a] days of gestation.

tum in F_2 generation (born to untreated offspring of virilized mothers) from 0.23 ± 0.01 to 0.17 ± 0.02 mm/g BW. The mechanisms involved in the effect were not clarified or confirmed. The authors suggested either that the gene had undergone transformation or that the concentration of receptor molecules was changed (which would also involve gene transformation). Possibly the observed change could have been the outcome of biological variation, not physiologically significant, despite the statistical validity.

Antiandrogens

Simultaneous administration of antiandrogens blocks the virilizing effects of androgens in the female. Elger *et al.* (1970) injected pregnant rats with methyl testosterone (MT) (1,3, or 10 mg daily) and 30 mg of cyproterone acetate (CA) from day 15 to day 21. Microscopic evaluation showed that CA prevented the virilizing consequence of 1 or 3 mg of MT but not 10 mg. Wolffian ducts retrogressed in most but not all female fetuses. In some rudimentary or even whole ducts persisted on one side. Goldman and Baker (1971) found that 30 mg of CA blocked the virilizing influence of 10 mg of DHT, or 5α-androstane-$3\alpha,17\beta$-diol, given between day 13 and 20 of gestation.

Influence on Human Babies

A female fetus, and the mother, may become virilized by exposure to an elevated androgen milieu during pregnancy. Increased androgen production can result from abnormal maternal adrenal function (adrenogenital syndrome), arrhenoblastoma, or thecal cell tumor (Murset et al., 1970; Fayez et al., 1974; Verkauf et al., 1977) or if the mother was treated with synthetic progestational agents (Section 3.2.2.2). Hayles and Nolan (1957; 1958) reported masculinization of a baby girl associated with the use of methyl testosterone during pregnancy.

Estrogens

Exposure of genetic females to estrogens induces teratological changes which persist into adulthood:

(i)   the vaginal epithelium proliferates abnormally leading to adenosis and ultimately to the development of adenocarcinoma of the vagina and cervix;
(ii) similar changes may develop in the mammary glands;
(iii) behavior patterns change; in humans personality aberration may develop.

In genetic females estrogen induces paradoxically virilizing changes (Greene et al., 1939c). Moore (1939) reported stimulation of Wolffian ducts by estrogens in the opossum of both sexes. A synthetic compound diethylstilbestrol (DES) is highly toxic (Chapter 5).

Estradiol

In mice, the vaginal epithelium is more sensitive to damage than is the reproductive capacity (Kimura, 1975). The reaction is rapid; the mitotic activity recedes within 2–3 days after exposure. Kimura et al. (1980) injected 50 µg of estradiol on day 17 of gestation and found in the offspring irreversible, ovary independent cornification, or stratification, of the vaginal epithelium. Estradiol had minimal effects on reproductive capacity; 8/12 animals had active corpora lutea. The sensitivity of the hypothalamus-pituitary axis to hormonal insult was greater earlier in fetal life. When estradiol was injected 2 days earlier (day 15 of gestation) only 1/6

Table 3.7 Influence of Estradiol Injected Subcutaneously from Day 16 to Day 19 of Gestation on the Anogenital Distance of 2 Day Old Male and Female Offspring.

Treatment	Daily dose µg	Anogenital distance, mm ± S. E.			
		Males	(No. of pups)	Females	(No. of pups)
0	0	3.7 ± 0.01	(28)	1.7 ± 0.02	(30)
Estradiol	100	3.6 ± 0.01	(15)	1.9 ± 0.03	(17)

Kincl (unpublished data).

Table 3.8 Influence of Various Steroids Injected Subcutaneously to Rats from Day 11 of Gestation on the Anogenital Distance of Male and Female Pups Removed by Cesarean Section.

Treatment	Daily dose mg	No. of treatment days	Anogenital Distance, mm ± SE					
			Males	(No. of pups)	Females	(No. of pups)	Intersex	(No. of pups)
0	0	0	3.08 ± 0.12	(18)	1.19 ± 0.07	(13)		
Cortisol	5 × 4	4	3.13 ± 0.07	(30)	1.24 ± 0.02	(38)		
	15 × 4	4	2.50 ± 0.09	(21)	1.21 ± 0.04	(34)	1.83 ± 0.01	(3)
Estradiol	0.1 × 5	5	2.42 ± 0.11	(19)	1.06 ± 0.04	(21)	0.0[a]	(1)
	1 × 5	5	2.38 ± 0.16	(8)	1.00 ± 0.0	(9)	1.38 ± 0.11	(16)
Progesterone	5 × 6	6	2.64 ± 0.06	(34)	1.02 ± 0.02	(30)		
	50 × 6	6	2.89 ± 0.06	(41)	1.20 ± 0.05	(28)		

[a] No apparent anogenital openings; Kincl (unpublished data).

females had ovaries containing CL at the age of 3–5 months. Injection of the same amount of estradiol after birth (day 3) provoked less severe changes in the vaginal epithelium but blocked completely the reproductive capacity.

The sex cords of rat embryos are sensitive to estrogen exposure in mid-pregnancy. Injection of 0.1 mg estradiol between days 16 and 19 had no effect on the formation of external genitalia or the anogenital distance in either sex (Table 3.7). Repeated dosing between days 11 and 16 resulted in a shortening of the distance in males and formation of intersexes (Table 3.8). Citti (1960) described endometrial changes (single cell layer with nuclei at the cellular base) in offspring of rats injected with 1 mg of alpha-estradiol (possibly meant estradiol − 17ß?).

Falconi and Rossi (1965) studied the teratological effects of estradiol benzoate.

3.2.2 Synthetic Steroid Hormones

Oral contraceptives are a mixture of a synthetic estrogen (ethinyl estradiol or its 3-methyl ether) mixed in varying proportions with a progestational agent, usually derived from an 4-estrene (19 nor testosterone); or estrogens are used alone followed by progestational agents. Injectable contraceptives are derivatives of 17-hydroxyprogesterone.

3.2.2.1 Ethinyl Estradiol (EE)

Exposure of pregnant mice (ICR-JCC strain) to EE, 0.02, 0.2 or 2 mg/kg BW (about 1 µg or more), prior to gonadal differentiation (days 11 through 17) produced in the adult offspring (10–14 weeks old) enlarged uteri, estrus- and proestrus-like proliferation of the vaginal epithelium (no diestrus smears) and cystic glandular hyperplasia with epidermization of the endometrium. The ovaries contained both growing follicles and CL. The number of primordial follicles decreased and the number of degenerating follicles, appearing as cysts, increased (Yasuda *et al.*, 1977). In males testes were smaller, and ovotestes were present in

Table 3.9 Influence of Testosterone Propionate and 17-Ethinyl Estradiol 3-Methyl Ether (MEE) Injected into Female Rats from Day 17 to Day 21 of Pregnancy on the Development of Female Offspring.

Treatment	Daily dose µg	Total no. born	With CL	Virilized	Virilization grade (number)	Prostate, mg $\pm$ S. E.
0	0	35	24	0	0	0.0 ± 0.0
Testosterone propionate	500	14	6	13	1 (13)[a]	8.8 ± 1.1
					2 (8)[b]	15.0 ± 3.0
MEE	10	10	6	9	1 (6)[a]	0.0 ± 0.0
					3 (1)[b]	0.0 ± 0.0

[a] Autopsy at 45 days of age; [b] autopsy at 80 days of age; for the description of virilization grade see Table 3.2; Kincl (unpublished data).

some. The gonads contained seminiferous tubes with spermatogonia, some with pachytene nuclei, Sertoli cells and Leydig cells in the interstitial tissue; in some animals testes were in the abdominal cavity. The ovarian portion had cells with pachytene nuclei (Yasuda *et al.*, 1985).

Kincl (Table 3.9) tested in rats the activity of the 3-methyl ether of EE (mestranol, MEE), another component of many contraceptive preparations. Daily injections of 4 µg between days 15 and 20 led to a significant decrease of the anogenital distance of male fetuses. A dose of 40 µg had no effect on females (Kincl and Dorfman, 1962). Most females became virilized (90%) by injecting 10 µg from day 17 through 22. The vaginal opening was present at the base of the urinary papilla in 60% and a common vaginal and urethra opening in 10%. No prostate was present in virilized females as was the case of malformations induced by testosterone. Estrogen handling had influenced the ovarian function of 45–80 day old offspring. Only 6/10 had active CL, and those were few.

A synthetic estrogen, RU 2858 (11β-methoxy-17α-ethinyl-1,3,5(10)-estratriene-$3,17\beta$-diol), injected (10 µg/day) from day 16 through 20 prevented in male offspring the development of Wolffian ducts, blocked the proliferation of prostate buds and induced hypospadia. In females the estrogen blocked the development of the lower vagina while the Wolffian ducts persisted. The compound was toxic: 59% of fetuses did not survive the treatment (Vannier and Raynaud, 1975).

3.2.2.2 Oral Contraceptives

Both types of progestational agents used, 4-estranes and 17-acetoxy derivatives, are teratogenic. Prior to the 1970's estrane derivatives were synthesized by reduction of phenolic ring A (estrone) and were always contaminated with varying amounts of EE or MEE. Only one group (Kincl and Dorfman, 1962) tested purified steroids.

Influence on Rodent Embryos

Injection of pregnant rats with either type of progestational agent induces masculinization of external genitalia (Pincus, 1956; Revesz *et al.*, 1960; Kincl and Dorfman, 1962; Whalen *et al.*, 1966; Saunders, 1967). Test compounds, dissolved in oil, were usually injected for several days from day 15, the period of the differentiation of sex organs. Progestational agents in high doses prevent birth. In such cases fetuses had to be removed by cesarean section on day 22. Most investigators examined the animals at birth and frequently later in life. The data of Revesz and coworkers (1960) are typical. 17α-Ethinyl-19-nor-testosterone (Norlutin) produced virilization when the daily dose was 1.5 mg or more. The anogenital distance was longer (2.5$\pm$0.03 mm vs 1.5$\pm$0.04 mm in controls). Malformations included enlarged clitoris and blind vaginae. In males the anogenital distance was shorter without other signs of feminization. Kincl and Dorfman (1962) observed a decreased anogenital distance only after a daily dose of 5 mg; the steroid they used was free of estrogenic contamination. A related steroid (17α-ethinyl-5(10)-17β-hydroxy-4-estren-3-one, norethynodrel), dose 12.5 mg, produced severe masculinization in females. The anogenital distance was long, 4.1$\pm$0.8 mm, and only 1.8$\pm$0.09 mm in controls. Many animals had a common vaginal-urinary opening and blind vagina. Whalen *et al.* (1966) used a 2.5 mg daily dose of Norlutin and found evidence of virilization. The group reported that the dose was enough to prevent parturition.

Progesterone Derivatives

Virilizing synthetic compounds with progestational activity include acetophenone derivatives obtained from 16α,17-dihydroxy progesterone (Lerner *et al.*, 1962), 17-hydroxy progesterone, and 19-nor-testosterone (Schöler and de Wachter, 1961; Suchowsky and Junkmann, 1961; Whalen *et al.*, 1966).

Three derivatives were studied extensively: medroxyprogesterone acetate (MAP), megestrol acetate (MGP), and chlormadinone acetate (CAP). MAP induces virilization in rats (Revesz *et al.*, 1960; Kincl and Dorfman, 1962; Sanwal *et al.*, 1970; Kimmel *et al.*, 1979) and cleft palate in rabbits but not in mice or rats (Andrew and Staples, 1977). A relatively low dose (0.15 mg daily) may influence proliferation of prostate buds; higher doses (1.5 mg or more) will induce the formation of blind vaginae, and the vaginal orifice will be absent. The closely related MGP also possesses virilizing properties (Sanwal *et al.*, 1970). Substitution of the methyl group on C-6 by chlorine (CAP) decreased the androgenic response. The steroid was not teratogenic up to a daily dose of 7.5 mg (Kincl and Dorfman, 1962; LePage and Guenguen, 1968). Mey (1963) found prostate development in females by increasing the dose to 10 mg.

Revesz and Chappel (1966) tested a 17-methyl progestational agent, medrogestenone (6,17-methyl-4,6-pregnadiene-3,20-dione). The steroid exhibited anti-androgenic properties. A daily dose of 10 mg induced hypospadias and cryptochridism in all the male offspring. There was no effect on females.

Influence on Human Embryos

Embryos became exposed to constituents of oral contraceptives either during a "pregnancy test", in support of a threat of a miscarriage, mistakenly without realizing that pregnancy existed, or when mothers used OCs during breast feeding. According to Shiono *et al.* (1979) over 70,000 fetuses were exposed to OCs in the US annually.

The first association between the use of OCs and masculinization was made by Wilkins *et al.* (1958). Other reports soon followed (Bongiovanni *et al.*, 1959; Grumbach *et al.*, 1959; Grumbach and Ducharme, 1960; Jones and Wilkins, 1960; Harlap *et al.*, 1975; Harlap and Eldor, 1980). Before that time there was no recognition that steroid hormones would masculinize girl babies. Physicians prescribed androgens (17α-ethinyl testosterone or methyl testosterone) as anabolic agents during pregnancy. The therapy often continued in the presence of frank masculinizing changes (deepening of the voice, hirsutism, acne, and clitoral enlargement) in the mother. Most of the baby girls were born with an enlarged clitoris (often to the point of differentiating as a phallus) with varying degrees of labioscrotal fusion. Virilization often took place without masculinization changes in the mother (see Diamond and Young, 1963). Wilkins (1960) reviewed 70 cases and concluded that OCs induced labioscrotal fusion only during the first trimester. Clitoral hypertrophy was influenced at any time during pregnancy. Jacobson (1961) reported masculinization in 18.5% of girl babies in Norlutin (NET) users. The incidence in unexposed pregnancies was only 1%. Fine *et al.* (1963) confirmed the teratological effect of NET.

Hagler *et al.* (1963) and Burnstein and Wasserman (1964) reported on the virilizing properties of Provera (MAP).

Bongiovanni *et al.* (1959), Neumann *et al.* (1970) and Lanier *et al.* (1973) described masculinization of female infants associated with estrogen therapy alone. Driscoll and Taylor (1980) tested the effect of maternal estrogens upon the male urogenital system.

In recent years the danger of teratogenic effects associated with the use OCs has significantly diminished. The danger was recognized, and the dose decreased; today preparations contain only 5%, or less, of the amounts used 30 years ago.

Earlier reports suggested a causal relationship between the use of OCs and congenital malformations in the offspring: heart defects (Levy *et al.*, 1973), limb reduction (Janerich *et al.*, 1974), facial deformation and, in boys, moderate to severe genital abnormalities (Lorber *et al.*, 1979); see also Nora and Nora (1975). Hines and Goldzieher (1968), Peterson (1969) and Vorherr (1973) questioned whether OCs cause malformations other than virilization. Rothman and Louik (1978) found no correlation in a large retrospective study of 7723 births. The frequency of congenital malformations, cleft palate, Down's syndrome, ear tabs, heart defects, limb malformations, neural tube, skin tags, abnormalities of upper alimentary tract, hydrocele, hypospadias and undescended testes was 4.3% for infants whose mothers had taken OCs during pregnancy. The incidence was 3.3% for infants born to mothers who did not use OCs for 3 years before conception. The authors concluded "...oral contraceptives present no major teratogenic hazard..." Ortiz-Perez *et al.* (1979) reached similar results in another retrospective study.

Prenatal OCs do not influence the sex of offspring (Crawford and Davies, 1973; Keserü *et al.*, 1974; Oechsli, 1974; Rothman and Liess, 1976). There is no danger of increased predisposition to cancer to the mother.

3.2.3 Diethylstilbestrol (DES)

DES, a nonsteroidal estrogen synthesized by Dodds (Dodds *et al.*, 1938), was in extensive use between 1945 and 1971 in pregnant diabetic women, in women with a history of an inability to carry to term, for treatment of postmenopausal syndrome and breast and prostatic cancer, and in beef cattle and poultry to achieve a more pleasing carcass fat distribution. In 1946 Smith (an obstetrician) and Smith (a biochemist) suggested giving increasing amounts of DES to all women in pregnancy to prevent, or decrease, the hazards of late complications. Smith (1948) summarized additional data to evidence the benefit of the *prophylactic* use of DES in threatened abortion, complications of late pregnancy, for women with known essential hypertension, with diabetes, and for premature delivery. The recommended dose regimen was 5 mg daily starting shortly after implantation, during the sixth or seventh week (counting from the last menstrual period) and increasing the dosage by 5 mg intervals to the 15th week (daily dose of 25 mg); thereafter the dose was increased by 5 mg at weekly intervals to 125 mg daily at 36 weeks. If the patient followed the suggested regimen she ingested a total dose of almost 10 g.

Smith and Smith believed DES stimulated estrogen production and augmented progesterone biosynthesis mediated by a better "utilization" of chorionic gonadotropin. The authors did not advocate a direct use of progesterone since the hormone "...markedly inhibited the pituitary stimulating properties of estrone..." An additional support for the use cited was that "small" amounts of estrogens increase pregnancy maintenance in rabbits. The authors believed "...stilbestrol is given not because it is estrogenic but because it stimulates the secretion of estrogen and progesterone...the dosages prescribed are not enough per se to raise the estrogen level above the physiologic norm of pregnancy...the level of estrogen of any patient on our schedule would still be within the range of normal..."

The hypothesis was deficient in two respects. The first error was to equate estradiol and diethylstilbestrol. The second was the design of the study which by present-day standards was poor: the results were based on 632 case histories comprising various complications during early-, mid-, and late pregnancy collected by 117 obstetricians; placebo was not used. Smith reported a 72% "success" rate (women that carried to term) and noted that the highest spontaneous cure rate reported in the literature was "only" 50%. Sommerville *et al.* (1949) challenged almost immediately the claim that diethylstilbestrol in some way increased the production of progesterone. The group reported a "...sharp decrease in urinary excretion of pregnanediol...withdrawal (of DES) was followed by a rise..."

The effectiveness of DES in preventing miscarriages was tested in 13 clinical trials between 1946 and 1955. In six tests the drug was without merit. For example, Dieckmann *et al.* (1953) evaluated more than 2000 users and concluded "...stilbestrol did not reduce the incidence of abortion, prematurity, or postmaturity..." In the discussion that followed Smith and Smith maintained "...our experience ...on the prophylactic value is very satisfactory..."

Many physicians were ardent prescribers. In one retrospective study in the Philadelphia area 72% of physicians that responded to the questionnaire (only 31.8% replied) had prescribed the drug during pregnancy (Mangan *et al.*, 1975). The Physicians' Desk Reference (PDR) for 1949 listed 25 mg DES tablet by

Squibb for "use in treatment of catastrophes of pregnancy caused by endocrine dysfunction". FDA published a recommendation against the use in pregnancy in 1971. E. Lilly & Co. carried in the 1972 edition of the PDR the following warning "...because of possible adverse reaction on the fetus, the risk of estrogen therapy should be weighed against the possible benefits when diethylstilbestrol is considered for use in a known pregnancy..." The 27th edition of PDR (1973) changed the warning to a contraindication reading "...a statistically significant association has been reported between maternal ingestion of diethylstilbestrol during pregnancy and the occurrence of vaginal carcinoma in the offspring..." The FDA banned the use of DES in farm animals in 1979.

During the "DES time" perhaps as many as 6 million pregnant women used the drug in the US alone. Perhaps 1 in 100,000 daughters may develop abnormalities which can lead to cancer (Section 3.3.1). The recognition of a link between DES use and cancer of the reproductive tract has resulted in many legal actions for damages against the manufacturing companies. The US Supreme Court affirmed the responsibility of the manufacturer for the product in 1980. It refused to interfere in a decision by the California Supreme Court that directed the manufacturers to share damages.

3.2.3.1 Effects in Animals

In mice, the test animal of choice, the vaginal epithelium differentiates after birth, and vaginal changes provoked by DES manifest more readily in the neonate (Chapter 5).

Females
The vaginal epithelium of untreated, 30 day old mice is stratified, composed of 5–10 columnar cell layers with or without superficial cornification. In DES exposed mice the epithelium develops poorly. It may consist of 2–3 cell layers interspersed with glandular structures lined by endocervical-like cells in the submucosa (adenosis). The lesions are most common in the fornix and upper vagina. The stroma may contain remnants of Wolffian ducts lined with cuboidal cells. The uterus is less sensitive to estrogen stimulation (Maier *et al.*, 1985). The epithelium proliferates and becomes stratified consisting of 3–6 cell layers.

Several groups have demonstrated the development of cervicovaginal adenosis-like lesions as a result of prenatal exposure (Plapinger and Bern, 1979; McLachlan *et al.*, 1980; Walker, 1980; Iguchi *et al.*, 1986a; Iguchi and Takasugi, 1987). Doses needed to develop the lesions in the ICR/JCL strain are between 200–2000 µg/day injected for 4 days beginning on day 14 of gravidity or 200 µg if continuously infused (iv), also for 4 days. The regimen induces lesions in the fornical and upper vaginal epithelium in 80–85% of females and stratification of the uterine epithelium in 38–70%. The appearance is dose dependent. The use of lower doses decreases the incidence (Nomura and Kanzaki, 1977; Newbold and McLachlan, 1982). The vaginal epithelium exhibits a mitotic activity at birth, and the activity persists in the portion derived from the urogenital sinus in mature females (Tanaka *et al.*, 1984).

In neonatal mice a dose of only 0.1 µg is often enough to reveal the changes in 36% of animals. In 10 day old ovariectomized mice a daily injection of 0.001 µg of estradiol is effective (Chapter 4). The observation suggests that ovarian steroids play a role in the induction of the DES syndrome (Iguchi et al., 1986b).

Effect on the Ovaries: Prenatal DES exposure alters the morphology of the ovaries and their steroidogenic capability (Haney et al., 1984). Polyovular follicles (PF), containing 2-9 oocytes occur spontaneously in the ICR/JCL strain, but the incidence is low. In DES handled mice the incidence increases 33 to 112 fold. PF appear already on day 5 of life in animals exposed to 2000 µg/day. Females injected with 20 µg/day develop PF when they reach 30 days of age (Iguchi and Takasugi, 1986).

Other Species: DES induced transplacental carcinogenicity was reported in mice, rats and hamsters (Section 3.3.1). A closely related triphenylethylene derivative, clomiphene citrate (used for induction of ovulation), injected into pregnant rats (2 mg/kg on day 0, 5, or 12) caused metaplasia in the uterine epithelium, cervix and the oviduct in offspring of both sexes, and mothers. Male offspring became azoospermic. The compound was toxic to embryos and reduced the number of offspring born (McCormack and Clark, 1979).

Males

Male mice (ICR/JCL strain), exposed *in utero* between days 9 and 16 to DES, retained both the Müllerian and Wolffian ducts. Some may exhibit sperm abnormalities, nodular lesions with squamous metaplasia of the coagulating glands and ampullae (McLachlan et al., 1975; McLachlan, 1977; Vorherr et al., 1979; Nomura and Masuda, 1980; Arai et al., 1983). A dose of 100 µg/kg BW daily (about 3 µg) produced lesions in 75% of adults 9-10 months old. Pathological findings included hypoplastic, fibrotic intraabdominal testes, often posterior to the urinary bladder or firmly fixed to the posterior pole of the kidney, nodular enlargement of the seminal vesicles and/or coagulating glands associated with squamous metaplasia. The treatment was toxic; the average number of male offspring born to DES mothers was 2.6 while in controls it was 4.6 (McLachlan et al., 1975). Injections between days 9 through 16 of gestation or continuous infusion (10 µg/h) from day 15 to 19 was teratogenic (Takasugi et al., 1983).

Chromosomal Changes: Several groups raised the possibility that DES may damage nuclear material. Ruddiger et al. (1979) observed increased or induced sister-chromatid exchange, Barrett et al. (1981) cell transformation and Tsutsui et al. (1983) increased chromosomal nondysjunction. Sawada and Ishidate (1978) reported cytostasis after DES treatment and Nickerson (1980) an induction of nuclear inclusions. Ivett and Tice (1981) noted inhibition of cellular proliferation and induction of aneuploidy and polyploidy.

Goldstein (1986) attempted to correlate the loss of reproductive viability in a nematode (*Caenorhabditis elegans*) to DES. The animal can utilize DES, instead of cholesterol, as a food source. A concentration of 5 µg/ml DES in the growth medium results in oocyte aging at the pachytene stage, and absence of synaptonemal complexes in the nuclei. The complexes are essential for pairing and segregation of chromosomes. A dose of 20 µg is toxic. It remains to be seen whether abnormalities observed in a fast maturing primitive hermaphroditic worm (development egg to adult takes 3½ days) are valid in explaining effects that occur in mammalian species.

3.2.3.2 Effect on Humans

Girl babies with masculinized genitalia were born to mothers who took DES during pregnancy (Bongiovanni *et al.*, 1959). Boys may be born with hypospadias (Kaplan, 1959; Yalom *et al.*, 1973).

Women
The most frequent changes include gross structural cervicovaginal abnormalities (vaginal and/or cervical ridges), incomplete cervical collar ("cocks comb") formation, transverse vaginal septa, lack of vaginal pars, and vaginal adenosis, the presence of glandular tissue of Müllerian origin (Sandberg and Hebard, 1977; Singer and Hochman, 1978). The incidence is highest in babies exposed to DES during vaginogenesis (weeks 12–13). Kaufman and Adam (1978) found a T-shaped uterus with bulbous cornual extensions in 23/66 women; in six the uterus was hypoplastic with a cavity of less than 2.5 cm. Early reports indicated a very high incidence of cervicovaginal malformations (almost 90%) in DES babies (Pomerance, 1973). Study of larger groups revealed a lower, yet still unfortunately very high frequency (Table 3.10). Vaginal adenosis is already present at birth. In stillborn babies the incidence (70%) is far above the 4% seen in unexposed babies (Johnson *et al.*, 1979).

Fertility: The age of menarche and the pattern of gonadotropin secretion are within normal limits. Dysmenorrhea, increased menstrual flow, cycle irregularities, primary anovulation and oligomenorrhea are frequent (Bibbo *et al.*, 1977; Siegler *et al.*, 1979; Cousins *et al.*, 1980; Schmidt *et al.*, 1980). Conception will take place in the absence of severe pathological changes in the reproductive tract. In individuals with extensive epithelial and anatomical changes sperm migration may be poor. In some patients cervical hypoplasia may develop into incompetence during pregnancy (Goldstein, 1978). Premature deliveries are more common (Barnes *et al.*, 1980; Cousins *et al.*, 1980; Rosenfeld and Bronson, 1980; Schmidt *et al.*, 1980).

Table 3.10 Incidence of Benign Cervicovaginal Changes in DES Exposed Women.

Number of patients	Incidence of abnormalities (%)			
	Cervicovaginal	Vaginal mucosa	Adenosis	Reference
110	22	56	35	Herbst *et al.* (1975)
84	39	78		Bibbo *et al.* (1975)
830	62	66	66	Mangan *et al.* (1979)

Men

In 1972 Linden and Henderson predicted there will be an increase of genitourinary abnormalities and cancer in men born to mothers who took DES during pregnancy. The projection was unfortunately correct. Bibbo *et al.* (1975; 1976) examined 42 DES men and found epididymal cysts in 4, hypoplastic penises and/ or testes in 2 and in one a testicular mass. Gill *et al.* (1977) examined 159 DES men (exposed on the average *in utero* to 12 g of DES) and found genital lesions in 25.8% and decreased sperm count, semen volume and sperm motility in 32%. The incidence of overall irregularities was only 7% in a control group of 161 men. Cosgrove *et al.* (1977), Whitehead and Leiter (1981) and Stillman (1982) reached a similar conclusion while Leary *et al.* (1984) did not observe deviations as a result of DES exposure in babies exposed to 1.5 g.

3.2.4 Adrenocortical Hormones and Stress

Elevated concentrations of cortical hormones during pregnancy influence adversely body growth and adrenal function in the offspring (Eguchi and Wells, 1965; Hansson and Angervall, 1966; Paul and D'Angelo, 1972), may induce cleft palate and produce changes in behavior. Differentiation of the genital apparatus is not transformed while reproductive function may be altered (see Table 3.8).

3.2.4.1 Effects in Animals

Prenatal ACTH and corticosterone decrease reproduction in mice of both sexes (Politch and Herrenkohl, 1984b). Maternal stress is equally damaging to the offspring. Herrenkohl and Politch (1978) stressed gravid rats in mid-gestation (day 14) by bright illumination and heat in restrained cages, three times for 45 min each time. Adult female descendants (100 days and older) exhibited abnormal estrus cycles; the estrus-metestrus stage was longer (7.7 ± 2.0 days) than in the control group (2.0 ± 1.0 days). The results were confirmed by Politch and Herrenkohl (1984a).

Daily stress (day 14 through 22) led to a postponement of birth by 2 days and reduced fertility and pup survival (Table 3.11). The poor survival was the result of decreased lactation in stressed mothers. Only 26–40% of pups were adequately nursed compared with "almost" 100% of litters born to non-stressed mothers (Herrenkohl, 1979). Pollard (1984) reported that maternal stress increased the fetal

Table 3.11 Fertility of Female Rats Born to Mothers Stressed During Pregnancy.

Group	Number used	Average number delivered	Pregnancy Failure %	Survival %
Nonstressed	72	11.7 ± 0.8 (SE)	17.9	100
Prenatally stressed	93	9.7 ± 0.1 (SE)	52.5	49

Survival % = percent surviving by day 10 postpartum; data from Herrenkohl (1979).

mortality rate and the growth of pup after birth. In males exposed to a similar regimen testosterone biosynthesis began 24 h earlier (Ward and Weisz 1980). Both sexes responded poorly to a stress stimuli; corticosterone concentration in blood was significantly lower in prenatally stressed animals (Pollard, 1984).

3.2.4.2 Effects on Humans

The influence of stress on the human fetus is poorly understood. Taeusch (1975) discussed the problems associated with evaluating the effect of stress on human babies.

3.3 Other Effects of Hormones

Triphenylethylene derivatives (DES) may induce cancer of the reproductive tract and mammary glands. Steroid hormones influence the growth of the fetus, may prevent palate closure, influence the development of the nervous system and induce behavioral and psychological changes.

3.3.1 Induction of Cancer

Triphenylethylenes (such as DES) are potent carcinogens in both sexes; synthetic steroid hormones are much less potent in this respect. The difference may be due to different biological half-lives and/or binding to hormone receptors and α-feto-protein.

DES in synergism with other carcinogens (nitrosoureas) provokes hepatic and pituitary cancer in castrated male rats (Sumi et al., 1980; Inoh et al., 1987).

3.3.1.1 Reproductive Tract

Cancer was induced by exposure to DES in utero in hamsters (Rustia and Shubik, 1976; Gilloteaux et al., 1982), mice (McLachlan, 1977; Barrett et al., 1981; Lamb et al., 1981; Newbold and McLachlan, 1982; Greenman and Delongchamp, 1986) and rats (Boylan, 1978).

In mice (CD-1 strain) a dose of 5-100 μg/kg BW injected between days 9 and 16 of gestation induced adenosis formation in 10-15% of animals. Two percent of exposed offspring developed cancer (Newbold and McLachlan, 1982).

In hamsters prenatal exposure to 100 μg (injected to mother) between days 8 and 11 increased the sensitivity to a later DES treatment (15 mg pellet of DES at the age of 50 days). Carcinoma developed 150-250 days later. The uteri became hyperplastic with papillae projecting into the lumen. Cystic glandular structures were inside the papillae with no opening into the luminal space. With time the glandular proliferations filled the uterine cavity leading to uterine swelling. In older animals the cystic stroma became hemorrhagic, invaded by inflammatory tissue and fibrous material, and sections of the uterus became hard. While some cysts became atrophic, others were hyperplastic, incorporating edematous hyperplasia including adenocarcinoma and carcinoma *in situ* (Gilloteaux *et al.*, 1982).

Humans

Cancer may develop 20-25 years after transplacental exposure to DES. The tumor is characterized by clear and hobnail cells originating from paramesonephric, rather than mesonephric, epithelium; hence the name clear-cell adenocarcinoma (CCA). This type of cancer was rare, limited to the reproductive and postmenopausal years (Ruffolo *et al.*, 1971). Only two cases were treated in the Massachusetts General Hospital between 1930 and 1965. Suddenly in a two year period (1969-1970) Herbst and Scully encountered four cases. The "surprising" appearance of so many cases in an adolescent population (15-23 years old) prompted the investigators to a retrospective study and a correlation between the use of DES in pregnancy and the appearance of CCA (Herbst *et al.*, 1971). Greenwald *et al.* (1971) reported vaginal cancer in daughters after maternal exposure to a related estrogen, dienestrol. Herbst *et al.* (1972) summarized the history of 91 girls (average age 17 years) with cervical and vaginal carcinomas. Adenosis was detected in 24/26 patients indicating a relationship between the development of CCA and adenosis formation. Sixty nine mothers were interviewed; 49 had taken DES, dienestrol, hexestrol, or a combination of the drugs. A connection between CCA and the use of triphenylethylenes has been confirmed by others (Folkman, 1971; Hatcher and Conrad, 1971; Yaffee, 1973; Noller and Fish, 1974; Tsukada *et al.*, 1972; Gilson *et al.*, 1973; Henderson *et al.*, 1973; Ulfelder, 1973; 1975; Scully *et al.*, 1974; Herbst *et al.*, 1975; Horwitz *et al.*, 1988). In older women benign vaginal adenosis may develop into squamous lesions similar to those found in the cervix and become neoplastic (Fetherston, 1975).

To help stricken patients the US Government established regional centers, under The National Cooperative Diethylstilbestrol Adenosis Project. The purpose of the clinics is to promote the awareness of potential DES damage, register women exposed *in utero* to DES, or other hormones, collect and evaluate information, and give, where necessary, advice and treatment. Two findings emerged from the collaborative studies:

(i) Daughters exposed prior to the second trimester, those who were born in the fall, or those whose mothers had a previous history of at least one spontaneous abortion are most at risk (Herbst *et al.*, 1986).

(ii) The risk of developing cancer is at least twice as high compared with unexposed women (Bornstein *et al.*, 1988) or 1 in 1000 afflicted women (Melnicks *et al.*, 1987).

The greatest problem encountered in the retrospective studies was to document the intrauterine exposure to triphenylethylenes from physicians, pharmacy or hospital records. In one study (Piver *et al.*, 1988) positive identification was possible in only 15.2% of the cases. The reasons for poor documentation are several: records going back to the 1940's are difficult to secure; in many cases the attending physician, if still alive, may be reluctant to provide such information, even if available.

Table 3.12 summarizes representative data obtained in two cooperative studies.

Table 3.12 The Incidence of DES-Associated Clear Cell Adenocarcinoma of the Vagina and Cervix in New York State (Years 1979–1986).

No. of patients	Age of patients	Incidence		
		Abnormalities[a]	Cancer	Reference
474	24 yrs. (average)	19.4%	2 patients	1
2000	16–31 yrs.		4 patients	2

[a] Gross vaginal and cervical abnormalities and adenosis;
references (1) Piver *et al.* (1988); (2) Kramer *et al.* (1987).

3.3.1.2 Mammary Glands

Females of BALB/cf C3H mice develop mammary tumors spontaneously. Neonatal exposure to androgens and estrogens (Chapter 4) carries a greater risk than prenatal exposure. Injection of DES (5 μg) on day 12 of gestation decreased, in 24 month old SLN mice bearing expressed mammary tumor virus, spontaneous tumor development to 42%. The incidence in controls was about 90%. The incidence did not decrease by injecting DES on day 17 (Nagasawa *et al.*, 1980).

Aged mice (C3H/HeN strain) and rats (ACI strain) are more susceptible to DES induced carcinogenesis than younger animals (Greenman *et al.*, 1986; 1987; Rothchild *et al.*, 1987).

Huseby and Thurlow (1982) induced mammary tumors by prenatal exposure to DES in male mice. Sumi *et al.* (1980) found DES synergism with nitrosoureas to produce mammary carcinoma in castrated male rats.

In humans prenatal DES increases the risk of developing mammary cancer but only slightly. Greenberg *et al.* (1984) found 134 breast cancers per 100,000 women years in a population of 3033 DES daughters. The incidence in a non-exposed group was 93 yielding a relative risk factor of 1.4 (95% confidence limits 1.1 to 1.9).

3.3.2 Other Teratological Effects

3.3.2.1 Mammary Glands

The developing mammary gland anlage is sex hormone dependent (Raynaud, 1947; 1949; Jean and Delost, 1964; Jean, 1971a,b). Elevated concentrations of androgens and estrogens have a teratological influence in rats and mice (Greene *et al.*, 1939a; 1940; 1941).

Androgens
Only females are sensitive to androgens; genetic males are insensitive. Androgens, present during the critical period of organogenesis, abolish the development of mammary buds in female rat fetuses. The anlage is about 10 times more sensitive to testosterone (1 mg dose is sufficient) than to dihydrotestosterone. A metabolite of testosterone (5α-androstane-3α,17β-diol) was more active than DHT; a 5 mg dose produced virilization (Goldman *et al.*, 1976).

In mice 5 mg TP will block mammary gland development if injected on day 12 of gestation. The effect is less if given later (Hoshino, 1965). The response is strain dependent. In neonatal animals androgens (and estrogens) stimulate mammary lobuloalveolar development in BALB/c, BALB/cfC3H, C3H, and ddY strains (Kimura and Nandi, 1967) but not in C57BL and R III strains (Mori *et al.*, 1976).

Antiandrogens
Cyproterone acetate (Elger and Neumann, 1966; Jost, 1972; Veyssière *et al.*, 1974) or inhibitors of 5α-reductase (Imperato-McGinley *et al.*, 1986) also restricted mammary gland genesis. The latter results suggest that DHT plays a double role in male fetuses; it stimulates the growth of external genitalia and inhibits nipple development.

Estrogens
In mice the sensitivity of the anlage is greatest on days 8 and 9 of gestation. Injection of estradiol benzoate (0.5 µg) to the mother blocks the growth of nipples and mammary buds in both sexes (Hoshino and Connolly, 1967). Greene *et al.* (1939b) tested large doses of DES (12-42 mg/day during days 12-21) in pregnant rats and reported the presence of nipples in both sexes. The external genitalia were of the female type. Raynaud (1961) described malformations (atrophy, micromastia, inverted nipple formation, epidermic invagination) in offspring born to mice injected with estrogens (Figure 3.4).

Boylan (1978) injected 120 µg (total dose) from day 15 to day 19 and found nipples developed in both sexes, but the epithelial hood was absent.

Antiestrogens
MER-25 (ethamoxytriphetol), 5 mg daily from day 13 to 18, inhibited nipple formation in Swiss strain mice (Iguchi *et al.*, 1979). Clomid caused reproductive tract abnormalities in both the mother and offspring (McCormack and Clark, 1979).

3.3.2.2 Growth and Bone Structure

Pharmacological amounts of estrogens bring on fetal death and reduce body growth (see previous sections). Adrenocortical hormones cause increases in postnatal growth in rats. Cortisone and corticosterone (0.75 mg twice daily) and a synthetic hormone dexamethasone (1 µg twice daily) injected from day 12 to day 22 of gestation stimulated the body, adrenal and pituitary weight during the second week of life (Angervall and Martinsson, 1969; Seeliger *et al.*, 1974). The use of higher doses (10 µg of dexamethasone twice daily from day 12 to day 18-21 of gestation) restricted body growth and the development of adrenal glands (Lemman *et al.*, 1977). Frank and Roberts (1979) reported retarded body, lung and liver development in rat fetuses exposed to a daily dose of 0.2 mg/kg maternal BW for 2 days before birth. Reinisch *et al.* (1978) state that in mice three daily doses of 100 µg of prednisone led to a poor body growth of 16 day old fetuses.

Body, brain, lung, liver (and placenta) development of rabbit fetuses is also sensitive to corticoid (betamethasone, 0.8 mg for two days) treatment (Barrada *et al.*, 1980).

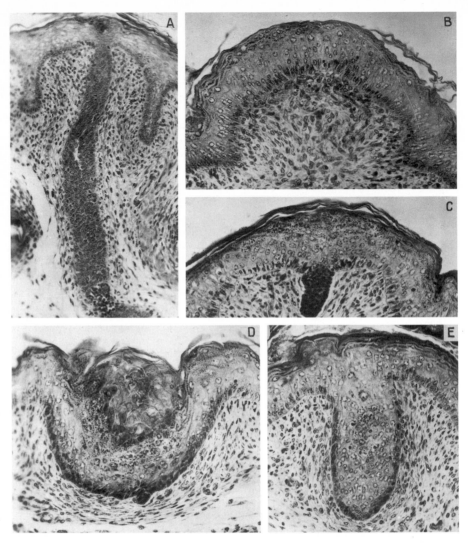

Figure 3.4. Malformation of Mouse Mammary Gland by Estradiol. **A** mammary gland of control newborn female mouse; **B** arrest of nipple formation; **C** micromastia; **D** inverted nipple formation; **E** epidermic invagination. Courtesy A Raynaud.

In humans adrenocortical hormones are used during the last trimester as a prophylactic measure to prevent respiratory distress syndrome (RDS) should infants be born prematurely (Liggins and Howie, 1972). RDS results from incomplete production of lung surfactant, increased surface tension, decreased lung compliance and decreased enzyme activity for phospholipid synthesis. Corticoids given to the mother promote morphological and biochemical maturation of fetal lungs (Taeusch, 1975; Gluck, 1976); the use carries a risk. The intrauterine growth of babies born to mothers who had taken 10 mg of prednisone may be significantly lower (Reinisch *et al.*, 1978).

Cleft Palate

Corticoids prevent palate closure in mice (Walker, 1965; Gabriel-Robez *et al.*, 1970), rabbits (Walker, 1967), and rats (Walker, 1971).

3.3.2.3 Nervous System

In fetal rats dopamine and epinephrine synthesizing enzymes begin to appear on day 12 of life. Fetal axon terminals connect in regions where synapses form, and the basic structure of the monoamine system develops (Olson and Seiger, 1972). The maturation of the system is relatively immune to large amounts of cortisol (56 mg/kg maternal BW) injected ip on day 12 or 15 (van Geijn *et al.*, 1979).

Gonadal hormones influence the development of sex dependent neural maturation and the formation of neural circuits in the fetus and the neonate (Arai, 1981; MacLusky and Naftolin, 1981; Torran-Allerand, 1978). In genetic males androgen production by the developing testes influences dimorphic evolution of the preoptic area (Raisman and Field, 1973; Greenough *et al.*, 1977; Gorski *et al.*, 1978), hypothalamus (Matsumoto and Arai, 1981), amygdala (Raisman and Field, 1973; Dibner and Black, 1978), spinal cord (Breedlove *et al.*, 1982) and sympathetic ganglia (Dibner and Black, 1978; Wright and Smolen, 1983).

Estradiol stimulated the growth of neural processes in dissociated cells obtained from the hypothalamus-preoptic area of 21 day old embryos of either sex. Cells from the cerebral cortex did not respond. Estrogens did not stimulate the growth of cells obtained from 17 day fetuses (Uchibari and Kawashima, 1985).

Sympathetic nerves originating in the hypogastric ganglion innervate organs derived from the Wolffian ducts and the urogenital sinus in male rats and mice. In females the ganglion supplies nerve fibers to the oviduct, uterus, and upper vagina (tissues derived from the Müllerian duct). In male neonate mice the average number of neurons is about 1800. The number increases to about 3800 in offspring born to mothers injected from day 9 to day 16 with TP (1 mg/day) or DES (4 µg/day). Female mice have fewer neurons, about 1300; hormonal treatment multiplied the cells to about 2500. Raising TP dose to 8 mg/day increased the number to 6000 in both sexes (Suzuki *et al.*, 1982; 1983). DHT (2 mg/day) proved equally active in this respect (Suzuki and Arai, 1985).

3.3.2.4 Immune System

Steroid hormones alter immune responses (Chapter 2). Administration of cortisone (250 µg/g BW; Angervall and Lundin, 1965), or cortisol hemisuccinate (Papiernik, 1977; Enzine and Papiernik, 1979), daily during the last 3 days of gestation to pregnant rats induced thymus involution in the neonate, albeit temporarily. The tissue recovered within the first week of life.

Luster *et al.* (1979) studied the influence of DES on immune responses.

3.4 Nonsteroidal Agents

Excesses of nonsteroidal hormones, vitamins, CNS active drugs, and other agents which cross the placenta can affect a variety of functions of the developing fetus (see Table 3.1).

3.4.1 Chorionic Gonadotropin

Elevated doses of human chorionic gonadotropin (hCG) or pregnant mares' serum (PMSG) interrupt gestation in hamsters (Yang and Chang, 1968). In gravid rats a single dose (50 IU/rat) will provoke fetal resorption if given between days 7 and 11 (Banik, 1975). Takasugi *et al.* (1985) infused daily 24 IU of hCG from day 15 to day 19. Adult (60–90 days) male offspring had abnormal testicular histology; spermatocytes detached from the basal layer of spermatogonia and Sertoli cells. In females (age 160 days) the gross structure of the ovaries appeared unchanged, albeit CL were present less frequently.

Kodituwakku and Hafez (1970) used PMSG (1500–2000 IU) followed by 2000 IU of hCG to induce multiple ovulation in cattle; the animals were then bred. Fetal survival was poor because of restricted placental development and poor vascular supply to each implantation site.

3.4.2 Prostaglandins

Prostaglandins used occasionally to induce abortion in women appear not to be teratogenic in mice (Persaud, 1974).

3.4.3 Thyroid Hormones

Adequate fetal thyroid function towards the end of pregnancy is essential for normal fetal development and brain maturation and critical for normal adaptation and growth after birth (see Chapter 5). Thyroid deficiency during gestation and the neonatal period causes impaired development of the central nervous system (see Smith *et al.*, 1957; Balazs *et al.*, 1968; Querido *et al.*, 1978) and alters maturation of many enzymatic systems in the liver (see Macho, 1979) and intestines (see Koldovsky *et al.*, 1975), activity of growth factors (somatomedins, erythropoietin, nerve growth factor and epidermal growth factor) and growth hormone (Fisher *et al.*, 1982), surfactant development in the lungs (Morishige *et al.*, 1982), and skeletal maturation (Hamburg, 1968).

In the rat the thyroid-pituitary axis becomes active on day 19 or 20 of gestation, 2 or 3 days before delivery (Eguchi *et al.*, 1980). Concentrations of thyroxine, reverse triiodothyronine, TRH and TSH are high in the human fetus. The concentration of TSH and thyroxine peak ½ h after delivery and then decline to adult levels (see Pasqualini and Kincl, 1985).

Insufficient thyroid function affects reproduction. The offspring of cretin rats have atrophied gonads: in males spermatogenesis is inhibited, and accessory sex organs remain atrophied; in females folliculogenesis is arrested, and the ovaries contain no CL (Leathem, 1961).

3.4.3.1 In Animals

Protein synthesis is abnormally low in the brain of thyroxine deficient monkeys (Holt *et al.*, 1975) and rats (Balazs *et al.*, 1968). In rats the thyroid begins to function on day 17 of fetal life (Hommes *et al.*, 1969). Maternal thyroid deficiency leads to fewer pregnancies and smaller litters (Stempak, 1962; Krohn and White, 1950; Varma *et al.*, 1978). The reports pertaining to the development of thyroidectomized rats are equivocal. Langman and van Faassen (1955) produced cleft palate and eye malformations by removing the thyroid during gestation. Christie (1966) could not obtain similar results. Varma *et al.* (1978) found no aberrations in offspring of rats made hypothyroid by treatment with triiodothyronine during pregnancy.

3.4.3.2 In Humans

Congenital hypothyroidism, both maternal and fetal, retards brain development and results in cretinism. Amniofetography (injection of an emulsion containing fat soluble radiopaque dye for the diagnosis of fetal malformations) may induce a transient impairment of fetal thyroid function (Radesch *et al.*, 1976). Theodoropolos *et al.* (1979) discussed the hazards of iodine induced hypothyroidism.

3.4.4 Insulin

Insulin action, modified by the presence of other hormones, nucleotides and electrolytes, promotes growth if adequate nutrition is available. During periods of feeding high insulin in the plasma promotes uptake and storage of glucose as glycogen, lipogenesis, and protein synthesis. When insulin in the blood is low (fasting) glycogenolysis, proteolysis and lipolysis take place.

Babies born to diabetic mothers tend to be overweight. Fetal hyperinsulinemia due to pancreatic cell tumor also leads to overweight. Intrauterine growth is retarded in the case of pancreatic agenesis, absence of insulin. Fetal growth is also retarded if the vascular supply is compromised as can be the case in toxemia, maternal hypertension, or diabetes mellitus with vascular disease (Hill, 1976).

3.4.5 Melatonin

Offspring born to pregnant rats injected with 500 µg of melatonin (days 17–20) had smaller pituitary and adrenal glands (Vaughan *et al.*, 1971).

3.4.6 Vitamins

The administration of large amounts of Vitamin A to pregnant rats (days 8-10) produces brain malformations and other teratological effects (Cohlan, 1954; Giraud and Martinet, 1956). The use of "sub-teratogenic" amounts during the same period (Malakhovskii, 1969; Butcher *et al.*, 1972) in mid-gestation, days 13-14 (Hutchings *et al.*, 1973; Vorhees *et al.*, 1978), and even during days 16-17 (Hutchings and Gaston, 1974) alters behavior and decreases the learning ability of the young.

3.4.7 CNS Active Drugs

In rats therapy with tranquilizers early in gestation alters postnatal activity and learning ability (Werboff and Dembicki, 1962; Werboff and Havlena, 1962; Werboff and Kesner, 1963).

The alkaloids of marijuana (cannabinoids) change both the reproductive ability and sex behavior. Perinatal exposure (50 mg/kg one day prior to expected parturition) of tetrahydrocannabinol to pregnant rats influences the reproductive capability of males. The adults are heavier with smaller testes, the concentration of LH in peripheral plasma is higher, and the copulatory behavior is suppressed (Dalterio and Bartke, 1979).

Phenobarbital (40 mg/kg BW during days 12-19) in gravid mice produced in daughters constant estrus and infertility. The concentration of estradiol and progesterone in peripheral plasma was higher whereas LH was lower (Gupta *et al.*, 1980). In mature male progeny (120 days old) testosterone biosynthesis and testosterone concentration in peripheral plasma decreased to about one-half (Gupta *et al.*, 1982).

References

Andersson T (1956) Intersexuality in pigs. Kungl Skogs-osh Lantg Tigskr Arg Gen 21: 136-153

Andonian RW, Kessler R (1979) Transplacental exposure to diethylstilbestrol in men. Urology 13: 276-279

Andrew FD, Staples RE (1977) Prenatal toxicity of medroxyprogesterone acetate in rabbits, rats and mice. Teratology 15: 25-32

Angervall L, Lundin PM (1965) Administration of cortisone to the pregnant rat. Effects on the lymphoid tissue of the offspring. Acta Endocrinol (Copenh) 50: 104-114

Angervall L, Martinsson A (1969) Overweight in offspring of cortisone-treated pregnant rats. Acta Endocrinol (Copenh) 60: 36-46

Arai Y (1981) Synaptic correlates of sexual differentiation. Trends Neurosci 12: 291-293

Arai Y, Mori T, Suzuki Y, Bern HA (1983) Long-term effects of perinatal exposure to sex steroids and diethylstilbestrol on the reproductive system of male mammals. Int Rev Cytol 84: 235-268

Balazs R, Kovacs S, Teichgraber P, Cocks WA, Eayrs JT (1968) Biochemical effects of thyroid deficiency on the developing brain. J Neurochem 15: 1335-1339

Banik UK (1975) Pregnancy-terminating effect of human chorionic gonadotrophin in rats. J Reprod Fertil 42: 67–76

Barnes AB, Colton T, Gundersen J, Noller KL, Tilley BC, Strama T, Townsend DE, Hatab P, O'Brien PC (1980) Fertility and outcome of pregnancy in women exposed in utero to diethylstilbestrol. N Engl J Med 302: 609–613

Barrada MI, Blomquist CH, Kotts C (1980) The effects of betamethasone on fetal development in the rabbit. Am J Obstet Gynec 136: 234–238

Barrett JC, Wong A, McLachlan (1981) Diethylstilbestrol induces neoplastic transformation without measurable gene mutation at two loci. Science 212: 1402–1404

Beach FA (1975) Bisexual mating behavior in the male rat: effects of castration and hormone administration. Psychol Zool 18: 390–402

Beach FA, Kuehn RE (1970) Coital behavior in dogs. X. Effects of androgenic stimulation during development of feminine mating response in females and males. Horm Behav 1: 347–367

Bibbo M, Al-Naqeeb M, Baccarini I, Gill W, Newton M, Sleeper KM, Sonek M, Wied GL (1975) Follow-up study of male and female offspring of DES-treated mothers. J Reprod Med 15: 29–32

Bibbo M, Gill WB, Azizi F, Blough R, Fang VS, Rosenfeld RL, Schumacher GFB, Sleeper K, Sonek MG, Wied GL (1977) Follow-up study of male and female offspring of DES-exposed mothers. Obstet Gynecol 49: 1-8

Bongiovanni M, DiGeorge M, Grumbach MM (1959) Masculinization of the female infant associated with estrogenic therapy alone during gestation: four cases. J Clin Endocrinol 189: 1004–1011

Bornstein J, Adam E, Adler-Storthz K, Kaufman RH (1988) Development of cervical and vaginal squamous cell neoplasia as a late consequence of in utero exposure to diethylstilbestrol. Obstet Gynecol Surv 43: 15–21

Boylan ES (1978) Morphological and functional consequences of prenatal exposure to diethylstilbestrol in the rat. Biol Reprod 19: 854–863

Breedlowe SM, Jacobson CD, Gorski RA, Arnold AP (1982) Masculinization of the female rat spinal cord following a single neonatal injection of testosterone propionate but not estradiol benzoate. Brain Res 237: 173–181

Brown-Grant K, Sherwood MR (1971) The early androgen syndrome in the guinea-pig. J Endocrinol 49: 277–291

Brunner JA, Witschi E (1946) Testosterone-induced modifications of sex development in female hamsters. Am J Anat 79: 293–320

Burstein R, Wasserman HC (1964) The effect of provera on the fetus. Obstet Gynecol 23: 931–934

Butcher RE, Brunner RL, Roth T, Kimmel CA (1972) A learning impairment associated with maternal hypervitaminosis-A in rats. Life Sci ll: 141–145

Cagnoni M, Famtini F, Morace G (1964) Studi sulla caratterizzazione sessuale delle strutture nervose deputate alla regolazione della funzione gonadica. Nota I. Gli effetti della somministrazione di testosterone in ratte proveniente da madri trattate nell'ultimo periodo della gravidanza. Ras Neurol Veget XVIII:275–284

Chaplin S, Smith JM (1987) Drugs and the fetus. IPPF Med Bull 21(no 4):1–2

Chrisman CL, Hinkle LL (1974) Induction of aneuploidy in mouse bone marrow cells with DES diphosphate. Can J Genet Cytol 16: 831–835

Christie GA (1966) Influence of thyroid function on the teratogenic activity of trypan blue in the rat. J Anat 100: 361–368

Citti V (1960) Studio comparativo dell'azione di sostanze ormonali sull'ovaio e sull'endometrio del feto e del neonato. Mon Ostet Gin Endocrinol Metab, Vol II, Modena, pp 1–18

Clarke IJ, Scaramuzzi RJ, Short RV (1976) Effects of testosterone implants in pregnant ewes on their female offspring. J Embryol Exp Morph 36: 87–99

Clarke IJ, Scaramuzzi RJ, Short RV (1977) Ovulation in prenatally androgenized ewes. J Endocrinol 73: 385–389

Clemens LG, Glade BA, Coniglio LP (1978) Prenatal endogenous androgenic influences on masculine sexual behavior and genital morphology in male and female rats. Horm Behav 10: 40–53

Cohlan SQ (1954) Congenital anomalies in the rat produced by the excessive intake of vitamin A during pregnancy. Pediatrics 13: 556–567

Cosgrove MD, Benton B, Henderson BE (1977) Male genitourinary abnormalities and maternal diethylstilbestrol. J Urol 117: 220–222

Courrier R, Jost A (1942) Intersexualité foetale provoqué par la pregneninolone au cours de la grossesse. Soc Biol (Paris) 136: 395–396

Cousins L, Karp W, Lacey C, Lucas WE (1980) Reproductive outcome of women exposed to diethylstilbestrol in utero. Obstet Gynecol 56: 70–76

Dalterio S, Bartke A (1979) Perinatal exposure to cannabinoids alters male reproductive function in mice. Science 205: 1420–1422

Dantchakoff V (1936) L'hormone male adulte dans l'histogenese sexualle du mammifère. C R Soc Biol 123: 873–876

Dantchakoff V (1937a) Sur l'edification des glandes annexes du tractus genital dans les free martins et sur les facteurs formatifs dans l'histogenése sexuelle male. C R Soc Biol 124: 407–411

Dantchakoff V (1937b) Sur l'obtention experimentale des freemartins chez le cobaye et sur la nature du facteur conditionnant leur histogenése sexuelle. C R Acad Sci 204: 195–196

Dantchakoff V (1938a) Sur les effets de l'hormone male dans un jeune cobaye male traité depuis un stade embryonnaire (production d'hypermales). C R Soc Biol 127: 1259–1262

Dantchakoff V (1938b) Sur les effets de l'hormone male dans une jeune cobaye femelle traité dupuis un stade embryonnaire (inversions sexuelles). C R Soc Biol 127: 1255–1258

Dantchakoff V (1938c) Effet du traitement hormonal de l'embryon de cobaye par la testostèrone sur ses facultés procréatrices et sur sa progéniture. C R Soc Biol 128: 891–893

Deanesly R (1939) Uterus masculinus of the rabbit and its reactions to androgens and estrogens. J Endocrinol 1: 300: 306

Diamond M (1967) Androgen-induced masculinization in the ovariectomized and hysterectomized guinea pig. Anat Rec 157: 47–52

Diamond M, Young WC (1963) Differential responsiveness of pregnant and nonpregnant guinea pigs to the masculinizing action of testosterone propionate. Endocrinology 72: 429–438

Dibner MC, Black IB (1978) Biochemical and morphological effects of testosterone treatment on developing sympathetic neurons. J Neurochem 30: 1479–1483

Dieckmann WJ, Davis ME, Rynkiewicz LM, Pottinger RE (1953) Does the administration of diethylstilbestrol during pregnancy have therapeutic value. Am J Obstet Gynecol 66: 1062–1081

Dodds EC, Goldberg L, Lawson W, Robinson R (1938) Estrogenic activity of certain synthetic compounds. Nature 141: 247–298

Driscoll SG, Taylor SF (1980) Effects of prenatal maternal estrogen on the male urogenital system. Obstet Gynec 56: 537–542

Eguchi Y, Wells LJ (1965) Response of the hypothalamic hypophyseal adrenal axis to stress-observations in fetal and caesarean newborn rats. Proc Soc Exp Biol Med 120: 675–676

Elger W, Neumann F (1966) The role of androgens in differentiation of the mammary gland in male mouse fetuses. Proc Soc Exp Biol Med 123: 637–640

Elger W, Steinbeck H, Cupceancu B, Neumann F (1970) Influence of methyltestosterone and cyproterone acetate on Wolffian duct differentiation in female rat foetuses. J Endocrinol 47: 417–422

Elsaesser F, Parvizi N, Ellendorff F (1978) Effects of prenatal testosterone on the stimulatory estrogen feedback on LH release in the pig. In: Dörner G, Kawashima M (eds) Hormones and brain development. Elsevier/North Holland Biomedical Press, Amsterdam, pp 61–67

Enzine S, Papiernik M (1979) Effet de l'hemisuccinate d'hydrocortisone injecté pendant la gestation et à la naissance sur le developpement du système immunitaire de la souris. INSERM 89: 105–112

Falconi G, Rossi GL (1965) Some effects of oestradiol 3-benzoate on the rat foetus. Proc Eur Soc Study Drug Tox 6: 150–156

Fayez JA, Brunch TR, Miller GL (1974) Virilization in pregnancy associated with polycystic ovary disease. Obstet Gynecol 44: 511–521

Fels E, Bosch LR (1971) Effect of prenatal administration of testosterone on ovarian function in rats. Am J Obstet Gynecol lll: 964–969

Fetherston WC (1975) Squamous neoplasia of vagina related to DES syndrome. Am J Obstet Gynecol 122: 176–181

Fine EH, Levin M, McConnell EL (1963) Masculinization of female infants associated with norethindrome acetate. Obstet Gynecol 22: 210–213

Flerkó B, Petrusz P, Tima L (1967) On the mechanism of sexual differentiation of the hypothalamus. Factors influencing the "critical period" of the rat. Act Biol 18: 27–36

Folkman J (1971) Transplacental carcinogenesis by stilbestrol. N Engl J Med 285: 404–405

Forsberg JG, Jacobsohn D, Norgren A (1968) Development of the urogenital tract in male offspring of rats injected during pregnancy with a substance with antiandrogenic properties (Cyproterone). Z Anat Entwickl Gesch 126: 320–331

Frank L, Roberts RJ (1979) Effects of low-dose prenatal corticosteroid administration on the premature rat. Biol Neonate 36: 1-9

Gabriel-Robez O, Clavert J, Roos M (1970) Periods of obtainment of palatine clefts with diethylstilbestrol dipropionate in mice. C R Soc Biol (Paris) 164: 2372–2375

Gill W, Schumacher GFB, Bibbo M (1976) Structural and functional abnormalities in the sex organs of mall offspring of DES (diethylstilbestrol) treated mothers. J Reprod Med 16: 147–153

Gill WB, Schumacher GFB, Bibbo M (1977) Pathological semen and anatomical abnormalities of the genital tract in human male subjects exposed to diethylstilbestrol in utero. J Urol 117: 477–480

Gilloteaux J, Paul RJ, Steggles AW (1982) Upper genital tract abnormalities in the syrian hamster as a result of in utero exposure to diethylstilbestrol. I. Uterine cystadenomatous papilloma and hypoplasia. Virchows Arch [A] 398: 163–183

Gilson MD, Dibona DD, Knab DR (1973) Clear-cell adenocarcinoma in young females. Obstet Gynecol 41: 494–500

Giroud A, Martinet M (1956) Teratogenèse par hautes doses de vitamine A en function des stades du developpement. Arch Anat Microsc Morphol Exp 45: 77–99

Gluck L (1976) Administration of corticosteroids to induce maturation of fetal lung. Am J Dis Child 130: 976–978

Goldman AS (1970) Virilization of the external genitalia of the female rat fetus by dehydroepiandrosterone. Endocrinology 87: 432–435

Goldman AS, Baker MK (1971) Androgenicity in the rat fetus of metabolites of testosterone and antagonism by cyproterone acetate. Endocrinology 89: 276–281

Goldman AS, Baker MK, Chen JC, Wieland RG (1972) Blockade of masculine differentiation in male rat fetuses by maternal injection of antibodies to testosterone-3-bovine serum albumin. Endocrinology 90: 716–721

Goldman AS, Shapiro BH, Neumann F (1976) Role of testosterone and its metabolites in the differentiation of the mammary gland in rats. Endocrinology 99: 1485–1490

Goldstein DP (1978) Incompetent cervix in offspring exposed to diethylstilbestrol in utero. Obstet Gynecol 52: 73 s

Goldstein P (1986) Nuclear aberrations and loss of synaptonemal complexes in response to diethylstilbestrol (DES) in Caenorhabditis elegans hermaphrodites. Mutat Res 174: 99–107

Gorski RA (1968) Influence of age on the response to paranatal administration of a low dose of androgen. Endocrinology 82: 1001–1004

Gorski RA, Gordon JH, Shryne JE, Southam AM (1978) Evidence for a morphological sex difference within the medial preoptic area of the rat brain. Brain Res 148: 333–346

Goy RW, Phoenix CH (1971) The effects of testosterone propionate administered before birth on the development of behavior in genetic female rhesus monkeys. In: Sawyer C, Gorski R (eds) Steroid hormones and brain functions. University of California Press, Berkeley, pp 193–202

Goy R, Bridson W, Young W (1964) Period of maximal susceptibility of the prenatal guinea pig to masculinizing actions of testosterone propionate. J Comp Physiol Psychol 57: 166–174

Greenberg ER, Barnes AB, Resseguie L, Barrett JA, Burnside S, Lanza LL, Neff RK, Stevens M, Young RH, Colton T (1984) Breast cancer in mothers given diethylstilbestrol in pregnancy. N Engl J Med 311: 1393–1398

Greene RR, Burrill MW, Ivy AC (1939a) Experimental intersexuality: the effects of antenatal androgens on sexual development of female rats. Am J Anat 65: 415–469

Greene RR, Burrill MW, Ivy AC (1939b) Progesterone is androgenic. Endocrinology 24: 351–357

Greene RR, Burrill MW, Ivy AC (1939c) Experimental intersexuality: modification of sexual development of the white rat with a synthetic estrogen. Proc Soc Exp Biol Med 41: 169–170

Greene RR, Burrill MW, Ivy AC (1939d) Experimental intersexuality. The paradoxical effects of estrogens on the sexual development of the female rat. Anat Rec 74: 429–436

Greene RR, Burrill MW, Ivy AC (1940) Experimental intersexuality: the effects of estrogens on the antenatal sexual development of the rat. Am J Anat 67: 305-345

Greene RR, Burrill MW, Ivy AC (1941) Experimental intersexuality: the effects of combined estrogens and androgens on the embryonic sexual development of the rat. J Exp Zool 87: 211-232

Greenman DL, Delongchamp RR (1986) Interactive responses to diethylstilboestrol in C3H mice. Food Chem Toxicol 24: 931-934

Greenman DL, Highman B, Chen JJ, Schieferstein GJ, Norvell MJ (1986) Influence of age on induction of mammary tumors by diethylstilbestrol in C3H/HeN mice with low murine mammary tumor virus titer. J Natl Cancer Inst 77: 891-898

Greenman DL, Kodell RL, Highman B, Schieferstein GJ, Norvell MJ (1987) Mammary tumorigenesis in C3H/HeN-MTV mice treated with diethylstilboestrol for varying periods. Food Chem Toxicol 25: 229-232

Greenough WJ, Carter CS, Steerman C, DeVoogd TJ (1977) Sex differences in dendritic patterns in hamster preoptic area. Brain Res 126: 63-72

Greenwald P, Barlow JJ, Nasca PC, Burnett WS (1971) Vaginal cancer after maternal treatment with synthetic estrogens. N Engl J Med 285: 390-392

Grumbach MM, Ducharme R (1960) The effects of androgens on fetal sexual development: androgen-induced female pseudohermaphroditism. Fertil Steril 11: 157-180

Grumbach MM, Ducharme R, Moloshok RE (1959) On the fetal masculinization action of certain oral progestins. J Clin Endocrinol Metab 19: 1369-1380

Gupta C, Sonawane BR, Yaffe SJ, Shapiro BH (1980) Phenobarbital exposure in utero: alterations in female reproductive function in rats. Science 208: 508-510

Gupta C, Yaffe SJ, Shapiro BH (1982) Prenatal exposure to phenobarbital permanently decreases testosterone and causes reproductive dysfunction. Science 216: 640-641

Hadziselimovic F, Guggenheim R (1980) Transmission and scanning electron microscopy of epididymis in male mice receiving estrogen during gestation. Acta Biol Acad Sci Hung 31: 133-140

Hagler SA, Schultz A, Hankin H, Kunstadter RH (1963) Fetal effects of steroid therapy during pregnancy. Am J Dis Child 106: 586-590

Hansson CG, Angervall L (1966) The parathyroids in corticosteroid-treated pregnant rats and their offspring. II. Effect of deoxycorticosterone acetate (DOCA). Acta Endocrinol (Copenh) 53: 553-560

Harlap S, Eldor J (1980) Births following oral contraceptive failures. Obstet Gynec 55: 447-452

Harlap S, Prywes R, Davies AM (1975) Births defects and oestrogens and progesterones in pregnancy. Lancet 1: 682-683

Hatcher RA, Conrad CC (1971) Adenocarcinoma of the vagina and stilbestrol as a "morning-after" pill. N Engl J Med 285: 1264-1265

Hauser H, Gandelman R (1983) Contiguity to males in utero affects avoidance responding in adult female mice. Science 220: 437-438

Hayles AB, Nolan RB (1957) Female pseudohermaphroditism: report of case in an infant born of a mother receiving methyltestosterone during pregnancy. Proc Staff Mayo Clin 32: 41-44

Hayles AB, Nolan RB (1958) Masculinization of female fetus, possibly related to administration of progesterone during pregnancy. Proc Staff Mayo Clin 33: 200-203

Henderson BE, Benton BDA, Weaver PT, Linden G, Wolan JF (1973) Stilbestrol and urogenital-tract cancer in adolescents and young adults. N Engl J Med 288: 354-356

Herbst AL, Ulfelder H, Poskanzer DC (1971) Adenocarcinoma of the vagina. Association of maternal stilbestrol therapy with tumor appearance in young women. N Engl J Med 284: 878-881

Herbst AL, Kurman RJ, Scully RE (1972) Vaginal and cervical abnormalities after exposure to stilbestrol in utero. Obstet Gynecol 40: 287-298

Herbst AL, Poskanzer DC, Robboy SJ, Friedlander L, Scully RE (1975) Prenatal exposure to stilbestrol: a prospective comparison of exposed female offspring with unexposed controls. N Engl J Med 292: 334-339

Herbst AL, Anderson S, Hubby MM, Haenszel WM, Kaufman RH, Noller KL (1986) Risk factors for the development of diethylstilbestrol associated clear cell adenocarcinoma: a case-control study. Am J Obstet Gynecol 154: 814-822

Herrenkohl LR (1979) Prenatal stress reduces fertility and fecundity in female offspring. Science 206: 1097-1099

Herrenkohl LR, Politch JA (1978) Effects of prenatal stress on the estrous cycle of female offspring as adults. Experientia 34: 1240-1241

Hill DE (1976) Insulin and fetal growth. In: Diabetes and other endocrine disorders during pregnancy. Alan R. Liss, New York, pp 127-139

Hill EC (1973) Clear cell adenocarcinoma of the cervix and vagina in young women. Am J Obstet Gynecol 116: 470-484

Hines DC, Goldzieher JW (1968) Large-scale study of an oral contraceptive. Fertil Steril 19: 841-846

Holt AB, Kerr GR, Cheek DB (1975) Prenatal hypothyroidism and brain composition. In: Cheek DB (ed) Fetal and postnatal cellular growth. John Wiley and Sons, New York

Hommes FA, Wilmink CW, Richters A (1969) The development of thyroid function in the rat. Biol Neonate 14: 69-73

Horwitz RI, Viscoli CM, Merino M, Brennan TA, Flannery JT, Robboy SJ (1988) Clear cell adenocarcinoma of the vagina and cervix: incidence, undetected disease, and diethylstilbestrol. J Clin Epidemiol 41: 593-597

Hoshino K (1965) Development and function of mammary glands of mice prenatally exposed to testosterone propionate. Endocrinology 76: 789-794

Hoshino K, Connolly MT (1967) Development and growth of mammary glands of mice prenatally exposed to estradiol benzoate. Anat Rec 157: 262-272

Hughes W (1929) The free martin condition in swine. Anat Rec 41: 213-247

Huseby RA, Thurlow S (1982) Effects of prenatal exposure of mice to "Low-Dose" diethystilbestrol and the development of adenomyosis associated with evidence of hyperprolactinemia. Am J Obstet Gynecol 144: 939-949

Hutchings DE (ed) (1989) Prenatal abuse of licit and illicit drugs. Ann NY Acad Sci 562: 1-388

Hutchings DE, Gaston J (1974) The effects of vitamin A excess administered during the mid-fetal period on learning and development in rat offspring. Dev Psychobiol 7: 225-233

Hutchings DE, Gibbon J, Kaufman MA (1973) Maternal vitamin A excess during the early fetal period. Effects on learning and development in the offspring. Dev Psychobiol 6: 445-457

Ichikawa S, Tamada H (1980) The effect of oestrogen on uterine plasticity in late pregnant rats. J Reprod Fertil 58: 165-168

Iguchi T, Takasugi N (1986) Polyovular follicles in the ovary of immature mice exposed prenatally to diethylstilbestrol. Anat Embryol 175: 53-55

Iguchi T, Takasugi N (1987) Postnatal development of uterine abnormalities in mice exposed to DES in utero. Biol Neonate 52: 97-103

Iguchi T, Tachibana H, Takasugi N (1979) Inhibitory effect of anti-estrogen (MER-25) on the occurrence of permanent vaginal changes in neonatally estrogen-treated mice. IRCS Med Sci 7: 579-583

Iguchi T, Takase M, Takasugi N (1986a) Development of vaginal adenosis-like lesions and uterine epithelial stratification in mice exposed perinatally to diethylstilbestrol. Proc Soc Exp Biol Med 181: 59-65

Iguchi T, Takei T, Takase M, Takasugi N (1986b) Estrogen participation in induction of cervicovaginal adenosis-like lesions in immature mice exposed prenatally to diethylstilbestrol. Acta Anat 127: 110-114

Imperato-McGinley J, Binienda Z, Gedney J, Vaughan ED Jr (1986) Nipple differentiation in fetal male rats treated with an inhibitor of the enzyme 5α-reductase: definition of a selective role for dihydrotestosterone. Endocrinology 11:132-137

Ingalls TH, Ingenito EF, Curley FJ (1963) Acquired chromosomal anomalies induced in mice by injection of a teratogen in pregnancy. Science 141: 810-812

Inoh A, Hamada K, Kamiya K, Niwa O, Yokoro K (1987) Induction of hepatic tumors by diethylstilbestrol in n-nitrosobutylurea initiated female rats. Jpn J Cancer Res 78: 134-138

Ivett JL, Tice RR (1981) DES-diphosphate induces chromosomal abberations but not sister chromatid exchange in murine bone marrow cell in vivo. Environ Mutagen 3: 445-452

Jacobson DB (1961) Abortion: its prediction and management: clinical experience with norlutin. Fertil Steril 12: 474-485

Janerich DT, Piper JM, Glebatis DM (1974) Oral contraceptives and congenital limb-reduction defects. N Engl J Med 291: 697–700

Jean C (1971a) Analyse des malformations mammaires du nouveau ne provoquées par l'injection d'oestrogènes à la mère gravide chez le rat et la souris. Arch Anat Microsc Morphol Exp 60: 147–168

Jean C (1971b) Développement mammaire post-natal de la souris issue de mère traitée par l'oestradiol pendant la gestation. Arch Sci Physiol 25: 145–185

Jean C, Delost P (1964) Atrophie de la glande mammaire des descendants adultes issus de mères traitées par les oestrogènes au cours de la gestation chez la souris. J Physiol (Paris) 56: 377–384

Jean-Faucher C, Berger M, de Turckheim M, Veyssiere G, Jean C (1977) Effects of an antioestrogen (MER-25) on sexual and mammary gland morphogenesis of the mouse fetus. J Reprod Fertil 51: 481–482

Johnson LD, Driscoll SG, Hertig AT, Cole PT, Nickerson RJ (1979) Vaginal adenosis in stillborns and neonates exposed to diethylstilbestrol and steroidal estrogens and progestins. Obstet Gynecol 53: 671–679

Jones HW, Wilkins L (1960) The genital anomaly associated with prenatal exposure to progestogens. Fertil Steril ll: 148–156

Josso N (1970) Action of testosterone on the Wolffian duct of the rat fetus in organ culture. Arch Anat Microsc Morphol Exp 59: 37–50

Jost A (1972) Use of androgen antagonists and anti-androgens in studies on sex differentiation. Gynecol Invest 2: 180–201

Kalter H (1978) Elimination of fetal mice with sporadic malformations by spontaneous resorption in pregnancies of older females. J Reprod Fert 53: 407–410

Kaplan NM (1959) Male pseudohermaphroditism: report of a case with observation of pathogenesis. N Engl J Med 261: 641–644

Kaufman RH, Adam E (1978) Genital tract anomalies associated with in utero exposure to diethylstilbestrol. Isr J Med Sci 14: 353–360

Kawashima K, Nakaura S, Nagao S, Tanaka S, Kuwamura T, Omori Y (1978a) Virilizing effect of methyltestosterone on female descendants in the rat. Endocrinol Jpn 25: 1-6

Kawashima K, Nakaura S, Nagao S, Tanaka S, Kuwamura T, Omori Y (1978b) Virilizing effect of testosterone and its metabolites on the urovaginal septum of female fetal rats. Endocrinol Jpn 25: 309–313

Keller K, Tandler J (1916) Über das Verhalten der Eihäute bei der Zwillingsträchtigkeit des Rindes. Mschr Ver Tierarztl Ost 3: 513–519

Keserü TL, Maráz A, Szabo J (1974) Oral contraception and sex ratio at birth. Lancet 1: 369

Ketchel MM, Pincus G (1964) In vitro exposure of rabbit ova to estrogens. Proc Soc Exp Biol Med 115: 419–421

Kimmel GL, Hartwell BS, Andrew FD (1979) A potential mechanism in medroxyprogesterone acetate teratogenesis. Teratology 19: 171–174

Kimura T (1975) Persistent vaginal cornification in mice treated with estrogen prenatally. Endocrinol Jpn 22: 497–502

Kimura T, Nandi S (1967) Nature of induced persistent vaginal cornification in mice. IV. Changes in the vaginal epithelium of old mice treated neonatally with estradiol or testosterone. J Natl Cancer Inst 39: 75–93

Kimura T, Kawashima S, Nishizuka Y (1980) Effects of prenatal treatment with estrogen on mitotic activity of vaginal anlage cells in mice. Endocrinol Jpn 27: 739–745

Kincl FA, Dorfman RI (1962) Influence of progestational agents on the genetic female foetus of orally treated pregnant rats. Acta Endocrinol (Copenh) 41: 274–279

Kincl FA, Maqueo M (1965) Secondary hermaphroditism in rats injected in utero with testosterone propionate. Excep Med Int Congr Ser 99: 415

Kiuturchiev B, Matrova T (1971) Comparative study on the enzyme changes in liver, kidney and placenta of rats and their fetuses after estrogen and carbon tetrachloride administration. Eksp Med Morfol 10: 164–171

Kodituwakku GE, Hafez ESE (1970) Prenatal degeneration following PMSG treatment in cattle. Cornell Vet LX(L970) 382–392

Korenchevsky V (1937) The female prostatic gland and its reaction to male sexual compounds. J Physiol 90: 371–376

Korenchevsky V, Dennison M (1936) The histology of the sex organs of ovariectomized rats treated with male or female sex hormone alone or with both simultaneously. J Pathol Bacteriol 42: 91–104

Kramer MS, Seltzer V, Krumholz B, Talebian F, Tedeschi C, Chen S (1987) Diethylstilbestrol-related clear-cell adenocarcinoma in women with initial examinations demonstrating no malignant disease. Obstet Gynecol 69: 868–871

Krohn PL, White HC (1950) The effect of hypothyroidism on reproduction in the female albino rat. J Endocrinol 6: 375–380

Lamb JC IV, Newbold NN, McLachlan JA (1981) Vizualization by light and scanning electron microscopy of reproductive tract lesions in female mice treated transplacentally with diethylstilbestrol. Cancer Res 41: 4057–4062

Langman J, van Faassen F (1955) Congenital defects in the rat embryo after partial thyroidectomy of the mother animal: a preliminary report on eye defects. Am J Ophthalmol 40: 65–76

Lanier AP, Noller KL, Decker DG, Elveback LR, Kurland LT (1973) Cancer and stilbestrol: a follow-up of 1,719 persons exposed to estrogens in utero and born 1943–1959. Mayo Clin Proc 48: 793–799

Leary FJ, Resseguie LJ, Kurland LT, O'Brien PC, Emslander RF, Noller KL (1984) Males exposed in utero to diethylstilbestrol. J Am Med Assoc 252: 2984–2989

Leathem JH (1961) Nutritional effects on endocrine secretions. In: Young WC (ed) Sex and internal secretion. The Williams and Wilkins Co, Baltimore, pp 666–704

Lemmen K, Maurer W, Trieb H, Ueberberg H, Seeliger H (1977) Morphologic changes in the adrenal glands of fetal and newborn rats following administration of glucocorticoids to the mother during pregnancy. Beitr Path 160: 361–380

LePage F, Gueguen J (1968) Results of a study on the possible teratogenic effects of chlormadinone and of its possible action on the course of pregnancy. Bull Fed Soc Gynecol Obstet 20 (Suppl): 313–314

Lerner LJ, Philippo M, Yiacs E, Brennan D, Borman A (1962) Comparison of the acetophenone derivatives of 16-17dihydroxyprogesterone with other progestational steroids for masculinization of the rat fetus. Endocrinology 71: 448–451

Levy EP, Cohen A, Fraser FC (1973) Hormone treatment during pregnancy and congenital heart defects. Lancet 1: 611

Liggins GC, Howie RN (1972) A controlled trial of antepartum glucocorticoid treatment for prevention of the respiratory distress syndrome in premature infants. Pediatrics 50: 515–525

Lillie FR (1916) The theory of the free-martin. Science 43: 611–613

Lillie FR (1917) The free-martin, a study of action of the sex hormones in the foetal life of cattle. J Exp Zool 23: 371–451

Linden G, Henderson BE (1972) Genital-tract cancers in adolescents and young adults. N Engl J Med 286: 760–761

Lorbber CA, Cassidy SB, Engel E (1979) Is there an embryo-fetal exogenous sex steroid exposure syndrome (EFESSES)? Fertil Steril 31: 21–24

Luster MI, Faith RE, McLachlan JA, Clark GC (1979) Effect of in utero exposure to diethylstilbestrol on the immune response in mice. Toxicol Appl Pharmacol 47: 279–285

MacLusky NJ, Naftolin F (1981) Sexual differentiation of the central nervous system. Science 211: 1294–1303

Maier DB, Newbold PR, McLachlan JA (1985) Prenatal diethylstilbestrol exposure alters murine uterine response to prepubertal estrogen stimulation. Endocrinology 116: 1878–1886

Malakhovskii VG (1969) Behavioral disturbances in rats receiving teratogenic agents antenatally. Bull Exp Biol Med 68: 1230–1232

Mangan CE, Giuntoli RL, Sedlacek TV, Rocereto T, Rubin E, Burtnett M, Mikuta JJ (1975) Six years experience with screening of a diethylstilbestrol exposed population. Am J Obstet Gynecol 134: 860–865

Marx L (1931) Versuche über heterosexuelle Merkmale bei Ratten. Arch Entwicklungsmech Organ 124: 584–612

Marx L (1932) Zur Anatomie der Prostata weiblicher Ratten. Z Zellforsch Mikroskop Anat 16: 48–62

Matsumoto A, Arai Y (1981) Effect of androgen on sexual differentiation of synaptic organization in the hypothalamic arcuate nucleus: an ontogenetic study. Neuroendocrinology 33: 166–169

McCormack S, Clark JH (1979) Clomid administration to pregnant rats causes abnormalities of the reproductive tract in offspring and mothers. Science 204: 629-631

McLachlan JA (1977) Prenatal exposure to diethylstilbestrol in mice: toxicological studies. J Toxicol Environ Health 2: 527-537

McLachlan JA, Newbold RR, Bullock B (1975) Reproductive tract lesions in male mice exposed prenatally to diethylstilbestrol. Science 190: 991-992

McLachlan JA, Newbold RR, Bullock BC (1980) Long-term effects on the female mouse genital tract associated with prenatal exposure to diethylstilbestrol. Cancer Res 40: 3988-3999

Melnick S, Cole P, Anderson D, Herbst A (1987) Rates and risks of diethylstilbestrol-related clear-cell adenocarcinoma of the vagina and cervix. An update. N Engl J Med 316: 514-516

Mey R (1963) Untersuchungen zur Frage einer intrauterinen Maskulinisierung durch 6-Chlor-6-dehydro-17α-acetoxyprogesteron. Arzneimittelforsch 13: 906-908

Moore CR (1939) Modification of sex development in the opposum by sex hormones. Proc Soc Exp Biol Med 40: 544-549

Murset G, Zachmann M, Prader A, Fischer J, Labhart A (1970) Male external genitalia of a girl caused by a virilizing adrenal tumour in the mother. Acta Endocrinol (Copenh) 65: 627-638

Nagasawa H, Mori T, Nakajima Y (1980) Long-term effects of progesterone or diethylstilbestrol with or without estrogen after maturity on mammary tumorigenesis in mice. Eur J Cancer 16: 1583-1589

Neumann F, Steinbeck H (1974) Antiandrogens. In: Eichler O, Farah A, Herken H, Welch AD (eds) Handbook of experimental pharmacology, vol XXXV/2. Springer, Berlin Heildelberg New York, pp 235-484

Neumann F, Elger W, Kramer M (1966) The development of a vagina in male rats by inhibiting the androgen receptors with an antiandrogen during the critical phase of organogenesis. Endocrinology 78: 628-633

Neumann F, Elger W, Steinbeck H (1970) Effects of oral contraceptives on the fetus. Lancet 2: 1258-1259

Newbold RR, McLachlan JA (1982) Vaginal adenosis and adenocarcinoma in mice exposed prenatally or neonatally to diethylstilbestrol. Cancer Res 42: 2003-2011

Nickerson P (1980) Effect of estradiol-17a and DES on advanced cortex and anterior pituitary gland of the mongolian gerbil. Tissue Cell 12: 117-123

Noller KL, Fish CR (1974) Diethylstilbestrol usage: its interesting past, important present and questionable future. Med Clin North Am 58: 793-810

Nomura T, Kanzaki T (1977) Induction of urogenital anomalies and some tumors in the progeny of mice receiving diethylstilbestrol during pregnancy. Cancer Res 37: 1099-1104

Nomura T, Masuda M (1980) Carcinogenic and teratogenic activities of diethylstilbestrol in mice. Life Sci 26: 1955-1962

Nora AH, Nora JJ (1975) A syndrome of multiple congenital anomalies associated with teratogenic exposure. Arch Environ Health 30: 17-21

Oechsli FW (1974) Oral contraception and sex ratio at birth. Lancet 1: 1004-1005

Olson L, Seiger A (1972) Early prenatal ontogeny of central monoamine neurons in the rat: fluorescence histochemical observations. Z Anat Entwickl Gesch 137: 301-308

Orotiz-Perez HE, De La Haba AF, Bangdiwala IS, Roure CA (1979) Abnormalities among offspring of oral and nonoral contraceptive users. Am J Obstet Gynecol 134: 512-517

Papiernik M (1977) Action de l'hemisuccinate d'hydrocortisone injecté au cours de la gestation sur l'immunogenèse de la souris. INSERM 73: 251-259

Pasqualini JR, Kincl FA (1985) Hormones and the fetus, vol I, Production, concentration and metabolism during pregnancy. Pergamon Press, Oxford

Paul DH, D'Angelo SA (1972) Dexamethasone and corticosterone administration to pregnant rats – effects on pituitary-adrenocortical function in the newborn. Proc Soc Exp Biol Med 140: 1360-1364

Payne DW, Katzenellenbogen JA (1979) Binding specificity of rat α-fetoprotein for series of estrogen derivatives: studies using equilibrium and nonequilibrium binding techniques. Endocrinology 105: 743-752

Persaud TVN (1974) The effects of prostaglandin F$_2$ on pregnancy and fetal development in mice. Toxicology 2: 25-29

Peterson WF (1969) Pregnancy following oral contraceptive therapy. Obstet Gynecol 34: 363-368

Phoenix CH, Goy RW, Gerall AA, Young WC (1959) Organizing action of prenatally administered testosterone propionate on the tissues mediating mating behavior in the female guinea pig. Endocrinology 65: 369–382

Pincus G (1956) Some effets of progesterone and related compounds upon reproduction and early development in mammals. Acta Endocrinol (Copenh) [Suppl] 28: 1-65

Piotrowski J (1968a) Experimental investigations on the effect of progesterone on embryonal development. II. Investigations carried out on rabbits. Folia Biol (Krakow) 16: 335–342

Piotrowski J (1968b) The effect of progesterone on the foetal development of rats of the Wistar strain. Part III. Folia Biol (Krakow) 16: 343–353

Piver MS, Lele SB, Baker TR, Sandecki A (1988) Cervical and vaginal cancer detection at a regional diethylstilbestrol (DES) screening clinic. Cancer Detect Prev ll: 197–202

Plapinger L, Bern HA (1979) Adenosis-like lesions and other cervicovaginal abnormalities in mice treated perinatally with estrogen. J Natl Cancer Inst 63: 507–518

Politch JA, Herrenkohl LR (1984a) Effects of prenatal stress on reproduction in male and female mice. Physiol Behav 32: 95–99

Politch JA, Herrenkohl LR (1984b) Prenatal ACTH and corticosterone: effects on reproduction in male mice. Physiol Behav 32: 135–137

Pollard I (1984) Effects of stress administered during pregnancy on reproductive capacity and subsequent development of the offspring of rats: prolonged effects on the litters of a second pregnancy. J Endocrinol 100: 301–306

Pomerance W (1973) Post-stilbestrol secondary syndrome. Obstet Gynecol 42: 12–17

Price D, Williams-Ashman HG (1961) The accessory reproductive glands of mammals. In: Young WC (ed) Sex and internal secretion. The Williams and Wilkins Co, Baltimore, pp 366–448

Raisman G, Field PM (1973) Sexual dimorphism in the neuropil of the preoptic area of the rat and its dependence on neonatal androgens. Brain Res 54: 1–29

Rauther M (1909) Neue Beitrage zur Kenntnis des Urogenitalsystems der Saugetiere. Deuschr Med-Natur-Wiss Ges Jena 15: 417–466

Raynaud A (1939a) Effect of estradiol dipropionate injected into mice during gestation. C R Soc Biol 130: 872–875

Raynaud A (1939b) Structure of the genital apparatus (gonads and ducts of Wolff and Muller) of male intersex mice obtained by the injection of estradiol dipropionate into the mother during gestation. C R Soc Biol (Paris) 130: 1012–1015

Raynaud A (1942) Modification expérimental de la differenciation sexuelle des embryons de souris, par action des hormones androgènes et oestrogènes. Actual Scient et Indus Nos. 925 et 926, Hermann Cie, Paris, pp 66–134

Raynaud A (1947) Effect des injections d'hormones sexuelles à la souris gravide, sur le développment des ébauches de la glande mammaire des embryones. L'action des substances androgènes. Ann Endocrinol (Paris) 8: 248–253

Raynaud A (1949) Nouvelles observations sur l'appareil mammaire des souris provenant de meres ayant reçu des injections de testostèrone pendant la gestation. Ann Endocrinol (Paris) 10: 54–62

Raynaud A (1957) Sur le développement et la différenciation sexuelle de l'appareil gubernaculaire du foetus de souris. CR Acad Sci 245: 2100–2103

Raynaud A (1958) Effects, sur l'appareil gubernaculaire des foetus, des hormones oestrogènes injectées à la souris gravide. CR Soc Biol CLII: 1461–1464

Raynaud A (1961) Les facteurs hormonaux de la croissance; la glande mammaire. G Doin et Cie and Masson and Co, Paris, pp 123–160

Reinisch JM, Simon NG, Karow WG, Gandelman R (1978) Prenatal exposure to prednisone in humans and animals retards intrauterine growth. Science 202: 436–438

Revesz C, Chappel CI (1966) Biological activity of medrogestone: a new orally active progestin. J Reprod Fertil 12: 473–487

Revesz C, Chappel CI, Gaudry R (1960) Masculinization of female fetuses in the rat by progestational compounds. Endocrinology 66: 140–144

Rodesch F, Camus M, Ermans AM, Dodion J, Delange F (1976) Adverse effect of amniofetography on fetal thyroid function. Am J Obstet Gynecol 126: 723–726

Rosenfeld DL, Bronson RA (1980) Reproductive problems in the DES exposed female. Obstet Gynecol 55: 453–456

Rothman KJ, Liess J (1976) Gender of offspring after oral contraceptive use. N Engl J Med 295: 859–861

Rothman KJ, Louik C (1978) Oral contraceptives and birth defects. N Engl J Med 299: 522–524

Rothschild TC, Boylan ES, Calhoon RE, Vonderhaar BK (1987) Transplacental effects of diethyl-stilbestrol on mammary development and tumorigenesis in female ACI rats. Cancer Res 47: 4508–4516

Rudiger HW, Haenisch F, Metzler M, Desch F, Glatt H (1979) Metabolites of DES induce sister chromatid exchanges in human cultured fibroblasts. Nature 281: 392–394

Ruffolo EH, Foxworthy D, Fletcher JC (1971) Vaginal adenocarcinoma arising in vaginal adeno-sis. Am J Obstet Gynecol lll: 167–172

Sandberg EC, Hebard JC (1977) Examination of young women exposed to stilbestrol in utero. Am J Obstet Gynecol 128: 364–370

Sanwal PC, Pande JK, Dasgupta PR, Kar AB, Setty BS (1970) Long term effect of a continuous low dose of megestrol acetate on the genital organs and fertility of female rats. Steroids 15: 711–722

Saunders FJ (1967) Effects of norethynodrel combined with mestranol on the offspring when administered during pregnancy and lactation in rats. Endocrinology 80: 447–452

Sawada M, Ishidate (1978) Colchicine-like effect of DES on mammalian cells in vitro. Mutat Res 57: 175–182

Schmidt G, Fowler WC Jr, Talbert LM, Edelman DA (1980) Reproductive history of women exposed to diethylstilbestrol in utero. Fertil Steril 33: 21–24

Scholer HF, de Wachter AM (1961) Evaluation of androgenic properties of progestational com-pounds in the rat by the female foetal masculinization test. Acta Endocrinol (Copenh) 38: 128–136

Schultz FM, Wilson JD (1974) Virilization of the Wolffian duct in the rat fetus. Endocrinology 94: 979–986

Scully RE, Robboy SJ, Herest AL (1974) Vaginal and cervical abnormalities including clear-cell adenocarcinoma related to prenatal exposure to stilbestrol. Ann Clin Labor Sci 4: 222–233

Seeliger H, Weigand R, Kosswig W, Von Seebach HB, Ueberberg H (1974) Wirkung von Östrogen und Glucocorticoidgaben auf die Nebennierenentwicklung des Feten. Teil 1. Wir-kungen von Ostradiolbenzoat (Untersuchungen an neugeborenen Ratten). Beitr Path 153: 18–34

Shiono PH, Harlap S, Ramcharan S, Berendes H, Gupta S, Pellegrin F (1979) Use of contracep-tives prior to and after conception and exposure to other fetal hazards. Contraception 20: 105–120

Siegler AM, Wang CF, Friberg J (1979) Fertility of the diethylstilbestrol-exposed offspring. Fertil Steril 31: 601–607

Singer MS, Hochman M (1978) Incompetent cervix in a hormone exposed offspring. Obstet Gynecol 51: 625–631

Slob AK, Goy RW, van der Werff ten Bosch JJ (1973) Sex differences in growth of guinea-pigs and their modification by neonatal gonadectomy and prenatally administered androgen. J Endocrinol 58: 11–19

Smith WNA (1970) Transplacental influence of androgen upon ovulatory mechanisms in the rat. J Endocrinol 48: 477–478

Smith WO (1948) Diethylstilbestrol in the prevention and treatment of complications of preg-nancy. Am J Obstet Gynecol 56: 821–834

Sommerville IF, Marrian GF, Clayton BE (1949) Effects of diethylstilbestrol on urinary excretion of pregnanediol and endogenous estrogens during pregnancy. Lancet 256: 680–682

Stempak JG (1962) Maternal hypothyroidism and its effects on fetal development. Endocrinology 70: 443–449

Stillman RJ (1982) In utero exposure to diethylstilbestrol: adverse effects on the reproductive tract and reproductive performance in male and female offspring. Am J Obstet Gynecol 142: 905–921

Stinnakre MG (1972) Etude de l'action "paradoxale" des estrogènes sur les canaux de Wolff des foetus femelles de rat, l'aide d'un anti-androgène l'acetate de "cyproterone". C R Hebd Seanc Acad Sci (Paris) D 275: 101–103

Stinnakre MG (1975) Period of sensitivity of rat fetal Wolffian duct to androgens. Arch Anat Microsc Morphol 64: 45–59

Suchowsky GR, Junkmann K (1961) A study of the virilizing effect of progestogens on the female rat fetus. Endocrinology 68: 341–349

Sumi C, Yokoro K, Kajitani T, Ito A (1980) Synergism of diethylstilbestrol and other carcinogens in concurrent development of hepatic, mammary, and pituitary tumors in castrated male rats. J Natl Cancer Inst 65(1):169–175

Suzuki Y, Arai Y (1977) Induction of estrogen-independent persistent vaginal cornification in cyproterone acetate (CA)-induced feminized male mice. Anat Embryol 151: 119–125

Suzuki Y, Arai Y (1985) Androgenic regulation of neuron number in the developing hypogastric ganglion of mice. Zool Sci 2: 809–811

Suzuki Y, Ishii H, Furuya H, Arai Y (1982) Developmental changes of the hypogastric ganglion associated with the differentiation of the reproductive tracts in the mouse. Neurosci Let 32: 271–276

Suzuki Y, Ishii H, Arai Y (1983) Prenatal exposure of male mice to androgen increases neuron number in the hypogastric ganglion. Dev Brain Res 10: 151–154

Swanson HE, van der Werff ten Bosch JJ (1964) The "early androgen" syndrome; differences in response to prenatal and postnatal administration of various doses of testosterone propionate in female and male rats. Acta Endocrinol (Copenh) 47: 37–50

Swanson HE, van der Werff ten Bosch JJ (1965) The "early androgen" syndrome; effects of pre-natal testosterone propionate. Acta Endocrinol (Copenh) 50: 379–390

Sweney LR, Shapiro BL (1975) Thyroxine and palatal development in rat embryos. Dev Biol 42: 19–27

Taeusch HW Jr (1975) Glucocorticoid prophylaxis for respiratory distress syndrome. A review of potential toxicity. J Pediatr 87: 617–623

Taguchi O, Nishizuka Y, Takasugi N (1977) Irreversible lesions in female reproductive tracts of mice after prenatal exposure to testosterone propionate. Endocrinol Jpn 24: 385–391

Takasugi N, Tanaka M, Kato C (1983) Effects of continuous intravenous infusion of diethylstil-bestrol into pregnant mice on fetus: testicular morphology at fetal and postnatal period. Endo-crinol Jpn 30: 35–42

Takasugi N, Iguchi T, Kurihara J, Tei A, Takase M (1985) Changes in gonads of male and female offspring of mice receiving a continuous intravenous infusion of human chorionic gonado-tropin during gestation. Exp Clin Endocrinol 85: 273–283

Tanaka M, Iguchi T, Takasugi N (1984) Early changes in the vaginal epithelium of mice exposed prenatally to diethylstilbestrol. IRCS Med Sci 12: 814–815

Theodoropolos T, Braverman LE, Vagenakis AG (1979) Iodine-induced hypothyroidism: a poten-tial hazard during perinatal life. Science 205: 502–503

Toran-Allerand CD (1978) Gonadal hormones and brain development: cellular aspects of sexual differentiation. Am Zool 18: 553–565

Tsukada Y, Hewett WJ, Barlow JJ, Pickren JW (1972) Clear-cell adenocarcinoma ("mesoneph-roma") of the vagina. 3 Cases associated with maternal synthetic nonsteroid estrogen therapy. Cancer 29: 1208–1214

Tsutsui T, Maizumi H, McLachlan J, Barrett JC (1983) Aneuploidy induction and cell transfor-mation by DES: a possible chromosomal mechanism in carcinogenesis. Cancer Res 43: 3814–3821

Uchibari M, Kawashima S (1985) Effects of sex steroids on the growth of neuronal processes in neonatal rat hypothalamus preoptic area and cerebral cortex in primary culture. Int J Dev Neu-rosci 3: 169–176

Ulfelder H (1973) Stilbestrol adenosis and adenocarcinoma. Am J Obstet Gynecol 117: 794–800

Ulfelder H (1975) Das Stilbostrol-adenosis-karzinom-syndrom. Geburtshilfe Frauenheilkd 35: 329–333

van der Werff ten Bosch JJ, Goldfoot DA (1975) Effects of various androgens administered to pregnant guinea-pigs on their female offspring. J Endocrinol 64: 35P

van Geijn HP, Zuspan FP, Copeland SJ, Vorys AS, Zuspan MP, Scott GD (1979) The effects of hydrocortisone on the development of the amine systems in the fetal brain. Am J Obstet Gynecol 135: 743–750

Vannier B, Raynaud JP (1975) Effect of estrogen plasma binding on sexual differentiation of the rat fetus. Mol Cell Endocrinol 3: 323–337

Vannier B, Raynaud JP (1980) Long-term effects of prenatal oestrogen treatment on genital morphology and reproductive function in the rat. J Reprod Fertil 59: 43–49

Varma SK, Murray R, Stanbury JB (1978) Effect of maternal hypothyroidism and triiodothyronine on the fetus and newborn in rats. Endocrinology 102: 24–30

Vaughan MK, Vaughan GM, O'Steen WK (1971) Pituitary adrenal pineal and renal weights in offspring of rats treated with testosterone and/or melatonin during pregnancy. J Endocrinol 51: 211–212

Verkauf BS, Reiter ED, Hernandez L, Burns SA (1977) Virilization of mother and fetus associated with lutema of pregnancy: a case report with endocrinologic studies. Am J Obstet Gynecol 129: 274–280

Veyssière G, Jean CH, Jean CL (1974) Evaluation quantitative de la croissance des ébauches mammaires des foetus de souris après injection de testostérone de dihydrotestostérone et d'acetate de cyprotérone à la mère gravide. Arch Anat Microsc Morphol Exp 63: 63–78

vom Saal FS, Bronson FH (1980) Variation in length of the estrous cycle in mice due to former intrauterine proximity to male fetuses. Biol Reprod 22: 777–780

vom Saal FS, Grant WM, McMullen CW, Laves KS (1983) High fetal estrogen concentrations: correlation with increased adult sexual activity and decreased aggression in male mice. Science 220: 1306–1308

Vorhees CV, Brunner RL, McDaniel CR, Butcher RE (1978) The relationship of gestational age to vitamin A induced postnatal dysfunction. Teratology 17: 271–276

Vorherr H (1973) Contraception after abortion and post partum. Am J Obstet Gynecol 117: 1002–1025

Vorherr H, Messer RH, Vorherr UF, Jordan SW, Kornfeld M (1979) Teratogenesis and carcinogenesis in rat offspring after transplacental and transmammary exposure to diethylstilbestrol. Biochem Pharmacol 28: 1865–1877

Walker BE (1965) Cleft palate produced in mice by human equivalent dosage with triamcinolone. Science 149: 862–863

Walker BE (1967) Induction of cleft palate in rabbits by several glucocorticoids. Proc Soc Exp Biol Med 125: 1281–1284

Walker BE (1971) Induction of cleft palate in rats with antiinflammatory drugs. Teratology 4: 39–42

Walker BE (1980) Reproductive tract anomalies in mice after prenatal exposure to DES. Teratology 21: 313–321

Werboff J, Dembicki EL (1962) Toxic effects of tranquilizers administered to gravid rats. J Neuropsychiatry 4: 87–91

Werboff J, Havlena J (1962) Postnatal behavior effects of tranquilizers administered to the gravid rat. Exp Neurol 6: 263–269

Werboff J, Kesner R (1963) Learning deficits of offspring after administration of tranquilizing drugs to the mother. Nature 197: 106–107

Whalen RE, Peck CK, Lopiccolo J (1966) Virilization of female rats by prenatally administered progestin. Endocrinology 78: 965–970

Whitehead ED, Leiter E (1981) Genital abnormalities and abnormal semen analyses in male patients exposed to diethylstilbestrol in utero. J Urol 47: 50

Wilkins L (1960) Masculinization of female fetus due to use of orally given progestins. J Am Med Assoc 172: 1028–1032

Wilkins L, Jones W, Holman G, Stempfel S (1958) Masculinization of the female fetus associated with administration of oral and intramuscular progestins during gestation: non-adrenal female pseudohermaphrodism. J Clin Endocrinol Metab 18: 559–585

Wilson JD (1973) Testosterone uptake by the urogenital tract of the rabbit embryo. Endocrinology 92: 1192–1199

Wilson JD, Lasnitzki I (1971) Dihydrotestosterone formation in fetal tissues of the rabbit and rat. Endocrinology 89: 659–668

Wright LL, Smolen AJ (1983) Effects of 17β-estradiol on developing superior cervical ganglion neurons and synapses. Dev Brain Res 6: 299–303

Yaffee S (1973) Stilbestrol and adenocarcinoma of the vagina. Pediatrics 51: 297–298

Yang WH, Chang MC (1968) Interruption of pregnancy in the rat and hamsters by administration of PMS or HCG. Endocrinology 83: 217–224

Yasuda Y, Kihara T, Nishimura H (1977) Effect of prenatal treatment with ethinyl estradiol on the mouse uterus and ovary. Am J Obstet Gynecol 127: 832–836

Yasuda Y, Kihara T, Tanimura T, Nishimura H (1985) Gonadal dysgenesis induced by prenatal exposure to ethynyl estradiol in mice. Teratology 32: 219–228

Young CW, Goy RW, Phoenix CH (1964) Hormones and sexual behavior. Science 143: 212–218

Zuckerman S (1940) The histogenesis of tissues sensitive to oestrogens. Biol Rev 15: 231–271

4 Effects of Steroid Hormones in the Neonate

A. Androgens and Estrogens

Estradiol, or testosterone, and a variety of C-19 and C-18 synthetic steroids with androgen- or estrogen-like biological activity given to the neonate induce sterility in the adult. Pregnanes are not active in this respect. Those competing with estrogens and androgens (progesterone) protect the neonate against the deleterious effect. These compounds are active in the neonate whether injected repeatedly or given only in one injection (Table 4.1). In most studies rodents (rats or mice) were the test animal of choice, but other mammals and even birds are sensitive.

Placement of small amounts of crystalline hormones into the hypothalamus induces sterilization; lower amounts are sufficient than when the compounds are given by injection. In the adult, continuous illumination or electrolytic lesions placed in the hypothalamus also produce sterility.

Female rats made sterile by a neonatal androgen treatment are referred to as "persistent estrus (PE)", "anovulatory (AN)", or "androgenized" animals. Animals made sterile by estrogens are classified as "estrogenized" and the process as "estrogenization".

4.1 Effects on Reproductive Function in Rats

In addition to reproductive changes neonatal sterilization induces growth abnormalities, changes in adrenal and thyroid functions, disturbances of behavioral patterns, and provokes, under specific conditions, the development of cancer.

4.1.1 Testosterone

Early workers (Shay et al., 1939; Burdick et al., 1940; Wilson et al., 1940; Huffman, 1941; Wilson and Wilson, 1943) used testosterone propionate (TP), estradiol dipropionate (EDP), or estradiol benzoate (EB) in oil solutions. The choice to use esters, instead of free steroids, whether intentional or due to serendipity, was fortunate. We know now that free steroids are less active. The developing organism must be exposed to the deleterious effects for at least two days. Arai and Gorski (1968a, b, d) used injections of barbiturates, or of an antiandrogen (cyproterone

Table 4.1 Induction of Sterility by Various Investigators in Adult Rodents by Injecting Neonatal Animals with Testosterone Propionate or Estradiol Esters.

Test animal	Injection sequence (days after birth)	Dose, μg Daily	Dose, μg Total	Reference
Female rats	*Testosterone propionate* from birth for number of weeks, 3 times a week	500–1000	5100–182,000	Shay *et al.* (1939)
	Alternate days from day 6 to day 38	250	3500	Bradbury (1941)
	12 injections from day 1 to 28	500–3000	6000–3600	Wolfe *et al.* (1944)
	3 successive days from day 1 to 7	100	300	Segal and Johnson (1959)
	One injection day 1, 5, 10, or 30		1250	Barraclough (1961)
	One injection day 5	1, 5, 10, 1250	1–1250	Gorski and Barraclough (1963)
	One injection day 2, 3 or 5		300, 500, 1000	Swanson and van der Werff ten Bosch (1963)
	One injection day 4		500	Tranezzani *et al.* (1963)
	One injection day 5		1250	Roy *et al.* (1964)
	One injection day 4 or 5		1250	Zeilmaker (1964)
	One injection day 4		3000	Van Rees and Gans (1966)
	One injection day 7, 10 or 13	10/g BW	100–200	Kikuyama and Kawashima (1966)
	One injection day 3		2500	Duluc and Mayer (1967)
	One injection day 2 or 3		500	Kurcz *et al.* (1967)
	Daily injection first 10 days	500		
	second 10 days	1000		
	last 10 days	2000	35,000	Arai and Masuda (1968)
	Same protocol as Arai and Masuda		35,000	Takewaki (1968)
	One injection day 4		1250	Colombo (1968)
	One injection day 5		100	Schiavi (1969)
	One injection day 5		1250	Everett *et al.* (1970)
	One injection day 5		50	Johnson (1971)
	One injection "early post partum"		500, 1000	Fels *et al.* (1972)
	One injection day 2		1250	Fuxe *et al.* (1972)
	One injection day 5		1000	Fujii (1973)
	One injection day 5		10, 50, 250, 1250	Malampati and Johnson (1974)
Female mice	*Testosterone propionate* One injection day 5, 10 or 20		1000	Barraclough and Leathem (1954)
	One injection day 1 or 5		5	Takasugi and Bern (1964)
	Estradiol dipropionate for 28 days, 3 times weekly beginning day 5, or 10	100	1200	Wilson (1943)

Table 4.1 (Continued)

Test animal	Injection sequence (days after birth)	Dose, µg		Reference
		Daily	Total	
	One injection day 10		1, 10	Merklin (1953)
	One injection day 5		10	Leathem (1956)
	One injection day 5		5	Gorski (1963 b)
	One injection day 5		5	Heinrich et al. (1964)
	One injection day 5		1250	Presl et al. (1966)
	One injection day 5		1250	Ošťádalová and Pařízek (1968)
	Daily from birth for 5 days	5	25	Takasugi and Bern (1962)
Female rats	*Estradiol benzoate*			
	One injection on day 4		250	Duluc and Mayer (1967)
	Daily injections			
	first 10 days	10		
	second 10 days	20	300	Hayashi (1975)
Female mice	*Estradiol*			
	Daily for 5 days beginning day of birth, 3, 5, 8, 11	20	100	Takasugi (1966)
	Daily from birth for 3 days	20	60	Mori (1979)
Female rats	*Estrone*			
	Daily from birth for 5 days	12.5	62.5	Takasugi (1958)
(PD females)	Daily injections			
	first 10 days	50		
	second 10 days	100		
	last 10 days	200	2500	Kawashima (1965)
Female mice (PE females)	*Estrone* Daily from birth for 5 days	25	125	Arai (1964 a)
(PD females)	Daily injections			
	first 10 days	25		
	second 10 days	50		
	last 10 days	100	1750	Arai (1964 b)
(PE females)	Daily from birth for 5 days	12.5	57.5	Hayashi (1969)
(PE and PD mixed)	Daily injections			
	first 10 days	50		
	second 10 days	100	1500	Hayashi and Kawashima (1971)
Male rats	*Testosterone propionate*			
	3 times a week for 4 weeks beginning day 5, 10, 15 or 20	63–3000	750–36,000	Wilson and Wilson (1943)
	One injection day 5		1250, 5000, 10,000	Kincl et al. (1962; 1963 a)
	One injection day 2, 3, or 5		300, 500, 1,000	Swanson and van der Werff ten Bosch (1963)
	Castrated day 1, injection day 5		2500	Morrison and Johnson (1966)
	One injection day 2 or 3		500	Kurcz et al. (1967)

Table 4.1 (Continued)

Test animal	Injection sequence (days after birth)	Dose, µg Daily	Total	Reference
Male rats	*Estradiol benzoate*			
	One injection day 5, 10 or 20		10, 30, 120	Kincl *et al.* (1962; 1963 a)
	From birth for 60 days	10–40	1800	Steinberger and Duckett (1965)
	One injection on day 5		100	Schiavi (1968)
	One injection day 4		250	Brown–Grant *et al.* (1975)
Male rats	*Estradiol dipropionate*			
	One injection on day 4 or 5		1250	Oštădalová and Pařízek (1968)

See text for additional references.

acetate), to block the activity of TP and found that unopposed androgen must be present for several days to produce sterility. Hayashi and Gorski (1974) used timed (12, 24, 48 and 72 h) implants of TP in the basal hypothalamus and demonstrated that an exposure of at least 48 h is necessary before more than 50% of the animals become anovulatory by 100 days of age. Shorter exposure times were not effective. The observations explain the low effectiveness of a single injection of nonesterified testosterone reported by Kincl *et al.* (1965) and Alklint and Norgren (1971). Presumably the metabolism and excretion of the free steroid is too rapid. Free testosterone has to be injected repeatedly (Mazer and Mazer, 1939; Bradbury, 1940; Selye, 1940; Selye and Friedman, 1940).

4.1.1.1 Females

Pathologies induced in female rodents vary with the dose employed and age given. Pups injected when born (day 1 of life) will be become anovulatory as adults with a lower dose; if older animals are injected a higher dose is needed (Table 4.2). The sensitivity is higher in pups delivered by caesarean section prior to expected birth (Butterstein and Freis, 1978).

Table 4.2 Influence of Various Doses of Testosterone Propionate Injected into Infant Female Rats During Various Days After Birth on Ovulation at the Age of 45 Days.

Dose, µg	Injection day	Ovarian response* 1	2	3	4	5
1		16/16*	8/8	15/15*	**	**
10		1/16	12/23	6/14	3/15	12/24
100		0/16	3/20	5/16	5/16	30/62

* Number of rats ovulating/total number of rats; ** not studied. After Rudel and Kincl (1966).

The Critical Period

Early workers invested considerable effort to define the period of life during which it is possible to induce sterilization in the belief that it would contribute to the understanding of the androgenization process. In rats the period was defined as occurring between days 1 and 10 after birth (Barraclough and Leathem, 1954; Barraclough, 1961; 1966; Gorski, 1963). Figure 4.1 illustrates typical data (Gorski, 1963; 1968). However, an accurate fixation of a "critical period" is beset with difficulties: the dose used, the age at which the adult was evaluated, and the end points have not always been uniform. Additional problems in interpreting and delimiting the "critical period" arise from the observation that some animals injected with a low dose of TP (5–10 µg) may ovulate for several cycles before becoming anovulatory (Swanson and van der Werff ten Bosch, 1964a; Gorski, 1968; Arai, 1971; Sheridan et al., 1976).

Persistent vaginal cornification can be also induced in young but sexually mature rats. Kawashima (1960) injected repeatedly small doses of estradiol (0.01 µg). The animals maintained a regular 4-day cycle for several months; the cycles then ceased, and the animals became persistently estrous or diestrous. Brown-Grant (1975) induced PE by injecting a high dose (2.5 mg) of estradiol benzoate and concluded "... the concept of critical period during development ... needs to be modified ..."

Brain Implants

Stereotaxic implantation of testosterone (Wagner et al., 1966; Nadler, 1972; Lobl and Gorski, 1974) in the dorsal preoptic-anterior hypothalamus induces a persistent estrus syndrome. The implants must be left in situ for about 48 h; shorter expo-

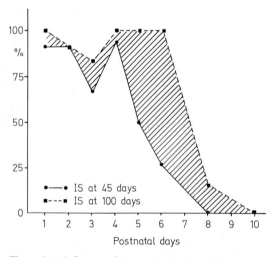

Figure 4.1. Influence of Age on Induction of Sterility in Female Baby Rats. The incidence of sterility (IS, percent) depends on the age (days after birth) at which the animals were injected. The dose of testosterone propionate was 10 µg. The presence, or absence, of corpora lutea was determined by laparotomy. The shaded area reflects those animals which exhibited the delayed anovulation syndrome.

sure will not be effective. To produce sterility by implants requires lower amounts than when systemic injection is the delivery route. For example an injection of 30 μg of TP will produce PE in one-half of females at 50 days of age (all will be anovulatory at 100 days). Implanting the same amount in the dorsal preoptic-anterior hypothalamus will sterilize all females by the age of 50 days (Hayashi and Gorski, 1974).

The areas of the hypothalamus which respond to testosterone are: the medial preoptic area, anterior hypothalamus, and paraventricular, dorsomedial, ventromedial or arcuate nuclei. Dihydrotestosterone is not effective in this respect. Several groups interpreted the lack of activity of DHT (an androgen which does not convert into estradiol) to indicate that estrogens, not androgens, are the active agents (Section 4.1.2).

Testosterone implanted in other areas of the brain besides the hypothalamus also produces sterility. Hayashi (1974) and Lobl and Gorski (1974) reported that implants in the cortical area will also induce the AN syndrome, provided the concentration of the hormone is higher. It is not known if the hormone influenced other centers in the brain, or if it diffused into the active center located in the hypothalamus.

Pathological Manifestations

Neonatal sterilization confers reproductive abnormalities in functions of the hypothalamus-pituitary axis, ovarian and uterine histology and end-organ response, the biosynthesis of ovarian hormones and their metabolism, and abnormalities of mating behavior. The severity of the changes depends upon the agent, dose used, and duration of the treatment. Low doses may produce only alterations in sexual behavior while large doses lead to malformations. Kawashima and Takewaki (1966) injected testosterone propionate, 0.5 mg during the first 10 days, 1 mg during the next 10 days and 2 mg during the last 10 days. This treatment has resulted in severe injury. The vaginal orifice was absent, the ovaries atrophied to the point where "... they were very small, consisting of only a few solid follicles and follicles at the beginning of antrum formation, and directly surrounded by adipose tissue ..." and the adrenal and thyroid functions were abnormal.

The Hypothalamus:

Androgenized females exhibit only discreet changes in the functions of the brain. The hypothalamic content of GnRH is lower in 70 day old females androgenized with 1.25 mg of TP on day 4 or 5 of life (Chiappa and Fink, 1977) but is still higher than in males. Sarkar and Fink (1979) report significantly lower concentrations in stalk plasma of GnRH (31 ± 4 pg/ml) and of LH (4.8 ± 1.2 ng/ml) than in controls (86 ± 12 pg/ml and 15.7 ± 1.9 ng/ml, respectively). Cyclic activity in the release of GnRH and of LH is absent. The lack of the activity is also apparent in the concentrations of neurotransmitting substances. In female rats dopamine concentration in the median eminence rises and falls during the cycle; a nadir occurs during proestrus and early estrus. In PE rats induced by 1.25 mg of TP on day 2 the cyclic turnover is abolished (Fuxe et al., 1972).

Biochemical differences exist between the brain of male and female rodents (see Raisman and Field, 1973), but androgenization does not produce any clear patterns. Salaman (1970) reported decreased synthesis of RNA. Packman *et al.* (1977) measured histochemically the activity of four dehydrogenases representative of major metabolic pathways (malic, lactic, glucose-6-phosphate and glutamic) in 100–120 day old female rats androgenized with 1 mg of TP on day 3: the treatment decreased the activity of malic dehydrogenase and increased the activity of the other three enzymes throughout the hypothalamic area.

The amount of AF-positive (Gomori's stain) neurosecretory material (NSM) may increase in the pars lateralis of the paraventricular nuclei, the pars dorsolateralis of supraoptic nuclei, and the median eminence, albeit slightly. The changes are more pronounced in animals bearing pituitary autotransplant under the renal capsule; the infundibular stem contains an increased number of glial cells and significantly greater amounts of NSM resembling pars nervosa. Oxytocin and vasopressor activity are higher (Takewaki and Kawashima, 1967).

The Pituitary: Changes in the pituitary cell population in androgenized females become manifested at the time of expected puberty, about day 30 of life, but not before (Kincl and Maqueo, 1972).

The Structure: The pituitaries of rats injected with a single dose of TP on day 5 of life are larger (Barraclough, 1961) particularly in animals injected with high doses (Gorski and Barraclough, 1963). The gland is atrophied in animals injected with repeated doses (Wilson *et al.*, 1940; Kawashima and Takewaki, 1966). Weight changes are accompanied by changes in cell population. Wolfe *et al.* (1944) injected female pups between 1st and 28th day of life with 36 mg of TP divided into twelve doses and found that the anterior lobe resembled that of male animals. The eosinophils were filled with granules, the Golgi apparatuses were usually small and applied close to the nucleus. The basophils were large and well granulated albeit less numerous. The chromophobes were small, often only a scant amount of cytoplasm was present and cell membranes poorly defined. Kawashima and Takewaki (1966) used repeated injection (0.5 mg to 2 mg of TP for the first 30 days of life) and found a significant decrease in the number of thyrotrophs and gonadotrophs. Self (1966) found differences between "LTH-", "LH-", and "FSH-" type of cells. Khan (1963) found a reduction in the number of gonadotrophs (hypertrophied with well marked Golgi zones), partial degranulation of acidophils and called the attention to ". . . a tumor-like condition of the anterior pituitary . . . well demarcated foci of intense hyperplasia involving acidophilic cells . . ." Maqueo *et al.* (1966) characterized a decrease in the number of gonadotrophs and acidophils and an increase in the number of castration cells. The changes were most explicit in females injected with a single dose of 10 mg of TP (Table 4.3).

The Function: Neonatally induced infertility is not the result of depressed concentrations of gonadotropins (Korenbrot *et al.*, 1975) but rather due to an abnormal hormonal milieu. Ovulation can be induced by transplanting ovaries into untreated donor females, or into males (under certain conditions), and by supplying exogenous gonadotropins. Estradiol injections stimulate LH release (Section 4.4).

Table 4.3 Pituitary Weights and Cell Population of 45 Day Old Female Rats Injected at the Age of 5 Days with Various Amounts of Testosterone Propionate. Cell Population in Each Pituitary Was Counted in Five Fields (8 animals per group).

Dose, μg	Body wt., g ± SE	Pituitary wt., mg ± SE	Pituitary cell count ± SE			
			Gonadotrophs	Basophils		Acido- phils
				Castration cells	Chromo- phobes	
0	144 ± 3	7.0 ± 0.4	96 ± 1	4 ± 1	930 ±	1104 ± 9
100	144 ± 6	7.8 ± 0.5	94 ± 5	6 ± 2	939 ± 24	NS
1000	150 ± 2	7.4 ± 0.3	87 ± 1	10 ± 3	955 ± 29	1106 ± 13
10000	174 ± 2	9.2 ± 0.3	67 ± 5	28 ± 3	985 ± 8	921 ± 9
Untreated males	224 ± 7	8.3 ± 0.3	98 ± 2	5 ± 1	910 ± 3	1108 ± 12

NS, not studied; data from Maqueo *et al.* (1966).

Table 4.4 Concentration of Gonadotropins in the Pituitary and Plasma of Androgenized Female Rats.

Treatment, TP, mg	Day of Observation	Pituitary concentration, μg/mg		Plasma concentration, ng/ml		Reference
		LH	FSH	LH	FSH	
1.25	60	0.8 (1.2)*				
	150	0.7 (1.1)	8.6 (15.2)		1420 (ND)**	Barraclough, 1968
0.05				36 (47)	50 (200)	Johnson, 1971
0.04, 0.05, 0.25, 1.25	40, 90 150	30–40	3.8–2.6	80	250–300	Mallampati and Johnson, 1974
1.25	60	0.8	~6	~1	~200	Chiappa and Fink, 1977

* ( ) values for controls; ** (ND) not detectable.

Comparison of data pertaining to gonadotropin content of androgenized rats measured by several groups is difficult since each used a different method and/or standard (Table 4.4). Barraclough measured LH with a bioassay, Mallampati and Johnson used a radioimmunoassay and LH RP-1 as a reference standard, and Chiappa and Fink also used RIA but with LH-S18 as a standard. Each standard has a different potency indicated by the large disparity in plasma LH concentrations.

LH: Sex-linked differences do not exist between male and female rats in the concentration of LH basal levels in blood, pulse intervals, and responses to GnRH. Only during the ovulatory peak is the amplitude higher in females. In untreated females the concentration of LH in the pituitary and in plasma begins to increase before sexual maturation, peaking when the rats are about 30 days old (Weisz and Ferin, 1970). The increase is abolished in PE rats, and the concentrations in the pituitary remain static (Barraclough, 1966; Weisz and Ferin, 1970; Chiappa and

Fink, 1977). Plasma concentration of LH is lower even than in males, and the intervals between pulse intervals are longer (Watts and Fink, 1984).

FSH: In PE rats concentrations of FSH in the pituitary gradually increase to about day 30 and then remains constant. In plasma no significant differences exist between the concentration of FSH in PE rats and controls from about day 40 of life (Weisz and Ferin, 1970; Johnson, 1971; Uilenbroek *et al.,* 1976; Chiappa and Fink, 1977).

Prolactin: The pituitaries of old female rats are larger than in young rats. Higher PRL content is reflected in higher amounts circulating in peripheral plasma (Clemens and Meites, 1971; Shaar *et al.,* 1975). Prolonged estradiol treatment induces elevated concentrations in young rats (Takahashi and Kawashima, 1981). PE females resemble old animals in pituitary morphology and PRL content (Kurcz *et al.,* 1967; Neill, 1972; Harlan and Gorski, 1977; Ruzsás *et al.,* 1977) or resemble old constant-estrus rats (Clemens and Meites, 1971). Johnson (1971) reported higher values (40 ng/ml on day 27 and 151 ng/ml on day 42) than in controls. Later investigators found higher values. Mallampati and Johnson (1974) determined prolactin values in several groups of PE rats (TP doses 0.01, 0.05, 0.25 or 1.25 mg) at 40, 90 and 150 days of age and found significantly higher values than in cycling rats. Danguy *et al.* (1977) and Gala *et al.* (1984) confirmed higher PRL secretion in androgenized females. The latter group found no effect on PRL values in response to blinding, olfactory bulbectomy, and pinealectomy.

Estimation of PRL in PE rats by earlier workers yielded contradictory results. Kikuyama (1963) and Kurcz *et al.,* (1967) could find no difference in PRL pituitary content between PE and control rats.

Altered ovarian steroid biosynthesis may be a cause of increased prolactin concentrations. In senile constant-estrous rats increased prolactin concentration in peripheral plasma is abolished by ovariectomy (Aschheim and Pasteels, 1963). Increased prolactin release could also be a result of altered opioid peptide concentrations and/or biosynthesis in the CNS (Bruni *et al.,* 1977; Dupont *et al.,* 1977; Shaar *et al.,* 1977).

Steroid Hormone Concentration in Blood: Estradiol plasma levels are within normal limits (Mallampati and Johnson, 1974; Kolena *et al.,* 1977). Falvo *et al.* (1974) found lower testosterone values in the plasma of PE rats than in control proestrus females (107 ± 10.8 pg/ml vs 172 ± 9.7 pg/ml). After castration, the concentrations decreased to 51 pg/ml in both preparations. Progesterone concentrations are low since corpora lutea are absent in PE rats.

The Ovary: Animals injected with "low" doses may undergo some ovulatory cycles, exhibit diestrus smears and form occasional corpora lutea. Animals injected repeatedly or with a single high dose of TP do not cycle (Swanson and van der Werff ten Bosch, 1964a; Kincl *et al.,* 1965a; Mayer and Duluc, 1970). The interstitial tissue is hypertrophied. In the ovaries are present numerous follicles in various stages of development. Self (1966) observed significant increases (200 – 400% from controls) in the number of granulosa, interstitial and theca interna cells. The presence of vesicular follicles and the absence of corpora lutea are prominent features though occasionally structures resembling "old" corpora lutea are seen (Figure 4.2). The changes become more severe with advancing age.

Hemiovariectomy or an exogenous LH source restore corpora lutea formation (Section 4.4).

Peripheral Organs: The uteri of PE rats are hypertrophied; the general structure of the myometrium is unchanged, while the endometrial stroma is compact, nonedematous, containing only few glands; the epithelium is highly folded, hypertrophied and composed of tall, columnar cells (Barraclough, 1961). When TP is injected repeatedly for 30 days after birth uteri atrophy (Wilson *et al.*, 1940). The development of lesions is dependent on the length of the treatment and the dose. Takewaki (1968) used repeated injections of TP and found in about 20 percent (12 or 54 animals) metaplastic areas of varying size. In few animals the whole uterine epithelium became metaplastic.

The uterine epithelium can respond to progesterone treatment and traumatization. Deciduomata form as in untreated females (Burin *et al.,* 1963).

Vaginae: Early vaginal introitus is the most visible result of neonatal androgen exposure. The age at which the vagina opens is dose dependent. Kincl *et al.* (1965a) reported vaginal introitus at 13 days in the group injected with 10 μg of TP while in the controls this was 37 days (Table 4.5). In markedly androgenized females (5 and 10 injections of TP from the day of birth) the vaginal opening as appears a minute entrance at the base of the clitoris. Or the vagina may end blindly under the perineal skin (Kikuyama, 1965). Direct androgen action may cause the early introitus. Butenandt and Kudszus (1935) noted that in 20 day old rats, treated for several days with testosterone or 4-androstene-3,17-dione, the vagina was opened at the end of the experiment. Tramezzani *et al.* (1963) recounted that in animals injected with 200 μg of TP at birth the vagina failed to open, while animals injected on day 4 or 5 of life with the same dose vaginal introitus occurred within a few days. Possibly, the failure of vaginal opening in this experiment was due to a transformation of genital skin into the scrotal type of skin (Martin *et al.,* 1961). The occurrence of an early introitus is independent of the presence of the ovaries (Colombo, 1968). Kimura and Takasugi (1964) ovariectomized PE rats (produced by daily injection of 0.0125 mg of TP from day 6 of life for 5 days) on day 13 of life. A single dosage of estradiol (1 mg) was injected into 94 – 144 day old animals. The vaginal epithelium became cornified 2 days after the injection, and the state persisted, interrupted by several stages of diestrus, for periods lasting more than 120 days. In the control group estradiol induced cornification lasted only 5–9 days.

Persistent cornification of the vaginal epithelium is another recognizable manifestation of androgenization. In animals injected with low doses of TP (5 or 10 μg) PE may become manifested after several ovulatory cycles have occurred (see above).

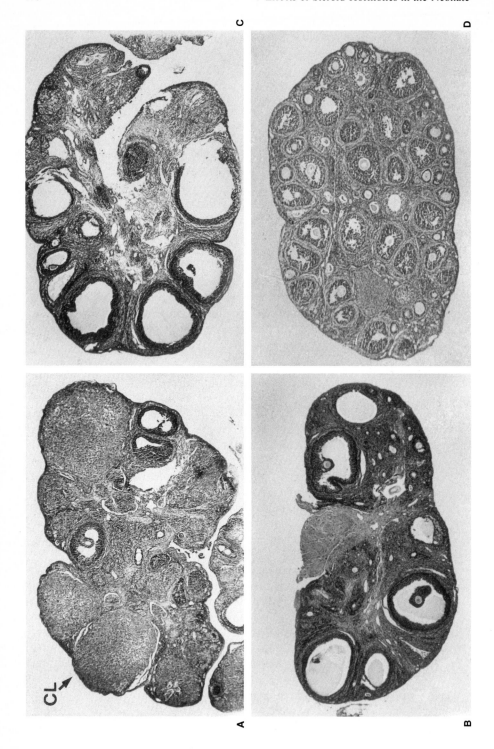

Table 4.5 The Influence of Graded Doses of Testosterone Propionate Given in a Single Injection to 5 Day Old Female Rats on Vaginal Introitus and Formation of Corpora Lutea at the Age of 45 Days.

Dose of TP, μg	No. of rats	Vaginal introitus days ± SE	No. of rats with corpora lutea	Mean no. of CL ± SE
0	43	37.3 ± 0.3	35	9.4 ± 1.0
10	10	30.3 ± 0.2	3	1.5 ± 1.4
100	22	19.6 ± 2.8	6	2.3 ± 1.2
500	31	13.0 ± 0.2	2	0.5 ± 0.4
1000	24	Not determined	0	0.0 ± 0.0

From Rudel and Kincl (1966).

The epithelium derived from the Müllerian ducts (the proximal zone) is longitudinally folded so it presents a rosette-shaped cross section. It may be stratified with mucified or cornified superficial layers (animals injected a few times with TP) or composed of a basal layer of cuboidal cells covered with a layer of columnar cells. This zone responds readily to estrogen treatment (Kikuyama, 1965). The epithelium of the distal portion is the one that contains detached cornified layers intermingled with leukocytes (Kimura and Takasugi, 1964; Takewaki, 1968).

4.1.1.2 Males

The early reports pertaining to the effects of androgens in prepubertal males include those by Moore and Price (1930, 1932, 1938), Biddulph (1939) and Greene and Burrill (1940). All have used repeated injections of TP for 20 to 30 days after birth and noted testicular atrophy and damage to seminiferous epithelium. Wilson and Wilson (1943) evaluated the influence of the dose and age on the response. TP (0.75 to 36 mg) was injected 3 times a week for 4 weeks, beginning on day 1, 5, 10, 15 or 20 after birth. Autopsy was at the age ranging from 125 to 190 days. The treatment produced a decrease in fecundity and sexual potency (only one male in ten sired a litter), atrophy of the testes, the prostate and seminal vesicles. The author found no microscopic abnormalities in the maturation of the seminiferous ep-

Figure 4.2. Ovary of Adult Rat Injected with Testosterone Propionate at the Age of Five Days. **A** Section from ovary of a control rat; note the presence of fresh corpora lutea (CL); × 10. **B** Rat was injected with 1.25 mg of testosterone propionate when 5 days old; growing and vesicular follicles are present; CL did not form; × 10. **C** Rat was injected with 60 μg of estradiol benzoate when 5 days old; note the formation of few Graafian follicles and absence of corpora lutea; × 10. **D** Rat was injected with estradiol benzoate, 120 μg, when 5 days old; the organ is small and only primary and secondary follicles are present; × 10. Note that the magnification of **A**, **C**, and **D** is the same.

ithelium. When treatment was begun on day 1 the damage was more severe than in groups injected on day 5 or 10 and was absent in the other groups. Despite the apparently normal microscopic appearance of the testes the function of Leydig cells was not normal as evidenced by accessory sex tissues atrophy. Wolfe *et al.* (1944) found significant decreases in the weights of testes, seminal vesicles and the prostate. Arai and Masuda (1968) confirmed that the degree of testicular patholo- gy was dependent on treatment length and the dose of TP used. Daily doses of 0.5 mg TP for the first 10 days of life, followed by 1 mg for the next 10 days and 2 mg daily from day 21 to day 30 of age resulted (rats 150 days old) in an atrophy of the testis (weight about 1/10th of controls) and of ventral prostate, seminal vesi- cles and coagulating glands. Maturation of the germinal epithelium was also af- fected: the diameter of seminiferous tubules was reduced, spermatids and sper- matozoa were almost totally absent, and "... only few spermatogonium-like cells and very few spermatocytes ... were found ... Leydig cells were hardly distin- guished from fibrocytes ...".

The responses to a single neonatal injection of TP are similar to those pro- duced by repeated dosing, albeit less severe. Kincl *et al.* (1962, 1965a) used graded doses of TP in 5 day old rats (0.25, 0.5, 1 and 10 mg) and examined the animals at the age of 45 days. In animals injected with the high dose the testes, ventral pros- tate, seminal vesicles and levator ani muscle atrophied. In a few animals spermat- ozoa and spermatids were absent, but otherwise the maturation wave was normal. In older animals injected with 1 mg (200 days) the testes atrophied (Kincl and Zbuzkova, 1971). This observation indicates that males may go through a delayed syndrome similar to the lag reported in females injected with low doses of TP.

Others reported similar findings: Johnson and Witschi (1963), Johnson *et al.* (1964), Swanson and van der Werff ten Bosch (1964b, 1965a), Jacobson and Nor- gren (1965), Ladosky and Kesikowski (1969), and Mayer and Duluc (1970). In Table 4.6 the "maximal tolerated dose" refers to the highest dose tested at which fertility was not inhibited as evidenced by successful mating and pregnancy. The

Table 4.6 Effects of a Single Dose of Testosterone Propionate in Male Rats on the Development of Gonadal Function in the Adult.

Dose μg TP	Age at injection, days	Age at autopsy	Maximal tolerated dose, μg	Threshold dose, μg	Reference
5, 10, 50, 500	1, 2, 4	77, 190	50*	50	Swanson and van der Werff ten Bosch (1964)
1500	5	30		1500	Jacobson and Norgren (1965)
500	1	150		500	Arai (1968 and Masuda, 1968)
250, 500, 1000 10,000	5	45	5000	1000	Kincl *et al.* (1962; 1965)
1,000	5	200		1000	Kincl and Zbuzková (1971)

* Higher dose not tested.

"threshold dose" refers to the minimal dose which produced a statistically significant atrophy of the testes and accessory sex tissue but not necessarily infertility. In studies with a short period of observation fertility (insemination of a female) was usually not tested. Delayed pathological manifestations could be missed in such an investigation.

Pituitary Function

Long treatment with TP (total dose 36 mg between days 1 and 28 of life) does not cause morphological changes in the gland (Wolfe *et al.*, 1944). FSH concentration is within normal variation in the peripheral blood in males given a single injection of 0.5 mg of TP on day 5 (Johnson, 1971) while Goldman and Gorski (1971) report a small decrease. Moguilevsky *et al.* (1977) found an increase in FSH serum concentrations in 90 day old males (397 ± 19.5 ng/ml ($\pm$ SE) vs 254 ± 11.4 ng/ml in controls) and no changes in LH values (37 ± 2.6 ng/ml and 33 ± 2.8 ng/ml, respectively). Prolactin concentration was higher (72 ± 10.0 ng/ml vs. 20 ± 3.6 ng/ml).

The ability of the gland to incorporate uridine, a measure of RNA synthesis, is impaired (Biró and Endröczi, 1977).

Steroid Hormone Concentration in Blood

Testosterone concentration in peripheral blood is within normal limits in 240 day old males injected neonatally with 250 µg TP. Average values were 5.3 ng/ml (± 0.7 SE) in plasma and 5 ± 0.6/ml in controls (Frick *et al.*, 1969). However, the biosynthesis of androgens is altered by neonatal TP treatment (Section 4.4).

4.1.2 Other Androgens

Dihydrotestosterone

In some organs reduction of testosterone to DHT is a prerequisite for the expression of androgenic response. It is the reduced form, not testosterone, that binds to cytosol and nuclear receptors in target cells (Chapter 2). The need for obligatory reduction in the adult is equivocal in the neonate. DHT and its propionate fail to induce the persistent syndrome in rats. Females injected with the hormone cycle, the ovaries contain CL, the vaginal epithelium undergoes cyclic variations, and sexual behavior is not altered (Whalen and Luttge, 1971; Arai, 1972; McDonald and Doughty, 1974; Ulrich *et al.*, 1972; Morishita *et al.*, 1975; van der Schoot *et al.*, 1976). However, the treatment advances the day of vaginal opening (Korenbrot *et al.*, 1975) and alters permanently the reactivity of the vaginal mucosa to progesterone and estradiol. Hormone treatment, which in controls supports the development and maintenance of deciduomata, causes cornification in DHT treated rats (Takewaki and Ohta, 1976). The observation suggests increased sensitivity of epithelium to estrogenic stimulation.

Implants placed into the hypothalamus do not produce sterility as does testosterone (Section 4.1.1.1). Since 5α-reduced androgens cannot aromatize, the results were interpreted to indicate that the activity of testosterone is the consequence of its conversion to estrogens in the brain (McEwen *et al.*, 1977). The findings in mice do not support this hypothesis. Iguchi and Takasugi (1983) reported that

DHT (50 μg) injected on day 5 into female mice induced permanent anovulation and estrogen independent, permanent proliferation of the vaginal epithelium. Yanai *et al.* (1977) found the β-isomer (17β-hydroxy-5β-androstan-17-one, 200 μg single dose) produced similar alterations.

Esters of 19-nortestosterone (NT) and 1-methyl-1-dehydro NT (for nomenclature see the Appendix) are active (Table 4.7). Booth (1976) found the propionate

Table 4.7 Induction of Persistent Estrus in the Infant Female Rodents by a Single Injection of Various Androstanes.

Steroid tested	Dose range tested, mg	Period of observation days	Minimum effective dose, mg	Reference
Rats				
Testosterone (T)	100–1000	45	1000	Kincl *et al.* (1965)
T dichloroacetate	10– 100	45	inactive	
Epitestosterone	500–5000	45	5000	
2 α-Methyl DHT	500–2000	45	inactive	
Oxymetholone	500–2000	45	inactive	
Androstanozol	500–2000	45	inactive	
1-Methyl DHT enanthate	3000	90–120	3000	Jacobson (1964)
5 α-Androstanediol	100	150	inactive	Brown-Grant *et al.* (1971)
5 α-Androstanediol propionate	250	40–180	250	Malampati and Johnson (1974)
AD enol propionate	50	110	50*	McDonald and Doughty
Androsterone propionate	50*	110	inactive	(1974)
5 α, 3 α AD dipropionate	50*	110	inactive	
5 α, 3 β AD dipropionate	50*	110	inactive	
19-HT propionate	50*	110	50*	
19-Hydroxy testesterone (19-HT)[a]	100	110	100	
5 α-19-HT	100	110	inactive	Booth (1976)
19-Nortestosterone (NT)[a]	100–1000	45	inactive	Kincl *et al.* (1965)
NT cyclopentylpropionate	100– 500	45	500	
NT phenyl propionate	1250		1250	Pařízek *et al.* (1963)
NT phenyl propionate	750	90–120	750	Jacobson (1964)
NT decanoate	1500	90–120	1500	
19-Norandrostenolone phenylpropionate	500, 2500	90–120	2500	Mayer and Duluc (1970)
Mice				
5 β DHT	250*	270	250*	Iguchi and Takasugi
5 β DHT	250*	270	250*	(1983)

* Injected 10 mg daily for 5 days beginning on the day of birth; [a] for nomencleture see Appendix; oxymetholone = 2-hydroxymethylene-17αmethyl-17β-hydroxy-5α-androstan-3-one; androstanozol = 17α-methyl-17β-hydroxy-5α-androstane-3.2-c pyrazole; 1-methyl DHT enanthate = 1-methyl-17β-hydroxy-1-5α-androsten-3-one 17-enanthate; 5α-androstanediol = 5α-androstan-3β, 17β-diol; AD enol propionate = 3β-hydroxy-3,5-androstadien-17-one 3-propionate; androsterone propionate = 3β-hydroxy-5α-androstane-17-one 3-propionate; 5α, 3α AD dipropionate = 5α-androstan-3α, 17β-diol 3,17-dipropionate; 5α, 3β AD dipropionate = 5α-androstan-3β, 17β-diol 3,17-dipropionate; 19-norandrostenolone phenylpropionate = 3β-hydroxy-4-estren-17-one 3-phenyl propionate; 5β DHT = 17β-hydroxy-5β-androstan-3-one.

of 19-hydroxy testosterone to be as active as testosterone propionate while the 5α-reduced analogue was not. Kincl *et al.* (1965) found that three "aromatizable" androgens (testosterone dichloroacetate, 100 µg), NT (1000 µg) and NT cyclopentyl-propionate (500 µg) did not inhibit luteinization at 45 days but caused precocious vaginal introitus. Another "non-aromatizable" androgen, 2α-methyl DHT (2000 µg) caused early vaginal opening. Gorski (1966) reported that 17α-methyl testosterone was inactive at 1000 µg dose.

Antiandrogens

Neumann and coworkers evaluated extensively the activity of antiandrogens in a battery of tests and showed that blocked endogenous androgen production favors in males the development of many female characteristics (Neumann and Steinbeck, 1974). Results obtained with cyproterone acetate are illustrative: male offspring born to rats treated with this compound during pregnancy develop mammary glands (Neumann *et al.*, 1966a) and a vagina (Neumann *et al.*, 1966b) and exhibit femalelike behavior patterns (Chapter 6).

4.1.3 Estradiol

Estradiol is more toxic in neonatal animals than are androgens. The treatment may lead to the development of genital tract cancer. Placement of microamounts of estradiol into the anterior hypothalamus of neonatal females produces a sterility syndrome (Döcke and Dörner, 1975; Hayashi, 1976). Hayashi (1979) induced sterility by implanting pellets of a weak uterotrophic estrogen, 16β-ethyl estradiol, into the anterior hypothalamus.

The presence of high amounts of estrogens during neonatal life induces a "wasting" disease similar to that seen after the removal of the thymus gland (Reilly *et al.*, 1967; Kincl *et al.*, 1970).

4.1.3.1 Females

Estrogenized females exhibit either a persistent estrus (PE) or a persistent diestrus (PD) depending on the dose used. Treatment with 12.5 µg of estrone for 5 days from the day of birth will produce PE females; the ovaries contain follicles of varying sizes but no corpora lutea (Takasugi, 1952; 1954a, b). Arai (1964a) obtained PE rats by injecting 25 µg of estrone daily from birth for 5 days. The nature of the dose also influences the nature of the cornification produced. Injection of "small" quantities results in ovary-dependent vaginal cornification, which disappears after ovariectomy in later life. Administration of "large" quantities generates cornification independent of the presence of gonads (Takasugi *et al.*, 1970).

Injecting higher amounts for longer periods (30 days) from birth produces PD rats; the ovaries contain only small follicles but no maturing follicles or corpora lutea. The animals are in a permanent diestrus. Kawashima (1965) and Kawashima *et al.* (1980a) used 50 µg of estrone for the first 10 days of life, 100 µg for the next 10 days and 200 µg from day 21 to day 30.

Greene and Burril (1941) gave a single injection of estradiol dipropionate (EDP) to rats at birth and found in the adult enlarged and inflamed oviducts, atrophied ovaries lacking corpora lutea and uteri showing squamous metaplasia. Turner (1941) furnished estrone daily for the first ten days of life and found prolonged periods of estrus, sterility and nonluteinized ovaries. Wilson (1943) gave EDP (100 µg every third day) for 28 days. The adults were sterile when injected on day 1 or 5 of life but not when injected on days 15, 20, 30, or 40 after birth.

Gorski (1963b) and Heinrich et al. (1964) used a single injection of estradiol benzoate on day 5. According to Gorski 5 µg is the minimal dose which will induce the sterility syndrome. Doughty et al. (1975) reported similarly.

Pathological Manifestations

The Hypothalamus: Peptidase activity, measured *in vitro* as oxytocin loss (a measure of LH secretion), is decreased in estrogenized rats (Griffiths and Hooper, 1973a). The finding was taken to indicate an increase in the secretion of LH. The argument was based on the observation that in male rats orchidectomy decreases the enzyme activity, whereas exogenous testosterone elevates the reaction (Griffiths and Hooper, 1973b).

The Pituitary: The pituitary of persistent estrus and persistent diestrus rats is atrophied. The number of gonadotrophs in PE rats is reduced; in PD females these cells are almost totally absent (Kawashima, 1965). PAS-positive AF-negative basophils may be completely absent, the chromophobes are small and shrunken and contain only small amounts of cytoplasm. PAS- and AF-positive basophils are more numerous (Arai, 1964a; Kawashima and Takewaki, 1966). Presl et al. (1966) reported the presence of an acidophil substance staining with PAS-orange 6-methylene blue which the authors believe may be related to the production and storage of LH.

In contrast to PE rats, castration of PD rats does not result in increases of castration cells indicating a more severe lesion. The low number, or complete absence, of gonadotrophs reflects a very low gonadotropin titer (Arai and Kasuma, 1968).

Prolactin producing cells are less numerous in aged PD rats, and PRL secretion is low (Table 4.8). No significant changes are seen in prolactin storage or plasma concentrations in younger animals (Kawashima et al., 1974); they are higher after adolescence (Nagasawa et al., 1973).

The Ovary: The gonads are atrophied and may contain growing follicles in various stages of development. In PD animals only primary follicles are usually present (Fig. 4.2). In 1½ year old PE rats hypertrophy of interstitial tissue may proceed to the formation of a hyaloid mass accompanied by a pituitary adenoma (Takasugi, 1958). Ovarian tumors may develop (Ostadalova et al., 1970).

Table 4.8 Prolactin Concentration in 18 Month Old Male and Female Rats and in Persistent Diestrus Females.

	Serum concentration μIU/ml	Concentration in pars distalis mIU ± SE	
		in whole gland	per mg
Control females	674 ± 113	701 ± 289	52 ± 23
PD females	364 ± 62	236 ± 92	30 ± 13
Control males	247 ± 19	88 ± 5	8 ± 0.4

Data from Kawashima *et al.* (1973).

Peripheral Organs: The uteri are atrophic, resembling organs obtained from untreated ovariectomized animals. In some animals the epithelium may have hyperplastic areas, occasionally infiltrated by leukocytes (Takewaki and Kawashima, 1967; Hayashi, 1968). Uterotrophic response to estrogens is minimal, related to a decrease in cytoplasmic estradiol receptors (Gellert *et al.*, 1977).

Estradiol is uterotrophic in the neonate. The uterus responds to treatment with estradiol (10 μg/day from day 1 through day 5) by increases (in 14 day old animals) in the dry weight of the uterus (an increase of 163% compared with controls), DNA content (193%) and proteins (211%); nuclear estrogen receptors and enzyme activity (ornithine decarboxylase) also increase (Sheehan *et al.*, 1981). The number of glands is decreased. In rats only luminal epithelial cells are differentiated at birth. The genesis of musculature and glandular epithelium occurs between day 9 and 15 of postnatal life. Exposure to estradiol (10 μg/day from day 1 to day 5) reduced by 30% the number of glands and induced premature differentiation (Branham *et al.*, 1985). The observed premature uterine growth and development led the authors to conclude "... neonatal toxicity of estradiol ... may arise from precocious and/or inappropriate induction of estrogen receptor-mediated biosynthetic and morphological events ..."

The uteri and vaginae of mice exhibit increased sensitivity to estrogens and develop readily neoplastic lesions (Section 4.3.1).

4.1.3.2 Males

Turner (1941) was one of the first to report permanent genital impairment in adult rats given estrogens during early life. Arai (1964b) used daily injection of estrone (25 μg first 10 days, 50 μg second 10 days, and 100 μg during last 10 days) beginning after birth. The treatment caused inhibition of spermatogenesis and marked atrophy of gonads and accessory sex tissue 7–11 months after the last injection. The pituitaries were small, gonadotrophs and castration cells were rare.

Kincl *et al.* (1962; 1963) used a single injection of graded doses of estradiol benzoate (EB), 10, 30, 120, or 250 μg, given on days 5, 10, or 15 after birth and evaluated the effect on fertility at the age of 140 days. Males injected with 10 μg were fertile while those injected on day 5 or 10 with 30 μg (or more) were infertile. Injections given later in life were less effective. Hendricks and Gerall (1970) injected 100 µg of EB into 4 day old rats and found delay in testes descent and de-creased body and testes weights. About 50% of the tubules were devoid of sper-matogenesis.

Pathological Manifestations

The Pituitary Gland and the Hypothalamus: Kawashima (1965) used repeated doses of estrone (50 µg during the first 10 days, 100 µg during the second 10 days and 200 µg during the last 10 days) and described in detail the histopathology of the neurosecretory cells in the hypothalamus, the anterior lobe of the pituitary, the testes, the thyroid, the adrenals, and the seminal vesicles. Increases were present in the neurosecretory material in supraoptical nuclei, the intermediate layer of the median eminence and the pars nervosa of the neurohypophysis. In the anterior hypophysis the gonadotrophs were almost totally absent.

An injection of 250 µg EB on day 5 leads to an increase in the number of cas-tration cells (5 ± 1, SE in controls vs 439 ± 6 in EB animals) and chromophobes (from 910 ± 8 to 1240 ± 4) while the number of gonadotrophs (98 ± 2 vs 66 ± 2) and acidophils (1108 ± 12 vs 980 ± 12) decrease (Maqueo *et al.,* 1966).

Prolonged treatment with estrogens leads to even more severe changes. Arai and Masuda (1970) injected 1 µg of EB for the first 10 days of life, 2 µg for the second 10 days, and 4 µg for the next 10 days. The treatment caused a significant atrophy of the pituitary gland. In a neonatally castrated estrogenized group chromophobes (castration cell) failed to develop; the authors ascribe the lack of response to a ... "long lasting malfunction in the hypothalamus-pituitary sys-tem ..."

Neonatal estrogenization does not influence the concentration of LH and FSH in the pituitary or plasma (Schiavi, 1969). Brown-Grant *et al.* (1975) reported sub-optimal levels of FSH during the time of puberty but no significant changes there-after.

The Gonads: Maqueo and Kincl (1964), Kincl and Maqueo, (1972) and Kincl *et al.* (1965a) analyzed the changes in the development of the seminiferous epitheli-um (Table 4.9) determined by random examination of about 200 tubules from

Table 4.9 Effect of Estradiol Benzoate Injected into 5 Day Old Pups on the Maturation of Semi-niferous Epithelium in 50 Day Old Rats; About 200 Tubules Were Examined in Each Testis and the Cell Count of Various Elements Was Recorded.

EB dose µg	No. of observations	Tubules with cells present, mean $\pm$ SE			
		Sperm	Spermatids	Spermatocytes	Spermatogonia
0	23*	87.3 ± 1.3	94.7 ± 0.9	99.9 ± 0.1	99.9 ± 0.1
120	15	35.1 ± 10.6	39.3 ± 10.0	76.3 ± 4.8	88.3 ± 2.9
240	18	4.7 ± 1.2	2.6 ± 1.2	35.0 ± 5.3	53.4 ± 6.1

* Total number of tubules examined: 4600.

each testis. The normal architecture of the testes had been preserved while the epithelium in many tubules was necrotic and consisted often of only spermatogonia and Sertoli cells. Leydig cells appeared decreased in number, and many underwent degenerative changes and were small with shrunken cytoplasm or without nuclei (Figure 4.3). Hendricks and Gerall (1970) confirmed the toxicity of 100 µg of EB. In 130 day old males testes were in the abdominal cavity, smaller than in controls, and had tubular abnormalities. Brown-Grant *et al.* (1975) described in detail the process of spermatogenesis in male rats injected on day 4 of life with graded doses of EB (3, 10, 30, 100, or 250 µg). Estrogen treatment induced irregular maturation of the seminiferous epithelium between the ages of 30 to 60 days. In older animals the testes of treated and control animals did not differ "... and the number of Leydig cells ... was not different at day 85 (from controls) ...". Yet, the fertility of experimental animals was significantly lower. Only 3/14 males mated and sired litters. In most animals active spermatozoa were absent; the epididymis contained only sperm heads or tails, together with cell debris and leukocytes.

Accessory Sex Organs: Kincl *et al.* (1962; 1963) examined the effects of graded doses of EB on organ weights in 45 day old animals (Table 4.10). The atrophy of testes, seminal vesicles, ventral prostate and levator ani muscle correlated with the dose used and the age at which it was given; the higher the dose, the greater was the atrophy. Others confirmed the dependence of age and dose on the severity of response (Mayer and Duluc, 1970; Harris and Levine, 1965; Brown-Grant *et al.*, 1975).

Table 4.10 Effect of a Single Injection of Estradiol Benzoate given to 5 Day Old Male Rats and Autopsied at the Age of 45 Days (8 animals/group).

Treatment, µg/pup	Organ weights, mg ± S.E.			
	Testes	Ventral prostate	Seminal vesicles	Levator ani
0	2190 ± 21	125 ± 6	88 ± 2	100 ± 3
30	1810 ± 53	110 ± 5	58 ± 4	112 ± 3
60	1430 ± 73	95 ± 4	41 ± 3	72 ± 6
120	1190 ± 30	80 ± 5	27 ± 6	65 ± 6
240	660 ± 260	34 ± 7	20 ± 4	40 ± 8

From Kincl *et al.* (1962; 1963).

Figure 4.3. Testis of Adult Rat Injected with Estradiol Benzoate at the Age of Five Days. A Section of one tubule of control rat; spermatogonia (large dark cells), primary and secondary spermatocytes (large lighter cells) and spermatids (including spermatozoa), small dark cells, with tails projecting into the lumen; × 164. **B** Dose 30 µg of estradiol benzoate (EB); well preserved germinal epithelium with minimal spermatozoa maturation; × 164. **C** Section of one tubule (same treatment as under **B**) showing an absence of spermatids but otherwise well preserved spermatocytes and spermatogonia; × 260. **D** Dose 120 µg EB; complete absence of spermatids, spermatocytes reduced in numbers, tubule diameter decreased, Leydig cells few in numbers; × 164. **E** Dose 240 µg EB; great damage to the germinal epithelium; absence of spermatids, spermatocytes and spermatogonia redued in numbers; tubule diameter reduced; × 164. **F** same treatment as **E**; × 260.

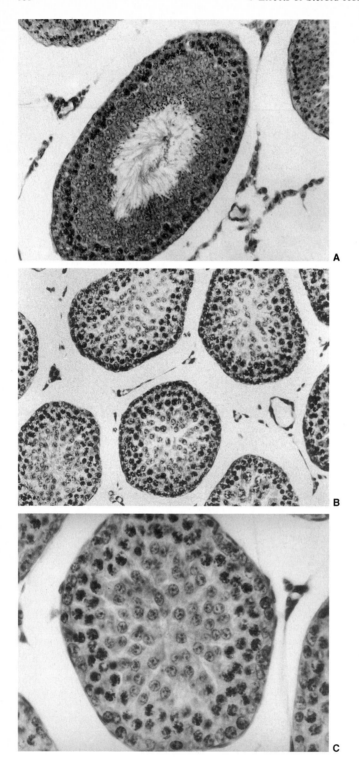

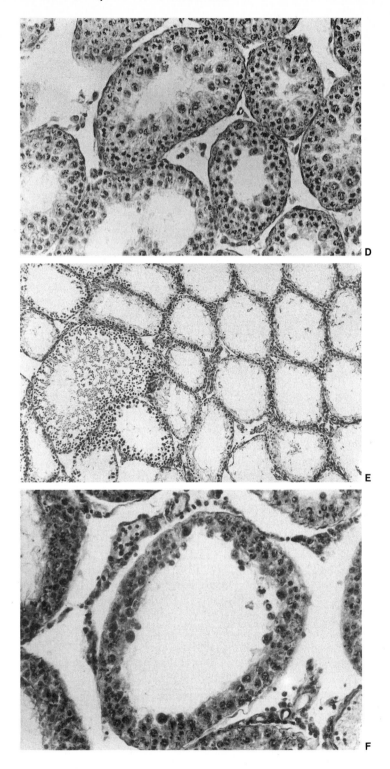

Table 4.11 Concentration of Testosterone in Peripheral Plasma and Organ Weights of 240 Day Old Male Rats Injected at the Age of 5 Days with Estradiol Benzoate (EB) or Testosterone Propionate (TP).

Treatment	Dose, µg	No. of rats	Testosterone ng/ml ± SE	Body wt., g ± SE	Organ weights, mg ± SE	
					Testes	Ventral prostate
Control	0	7	5.0 ± 0.7	422 ± 20	3430 ± 145	618 ± 21
EB	50	11	2.6 ± 0.3*	469 ± 11	2660 ± 148*	525 ± 31*
	250	4	1.9 ± 0.3*	526 ± 18*	2450 ± 132*	343 ± 80*
TP	250	6	5.3 ± 0.7	494 ± 26	3620 ± 46	758 ± 47

* $P < 0.05$; after Frick *et al.* (1969).

In males exposed to repeated injections of EB (from day 1 to day 30) the epithelium in the coagulating glands may undergo squamous stratification and cornification (Arai, 1968); the metaplasia is prevented by concurrent injections of TP (Arai, 1970). Squamous cell carcinoma develops in offspring born to mothers exposed to DES (Chapter 3).

The atrophy of accessory sex glands is the result of decreased testosterone plasma levels. Frick *et al.* (1969) reported about one-half the value of controls (Table 4.11); Brown-Grant *et al.* (1975) confirm the findings and report plasma concentrations depressed from day 6 onwards.

4.1.4 Other Estrogens

Females: Exposure to diethylstilbestrol either *in utero* or postnatally induces permanent changes in the accessory sex organs. The lesions often convert into a precancerous or cancerous stage. A 50 µg dose of diethylstilbestrol (DES) suppressed luteinization in all the treated animals and produced marked atrophy of the ovaries (Table 4.12). Gorski (1963) found DES active at a dose of 10 µg and hexoestrol at a dose of 100 µg (he did not study lower doses). Presl *et al.* (1967) confirmed the toxicity of DES.

The toxicity of diethylstilbestrol is covered in Chapter 5.

Table 4.13 shows an assessment of the effectiveness of various C-18 aromatic steroids in female animals, based on the degree of luteinization, vaginal introitus, and tissue weights. Mestranol (3-methyl ether of 17α-ethinyl estradiol) was most toxic. The minimum effective dose, 0.1 µg, inhibited completely luteinization and produced marked ovarian and uterine atrophy. Ethinyl estradiol, estradiol, and EB were active at a dose of 30 µg. Introduction of a 3-methyl ether group into estradiol decreased the activity; the 3-methyl ether of estradiol was only moderately active at a dose of 500 µg. Estriol, an "inactive" (uterotrophic) estrogen, was active at a dose of 100 µg. A dose of 1 µg induced vaginal opening in 27 day old animals. At this dose the ovarian function was normal in four of six animals. Several 2, 4, 6 and 7 substituted estrogens were active in both sexes albeit only when tested at high doses.

Gorski (1966) used the status of vaginal smears, the presence or absence of corpora lutea and tissue weights (ovaries, uterus) as the end points. Estradiol benzoate was active at a dose of 5 µg.

Table 4.12 Inhibition of Sexual Development in 45 Day Old Female Rats Produced by Various Phenolic Steroids When Injected at the Age of 5 Days.

Steroid used	Total no. of rats used	Dose range studied, μg	Minimum effective dose, μg	Response at minimum effective dose				Tissue weight mg ± SE	
				Ovarian response*	Mean corpora lutea ± SE	Vaginal introitus days ± SE	Body wt. g ± SE	Ovaries	Uterus
0	30	0	0	26/30	9.0 ± 0.3	40.1 ± 0.5	146 ± 2	41.9 ± 2.0	164 ± 11
Mestranol	50	0.01–30	0.1	0/8	0.0 ± 0.0	13.0 ± 0.0	137 ± 3	19.2 ± 0.7	67 ± 5
Ethinyl estradiol	16	0.1–1	1	1/8	1.0 ± 0.9	34.9 ± 2.9	145 ± 3	24.3 ± 1.6	74 ± 8
Estradiol benzoate	54	30–120	30	0/6	0.0 ± 0.0	12.7 ± 0.4	140 ± 3	12.4 ± 1.2	91 ± 19
Diethylstilbestrol	16	50–200	<5	0/0	0.0 ± 0.0	–	140 ± 3	18.7 ± 2.9	128 ± 12
17-Deoxyestrone	16	100–1000	100	0/8	0.0 ± 0.0	–	142 ± 2	16.4 ± 0.9	54 ± 8
Estriol	38	1–1000	100	2/8	1.9 ± 1.3	22.0 ± 1.4	144 ± 3	28.0 ± 1.9	98 ± 17
Estradiol	48	10–500	120	2/8	0.5 ± 0.1	13.5 ± 0.6	125 ± 5	15.5 ± 2.4	63 ± 16
Estradiol 3-methylether	14	100–500	500	4/6	5.7 ± 1.6	23.1 ± 1.1	147 ± 5	33.5 ± 5	160 ± 34

* Number with corpora lutea/total number observed.

Table 4.13 Inhibition of Sexual Maturation in 45 Day Old Male and Female Rats by Various Phenolic Steroids Injected at the Age of 5 Days.

Steroid tested	Doses used, µg	No. of animals used	Effective dose, µg Males	Females
17-Estradiol	100, 500, 1000	44	500	500
Estrone	60, 100, 120, 300, 1000	80	1000	1000
2-Allyl estradiol	3000	16	3000	3000
2-Methyl estradiol	1000	16	>1000	>1000
2,4-Diallyl estradiol	3000	8	NS	>3000
6α, 7α-Dihydroxyestrone triacetate	3000	8	NS	>3000
1-Methyl-3-hydroxy-1,3,5(10),6-estratetraen-17-one	3000	8		>3000
1-Methyl-e-acetoxy-1,3,5(10),6-estratetraen-17-one	100, 1000	32		1000
17α-Methyl estradiol 3-methyl ether	0.1, 1.0, 10	48	10	0.1
17α-Methyl estradiol 3-butyl ether	0.1, 1.0, 10	23	NS	10
EE[a] 3-methyl ether 17-acetate	0.1, 1.0, 10	40	NS	1
EE 3-methyl ether 17-caproate	0.01, 0.1, 1	48	>1.0	0.1
EE 3-butyl ether	0.1, 1.0, 10	48	1.0	1
EE 3-butyl ether 17-acetate	0.1, 1.0, 10	48	10	10
3-Methoxy-17β-cyanoethoxy-1,3,5(10)-estratrien	1, 10, 30, 120	43	30	30
3-Methoxy-17β-fluorethoxy 1,3,5(10)estratriene	1, 10, 30, 120, 100, 500, 96	96	100	
3-Methoxy-1,3,5(10),9(11)-estratetraen-17-one	30, 100, 300	24	NS	300
Estra-1,3,5(10)-triene	100, 300, 500, 1000	80	>1000	>1000
6α-Methyl-1,3,5(10)-estratriene	100, 1000	8	NS	>1000

[a] EE = Ethinyl estradiol; NS = not studied. Data from Kincl *et al.* (1965) and Kincl and Dorfman (1967).

Table 4.14 Inhibition of Sexual Maturation in 45 Day Old Male Rats Produced by Various Phenolic Steroids Injected at the Age of 5 Days.

Steroid used	Total no. of rats used	Dose range studied, µg	Minimum effective dose µg	Maximum inhibition (in percent on a weight basis) Body	Testes	Ventral prostate	Seminal vesicles	Leva-tor ani
Mestranol	48	0.01–30	0.1	15	41	NS	73	54
Ethinyl estradiol	15	0.1–1	1	NS	NS	34	25	NS
Estradiol benzoate	102	30–240	30	15	70	73	77	50
Stilbestrol	16	50–200	50	NS	39	33	75	39
Estriol	37	1–1000	100	18	64	51	83	49
17-Deoxyestrone	14	100–1000	100	NS	72	50	72	20
Estradiol	56	10–500	120	26	50	45	77	40
Estradiol 3-methylether	16	100–500	500	NS	28	NS	52	NS
3-Deoxyestrone	11	100–1000	1000	18	41	38	47	25

NS indicates P > 0.01; from Kincl *et al.* (1965).

Males: Takasugi (1954a, b) used a total dose of 2.5–5 mg of estrone, divided into daily amounts of 12.5 μg, for 20 to 40 days beginning on the day of birth. Arai (1964b) employed graded doses of estrone (25–100 μg) from birth for 30 days. Both investigators found azoospermia and decreased weights of testes and accessory sex organs. Kincl *et al.* (1965a) studied the effects produced by treating 5 day old rats with graded doses of various ring A aromatic steroids; suppression of the growth of gonads and accessory sex tissues at 45 days of age was the endpoint. Mestranol was the most active compound tested; the minimum effective dose was 0.1 μg. The free compound, ethinyl estradiol, was ten times less active. The minimum effective dose of estradiol was 120 μg. Esterification (EB and estradiol 3-methyl-17-cyanoethyl bis ether) increased the activity about four fold. Table 4.14 lists the activity of other estrogens.

4.2 Effects of Estrogens During Lactation

Weichert and Kerrigan (1942) injected lactating rats on the day of parturition with estradiol (Theelin, 50–500 μg) or DES (doses 25–100 μg daily) to investigate transfer of hormones to milk. DES was more toxic than estradiol. In female pups the treatment produced (12 days later) frank estrogenic effects to the urogenital systems. The gonads were small and developed only slightly beyond the sex cord stage. The growth of the oviducts was inhibited, and the epithelium was hyperplastic. The uteri and the vaginae were greatly changed. The myometrium became thickened, the epithelium stratified, and secretory glands were absent. Vaginal epithelium became cornified and atypical; it did not show the characteristic folding, the lining was smooth and the lumen decreased. In males DES even in the lowest dose produced decreases in the diameter of seminiferous tubules and ductus deferens; the rete tubules were enlarged. The height of the cells lining the epididymis, seminal vesicles and the prostate gland was decreased. Hale (1944) gave daily injections of DES (60 mg) or estrone (0.5 mg) during the first 14 days of lactation and studied the consequence in female pups. In most animals the vagina opened early, the cycles were irregular and estrus cycles prolonged. Pathological findings included pyometra, cystic glandular hyperplasia and squamous metaplasia of the epithelium in the uterus and the oviducts. The ovaries were small and contained a few developing and large vesicular follicles; wide areas of interstitial tissue were present. Hale also noted in the adult an increased sensitivity of the vaginal epithelium to estrogen treatment. Lunaas (1963) confirmed the transfer of estradiol to milk in cattle and McKenzie *et al.* (1975) the transport of prednisolone to breast milk in humans.

The wide use of contraceptives containing steroid hormones, and their use during lactation, combined with toxic effects seen in neonatal animals, led several groups of investigators to examine the possibility of estrogen transmittal through mother's milk to the offspring.

Rudel and Kincl (1966) gave mestranol orally to lactating rats, or mice, for five days, beginning on the day of parturition and evaluated the effects in adult female

(Table 4.15) and male offspring (Table 4.16). In female rat offspring a daily dose of 10 µg given to mothers produced a statistically significant inhibition of luteinization. The dose of 100 µg resulted in a complete cessation of reproductive function in all the daughters. Mice reacted in a similar manner. A daily dose of 10 µg for five days to lactating mothers decreased significantly the incidence of ovulation in female offspring. Daily doses of 1 and 3 µg were not effective.

Male offspring were less sensitive. Only the dose of 100 µg provoked a significant decrease in the weights of testes and accessory sex organs, indicating a decline in the endogenous testosterone production. Abdel-Aziz et al. (1969) confirmed the transfer of mestranol to milk and observed similar toxic effects.

Saunders (1967) tested the activity of norethynodrel combined with mestranol, or mestranol alone, during pregnancy and for 21 days after parturition. Females born to mothers fed 100 µg/kg weight of the combination were sterile. Those fed a dose of 10 µg/kg weight (about 2 µg of mestranol) were also sterile. The ovaries were small, corpora lutea were absent, occasionally oviducts were dilated and inflamed; in some abscesses had formed. In others the parovarian space was absent. The males were apparently normal. Clancy and Edgren (1968) used ethynyl estradiol (EE) and EE in combination with norgestrel (10:1). The highest dose of EE used was 6.25 µg/day. The authors only measured organ weights and reported no differences between treated and control groups. The gonads were not examined for pathology. Holzhausen et al. (1984) injected medroxyprogesterone acetate

Table 4.15 Effect of Mestranol Given by Gavage to Lactating Rats for the First 5 Days of Lactation on the Development of 45 Day Old Female Offspring.

Daily dose µg	Body wt., g ± SE	Ovarian response[a] %		Average no. of CL ± SE	Tissue weights, mg ± SE		Vaginal introitus days ± SE
					Ovaries	Uterus	
0	143 ± 5	37/42	88	12.5 ± 0.8	43.3 ± 2.8	180 ± 10	38 ± 0.6
3	164 ± 3*	15/16	94	13.2 ± 0.9	53.9 ± 0.9	214 ± 15	37 ± 0.4
10	145 ± 3	24/35	69*	7.2 ± 0.7*	39.8 ± 2.5	164 ± 12	27 ± 2.0
30	131 ± 5	15/31	48*	5.5 ± 0.8*	35.3 ± 3.0*	150 ± 2	32 ± 2.0
100	130 ± 8	0.10	0*	0.0 ± 0.0*	12.0 ± 1.3*	81 ± 23*	15 ± 0.5*

* P < 0.05
[a] Ovarian response: number of rats with corpora lutea/total number of rats. Data from Rudel and Kincl (1966).

Table 4.16 Effect of Mestranol Given by Gavage to Lactating Rats During the First 5 Days of Lactation on the Development of 45 Day Old Male Offspring.

Daily dose µg	Total no. of rats	Body wt., g ± SE	Tissue weights, mg ± SE			
			Testes	Ventral prostate	Seminal vesicles	Levator ani
0	6	169 ± 3	1990 ± 74	72.0 ± 13.6	74.8 ± 3.1	67.4 ± 8.8
10	30	145 ± 5*	1800 ± 43	85.6 ± 4.1	68.2 ± 3.8	67.4 ± 5.2
100	11	153 ± 5	1190 ± 125*	69.8 ± 4.2*	23.5 ± 3.8*	26.0 ± 4.1*

* P < 0.05. Data from Rudel and Kincl (1966).

(5 μg/g BW) into lactating rats from the day of delivery for 20 days. In the adult female offspring the exposure caused a significant reduction in the magnitude (45%) and amount (by 27%) of the proestrus LH surge.

Other steroids given to lactating mice did not influence the reproductive capability of the offspring. Methyl testosterone given by gavage in doses of 500 and 2500 μg and norethindrone (100 μg per day) produced no effect. Injected testosterone (300 and 1500 μg), and estradiol benzoate (100 and 500 μg), were also not effective (Rudel and Kincl, 1966).

In lactating monkeys a daily dose (for 12 weeks) of 100 μg of ethynyl estradiol did not produce adverse effects on the offspring (Bernard, 1977).

4.2.1 Transfer of Steroids to Milk

An *indirect evidence* for the transfer of estrogens to mother's milk was revealed by measuring the uterotrophic effect of the milk. Laumas *et al.* (1970) found discrete estrogenic activity in human and goat milk after oral dosing with 19-nor steroids, but not after 17-acetoxy progestational agents which cannot aromatize.

Abdel-Aziz *et al.* (1969) used ^{3}H mestranol to show steroid transport to milk in rats, guinea pigs, and sheep. Rats were given five oral doses (100 μg/day) during the lactation period. The milk contained about 1% of given radioactivity recovered during the period of study. This means that each pup in a litter of 10 might ingest about 0.5 μg of mestranol, a sufficiently toxic dose. The transfer to milk in guinea pigs was lower than in rats. Only 0.05% was found after subcutaneous injection, and only 0.025% after gavage. The milk of sheep (one ewe) injected via the jugular vein contained 0.5% of MEE within the first 24 h.

4.2.1.1 Effects on Humans

Contraceptive steroids are used widely during lactation. The use of oral contraceptives (OC) leads to a decrease in the duration of lactation (Chopra, 1972; Buchanan, 1975; Nilsson *et al.*, 1986) and may result in a delay of the child's weight gain (Jimenez *et al.*, 1984) or have no effect on the growth of the children (Kamal *et al.*, 1969; Vorherr, 1973; Nilsson *et al.*, 1986). Steroids are transferred to breast milk and appear in the infants' blood. The transferred amounts are small, about 0.1% of the dose taken by the mother; this correlates with data obtained in animals (Table 4.17). Nilsson and Nygren (1979) used radioimmunoassay to analyze the concentration of ethinyl estradiol (EE), found to be 0.015 ng/ml in milk. Assuming a daily consumption of 600 ml of milk an infant would be exposed to about 10 ng of EE a day, or 0.3 μg a month. A minipill containing 30 μg of D-norgestrel could provide a daily dose of 1 μg per month.

Breast feeding babies, whose mothers were taking an oral contraceptive, developed gynecomastia (Curtis, 1964; Marriq and Oddo, 1974). Admittedly the cases are few, and the mothers were taking OC containing high amounts of estrogens. These amounts might or might not have produced adverse effects on the developing hypothalamus-pituitary-gonadal axis of the infant. A partial answer may be-

Table 4.17 Transfer of Ingested Steroids to Milk in Women.

Steroid taken	Transferred to milk, %	Plasma:milk	Concentration in milk, mg/ml	Reference
Norethynodrel	1			(1)
Norethynodrel	0.13			(2)
Ethynodiol diacetate	0.13			(2)
Lynestrenol	0.14			(3)
Mestranol	0.14			(3)
Ethynyl estradiol	0.1	100:25	0.015	(4)
D-Norgestrel		100:15	0.25–0.5	(4)
Megestrol acetate		100:80	3.2	(4)

References: (1) Laumas et al. (1967); (2) Pincus et al. (1966); (3) van der Molen et al. (1969); (4) Nilsson and Nygren (1979).

come known in a few years. Nilsson et al. (1986) follow 48 breast fed children whose mothers had used OC containing 50 μg of EE while lactating. The group found no adverse effects upon intellectual or psychological advancement or development of diseases up to the age of 8 years. At this age any adverse effects upon the development of sexual activity would not become manifested. Without definite knowledge the circumstantial evidence is sufficient, in my opinion, to contraindicate the use of oral contraceptives (and indeed of any other drug taken for convenience) in women breast feeding their children (Kincl, 1980). If oral contraceptives must be used low dose estrogen-free preparations are the agents of choice (Nilsson and Nygren, 1979; 1980). Those containing the most toxic estrogen (in this respect), mestranol, should be avoided.

4.3 Effects of Androgens and Estrogens on Reproduction in Other Species

Gonadal hormones affect reproduction in other species besides rats. Some mammalian species that are born well developed, guinea pigs and primates, did not respond to neonatal androgens or estrogens. The observations led some investigators to believe that in these species the brain is so well developed that it cannot be "damaged" by the treatment. Equally, treatment failure could have resulted from insufficient exposure time or dose of the steroid tested.

The phenomenon is not restricted to mammalian species. In one study Kincl et al. (1967) found that estrogens, injected into 1 day old Japanese quail (Coturnix coturnix japonica), inhibited egg laying in adult birds.

4.3.1 Mice

The peripheral organs (vagina and uterus) of neonatal mice are more sensitive to steroid treatment than are the organs of infant rats.

4.3.1.1 Females

Persistent changes in the vaginal epithelium (cornification or diestrus) are the most recognizable manifestations of androgenization or estrogenization. Animals injected with low doses may go through several cycles before the development of the anovulatory syndrome. The evidence is strong that neonatal steroid treatment affects vaginal epithelium directly by selecting cell lines having the potential to differentiate (cornify) without the need of estrogen stimulation (Takasugi et al., 1962; Takasugi, 1963; 1971; Takasugi and Bern, 1964; Takasugi and Kamishima, 1973).

Androgens

Female mice respond to neonatal androgens as do rats. An injection of 1 mg of TP to 5 day old pups renders all females infertile when tested at 100 days of age. The same dose given to 20 day old animals is not active, and the females are fertile (Barraclough and Leathem, 1954).

Iguchi and Takasugi (1983) found that DHT (50 μg) injected on day 5 induced permanent anovulation and estrogen independent, permanent proliferation of the vaginal epithelium; β-DHT was less active. Yanai et al. (1977) confirmed the activity of 200 μg (single dose).

Estrogens

One injection of 10 μg of estradiol dipropionate reduced the fertility of Swiss albino females when tested at the age of 70–90 days. A dose of 12 μg was not active (Merklin, 1953). Estradiol injections (5 or 20 μg for 5 days) begun within 6 days after birth were enough to induce vaginal opening by the time of the last dose. After sexual maturation, episodes of cornification were interspersed by diestrus smears lasting from 1 to 15 days. Parakeratotic and inflammatory areas were present, and the epithelium (stratum germinativum) grew into the tunica propria (Takasugi, 1963). Beginning at about 50 days of age crystalline material appeared which was later (6 to 7 months) transformed into large, white concretions (Takasugi and Bern, 1964; Kimura and Nandi, 1967; Mori, 1967; 1968a; 1969a).

The Pituitary

Five daily injections of estradiol (days 1–5) did not change prolactin secretion in aged (18–26 months) castrated mice (Nagasawa et al., 1978).

Ovaries

Iguchi (1985; Iguchi et al., 1987a, b) reported increased incidence of polyovular follicles in DES injected mice (Chapter 5).

Peripheral Organs

The epithelial lining of the vagina and cervix, derived from Müllerian ducts, is more sensitive to estrogens in early life than the urogenital sinus derived portion of the vagina (Figure 4.4). This was demonstrated by exposing vaginal grafts, obtained from females within 1 day after birth, to continuous exogenous estradiol (in

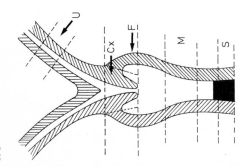

Figure 4.4. Diagram of a Frontal Section of the Mouse Reproductive Tract. The sensitivity of the epithelial lining of the neonatal mouse reproductive tract to estrogen injury depends on the tissue from which the section is derived. The vagina and the cervix (**M** , **Cx**, **F**), derived from Müllerian ducts, are more sensitive than the portion of the vagina derived from the urogenital sinus (**S**). Redrawn from Iguchi et al. (1985a) with permission.

PDS implant). The grafts developed adenosis after a 1 month exposure while adenosis occurred less frequently in tissue obtained from 7 or 10 day old donors (Iguchi et al., 1985a). The formation of vaginal adenosis is considered a premalignant stage in the development of adenocarcinoma (Takasugi, 1972; Forsberg, 1979; Forsberg and Kalland, 1981). When a tumor develops it becomes independent and can be transplanted (Jones and Pacillas-Verjan, 1979).

The Uterus: In rats estradiol (10 μg daily from birth) stimulates rapidly the growth of the tissue (Sheehan et al., 1981). Within 5 days the weight is doubled (230% increase). Increases in protein content (210%), DNA content (190%), and nuclear estrogens receptors (890%) accompany the development. The treatment does not influence the development of the endometrium and gland (Branham et al., 1985).

In mice the uterine epithelium undergoes irreversible proliferation, independently of the presence of ovaries, sometimes resulting in squamous metaplasia and cornification (Takasugi and Bern, 1964; Kimura and Nandi, 1967; Nori, 1975). Mori (1977) noted that the epithelial cells possessed a few microvilli, regardless of the absence or presence of estrogens. Stratified and squamous cells, resembling those appearing in cancerous vagina, made their appearance 13 months later more frequently than in controls (Hayashi, 1975; Takasugi, 1964; 1966; 1972; 1976; 1979; Takasugi et al., 1970; Forsberg, 1969; Jones and Bern, 1979; Bern et al., 1984).

The Vagina: Neonatal injection of 20 μg of estradiol, injected for 5 days beginning within 24 h of birth, induces persistent, estrogen independent cornification of the vaginal epithelium in the C57BL/Tw strain. The source of cornified cells B-cell nodules appearing within 3–5 days after initiating is treatment (Takasugi, 1971; 1976; Iguchi et al., 1976). Nodules comprised of polygonal cells (B cells) form within a few days within the epithelial lining. Within about 10 days the nodules form a continuous sheet culminating in an irreversible proliferating epithelium

(Takasugi and Kamishima, 1973; Mori *et al.*, 1983). Figure 4.5 illustrates typical changes.

The cornification of the vaginal epithelium is independent of endogenous estrogen stimulation (Kimura *et al.*, 1967). Iguchi (1984) exposed *in vitro* vaginal tissue obtained within 12 h of life to various steroids (testosterone, dihydrotestosterone, estradiol, or diethylstilbestrol) for 72 h and then transplanted the explant under the kidney capsule of 2 month old females. The recipients were oophorectomized 1 month after the operation. The transplants were harvested for histological evaluation 30 days later. In controls vaginal epithelium was atrophic with some isolated 1–2 cell layers of cornified cells. Epithelium obtained after exposing the tissue to estradiol, or DES, consisted of 4–10 cell layers frequently undergoing either parakeratosis or cornification. DHT induced cell proliferation and occasionally parakeratosis but not cornification. The incidence of epithelial proliferation was much lower in tissue exposed to testosterone prior to transplantation.

Vaginal cancer in 10 month old C57 Black/Tw strain mice as a result of 5 daily injections of 20 µg of estradiol during the first five days of life was reported by Mori *et al.* (1983). The adult vaginal epithelium exhibits increased sensitivity to estrogens (see Section 4.4).

4.3.1.2 Males

Androgens

Barraclough and Leathem (1959) injected 1 mg of TP to mice aged 5, 10 and 20 days. The group found suppression of testis growth but normal histological appearance and well developed spermatozoa in 40–60 day old animals.

Estrogens

Takewaki and Kawashima (1967) used repeated injections of estrone (10–30 days) from the day of birth; the treatment produced a long-lasting suppression of spermatogenesis. Takasugi (1970) and Ohta and Takasugi (1974) contributed detailed description of testicular damage.

The damage caused by estrogens is prevented by concomitant androgens. Takasugi and Furukawa (1972) gave daily injections of estradiol (20 µg) for the first 10 days of life and 40 µg for the next 5 days and testosterone propionate (4–40 µg)

Figure 4.5. Changes in the Vaginal Epithelium of Adult BALB/c Mice Exposed to Estradiol at Birth. **A** Extensive epithelial downgrowth of vagina in a 10 month old mouse treated with 5 µg of estradiol for 5 days from the day of birth; × 140. **B** Hyperplastic lesions in vaginal fornix of a 12 month old mouse with 5 µg of estradiol for 5 days from the day of birth; × 140. **C** Epithelial downgrowth of vagina in 10 month old mouse treated with 5 µg of estradiol for 5 days from the day of birth; cyst bearing pearls, masses of epithelial cells, cellular debris and accumulation of leukocytes and erythrocytes are visible; × 140. **D** Epithelial downgrowth of vagina in a 15 month old mouse treated with 20 µg of estradiol for 10 days from the day of birth; the mouse was ovariectomized when 40 days old; × 140. **E** and **F** Adenosis in hyperplastic lesions of cervix in a 20 month old mouse treated with 20 µg of estradiol for 5 days from the day of birth. Courtesy T Mori (Tokyo, Japan).

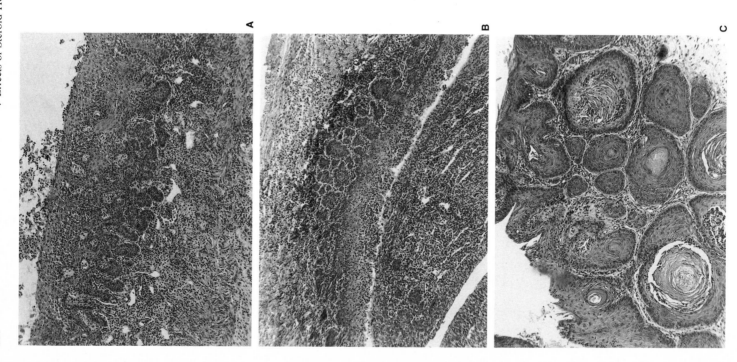

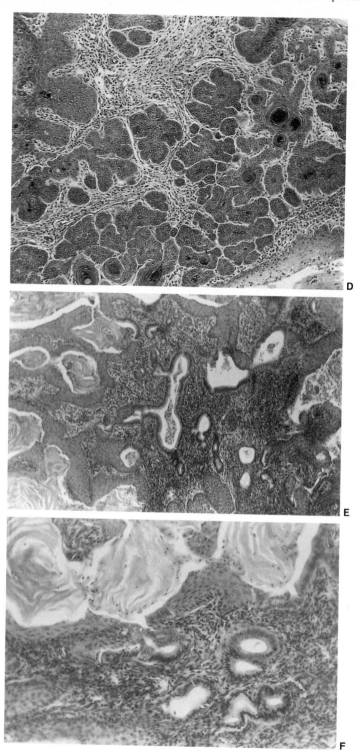

for the first 10 days and twice the dose for the following 5 days. Unopposed estradiol produced inhibition of spermatogenesis and atrophy of seminal vesicles and
the prostate (60 day old animals). In males given TP in addition to estradiol spermatogenesis was normal, yet the weights of the gonads and of the sex tissues were
decreased. Since the period of observation was short, it is not possible to conclude
whether the protective action of testosterone was permanent or only transitory.

4.3.2 Guinea Pigs

Dantchakoff (1938a, b, c) used a combined pre- and postnatal exposure to androgens, or estrogens, and found reproductive and behavioral disturbances in
both sexes. In newly born male guinea pigs daily injection of TP (2 mg) for 3 days
caused in 30 day old animals "... d'une puberté précoce ...": accessory sex tissue
(seminal vesicles, prostate and Cowper's glands) were enlarged, secretory activity
was apparent, and spermatogenesis was active.

Kincl and Folch Pi (1963) injected 5 mg of TP into female offspring, and
120 µg of EB in male pups, within 2 h after birth and into 5 day old pups and
evaluated fertility (mating), the status of the gonads and effects on accessory sex
organ weights in 100–150 day old animals. Injection after birth produced sterile
females. The ovaries were atrophied and not luteinized. When the injection was
given on day 5 of life one-half of females were fertile, and their ovaries were fully
developed. Likewise only the males injected immediately after birth were sterile,
their testes and accessory sex organs were atrophied, and spermatogenesis was abnormal (Figure 4.6). Males injected at the age of 5 days developed normally.

Carlevero et al. (1969) used 10 mg of TP within 24 h of birth and found no effect on females. D'Albora et al. (1974) implanted a pellet of testosterone enanthate
(10 mg) one day after birth and observed precocious vaginal opening but no effect
on ovulation. Brown-Grant and Sherwood (1971) injected 5 mg of TP on day one
and observed a similar effect on vaginal introitus and some influence on fertility:
only 4 or 7 animals mated and became pregnant.

TP treatment of pregnant guinea pigs masculinized female offspring. As adults
the animals did not mate and did not ovulate (Chapter 3).

Schulz et al. (1972a, b) studied the influence of estradiol (0.1 or 10 µg/100 g of
body weight) on the functions of the central nervous system, liver, adrenals, ovaries and the uterus by measuring the RNA and protein synthesis in these organs.
According to this group only the uterus and the liver are sensitive to neonatal
damage.

4.3.3 Hamsters

Hamsters react similarly to rats and mice. Females injected with 300 µg of TP or
150 µg of EB on the second day after birth exhibit persistent cornification of the
vaginal epithelium, fail to cycle and to form corpora lutea. The adrenals are hypertrophic (Swanson, 1966; Schwartz and Gerall, 1979).

In males TP treatment results in moderate decrease of the weight of testes and

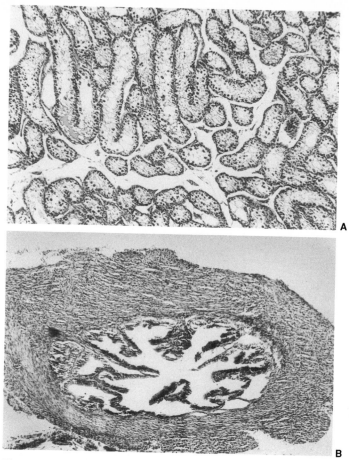

Figure 4.6. Testis and Prostate of an Adult Guinea Pig Injected with Estradiol Benzoate when Born. **A** Testis of 60 day old guinea pig injected with 120 µg of estradiol benzoate on the day of birth; the diameter of the tubules is decreased, with almost a complete absence of the germinal epithelium; ×25. **B** Decrease of the epithelial height in the prostate gland (ventral lobe) of the same guinea pig; ×25.

azoospermia in some, but not all, animals (Whalen and Edgren, 1978). The effects of estradiol benzoate are more intense. Gonads and seminal vesicles atrophy and body weights sink (Swanson, 1966).

4.3.4 Rabbits

Campbell (1964) injected 1 mg of TP (or 200 µg of EB) on the second or third day of life and noted no adverse influence on either the females or the males. Anderson *et al.* (1970) reported that androgen treatment during gestation inhibited female behavior (nest building) in the offspring (Chapter 6).

4.3.5 Ferrets

Neonatal testosterone given in a released form (PDS capsule) induces masculine behavior (neck grip, mount and pelvic trusts) in females. Dihydrotestosterone given in a similar form was not effective (Baum *et al.*, 1982). The diffusion rate through polydimethylsiloxane membranes varies from steroid to steroid; the authors used implants of the same size and thus did not provide the same dose and may, or may not, have detected differences between the two hormones.

4.3.6 Sheep

Short (1974) and others described masculinization of ewes as a result of *in utero* masculinization (Chapter 3). There are no studies about the effect of steroid hormones during the neonatal period.

4.3.7 Pigs

Kincl (unpublished results) injected male piglets with 1 mg/kg body weight of EB within 24 h of birth. At the age of 5 months testis and accessory sex tissues were atrophic, and the characteristic odor of male pigs was absent. Dörner *et al.* (1977) found that males exposed prenatally to EB exhibited decreased sexual behavior as adults. Elsaesser and Parivizi (1979) found decreased ovulation in gilts born to mothers injected with TP (Chapter 3).

4.3.8 Dogs

Gustafsson and Beling (1969) injected beagle pups with EB (200 µg/kg body weight) 5 times weekly for the first 6 weeks of life. The treatment caused skeletal malformations: femoral heads were smaller, leading to point laxity. No information was included on the endocrine status of the animals. Zimbelman and Lauderdale (1973) used TP (25 or 50 mg) in females within 2 days of birth and found no influence on the occurrence of the first estrus, length of later cycles, and fertility. The clitoris was large, and os clitoridis was present.

4.3.9 Bovine

Neonatal treatment of heifers with TP (50 or 100 mg) produced in some animals congenital abnormalities of the reproductive tract (absence of the uterus). The animals did not cycle during the period of observation which was 18 months (Zimbelman and Lauderdale, 1973).

4.3.10 Primates

It has been occasionally argued that rodents are born with an undeveloped brain and endocrine systems vulnerable to steroid manipulation while this is not the case in primates. The argument is based on insufficient evidence. The similarity of dysfunctions caused *in utero* by teratological agents, including steroids, would indicate otherwise (Chapter 3).

Treloar *et al.* (1972) attempted to "masculinize" female rhesus babies by injecting testosterone (35 mg/kg body weight) within 24 h after birth. The treatment had no effects on the menstrual cycle, gross ovarian morphology, or plasma progesterone concentrations. The authors elected to use non-esterified testosterone suspended in an aqueous medium and studied only one dose. In view of design limitations the results may, or may not, indicate that the rhesus monkey is insensitive to "neonatal" damage.

On rare occasions human babies were exposed to contraceptive steroids transmitted in the milk during breast feeding. In a few estrogenic effects (gynecomastia) became apparent (Section 4.2.1.1). Reports of other detrimental effects, if any, are not known.

Chapter 3 covers transplacental toxicity in children born to mothers treated for threatened abortion with estrogens (DES).

4.4 Tissue Responses

Pharmacological doses of estradiol and testosterone during neonatal period (and *in utero*) influence in the adult the sensitivity of target tissues, hormone biosynthesis and clearance rates, and the metabolism and biological half-life of the same hormones. Changes in responsiveness of various organs depend upon the sex and the agent used. In the female early treatment with estrogens (but not androgens) predisposes uterine and vaginal epithelium to estrogens. Accessory sex organs of androgenized males are more sensitive to androgens while estrogenized males respond only poorly.

4.4.1 Hypothalamus-Pituitary Axis

4.4.1.1 Androgenized Animals

Females
Several groups (Petrusz and Nagy, 1967; Napoli and Gerall, 1970; Dunlap *et al.,* 1972) measured ovarian compensatory hypertrophy after unilateral castration, noted a decreased response, and speculated that the hypothalamus of androgenized animals exhibited diminished sensitivity to estrogens. The response to GnRH stimulation is reduced (Turgeon and Barraclough, 1974). Clark and Zarrow (1968) reported decreased ovarian depletion of cholesterol in response to LH stimulation.

The pituitaries of rats which develop the delayed AN syndrome respond at the time of sexual maturation (day 30) to stimulation with estradiol (EB), and LH is released. Females sterilized with a high dose of TP (30 µg or more) do not respond (Harlan and Gorski, 1977).

Prolactin

Pituitary response in respect to prolactin release is normal; Johnson (1979) reported release of prolactin following the treatment with pregnant mares' serum gonadotropin (PMSG). Pantic and Genbacev (1972) speculated that in males altered prolactin release may account for the sensitivity to estrogens.

4.4.2 Restoration of Gonadal Function

Ovulation is induced in PE rats by electrical stimulation of specific areas in the hypothalamus, by providing exogenous luteinizing hormones, by steroid hormone treatment, by pinealectomy, by superior cervical sympathetic ganglionectomy (Ruzsas et al., 1977) or by intraventricular infusion of norepinephrine (Tima and Flerkó, 1974; 1975).

4.4.2.1 Androgenized Animals

Females

Electrical Stimulation: Placing the electrodes either in the medial preoptic area, the arcuate nucleus, or the median eminence induces ovulation. Barraclough and Gorski (1961) induced ovulation in TP females sterilized with a "low" dose by a monophasic, rectangular pulse delivered through bilateral concentric bipolar electrodes delivered to the arcuate-ventromedial region of the hypothalamus. Animals rendered anovulatory with a "high" dose did not respond unless they were pretreated with progesterone (Illei-Donhoffer et al., 1970). Terasawa et al. (1969) demonstrated that in androgenized rats the threshold current needed to induce ovulation is lower (100 µA) than in old, spontaneous PE rats (400 µA). Everett et al. (1970) have shown that PE rats have higher refractiveness to electrical stimulation than untreated animals.

The release of LH after electro-stimulation of the forebrain is similar in normal and androgenized rats (Kuvo et al., 1975).

Exogenous Gonadotropins: The pituitaries of androgenized rats hold sufficient LH to induce ovulation. Autologous pituitary extracts injected into the donors will induce ovulation in ⅔ of the animals (Tima and Flerkó, 1967; 1968). Treatment with synthetic GnRH (Borvendeg et al., 1972; Mennin et al., 1974; Uilenbroek and Gribling-Hegge, 1977) or brain (sheep) extracts (Johnson, 1963) will also induce ovulation. Ovaries of androgenized rats also respond to a treatment with PMSG (to induce adequate follicle formation) followed by hCG stimulation to bring about ovulation (Brown-Grant et al., 1964; von Fels et al., 1972), PMSG alone (Uilenbroek and van der Werff ten Bosch, 1972), hCG alone (Duluc and Mayer,

1967), or LH alone (Barraclough and Fajer, 1968). The corpora lutea secrete progesterone (Barraclough and Fajer, 1968). Barraclough (1968) induced ovulation with hCG (plus reserpine to stimulate PRL secretion) and measured the secretion rate in an effluent ovarian vein. The rate was 15 μg/hr/left ovary, a rise from 0.3 μg/hr/left ovary prior to stimulation. Ova were present in the oviduct.

The corpora lutea which are formed have the ability to incorporate acetate into progesterone and 20-OHP (20α-hydroxy-4-pregnen-3-one). Both hormones are undetectable in androgenized animals. Cortes *et al.* (1971) detected 620 ng/ml of progesterone in ovarian blood 10 h after the LH pulse; 20-OHP rose to 260 ng/ml. The surge resembled increases seen in untreated rats during spontaneous ovulation. The concentration of progesterone, 20-OHP, and prolactin are comparable to levels seen in pregnant rats which is suggestive that the CL are "... functional and maintained ..." (Johnson, 1979).

The competence of the ovary to ovulate does not assure successful gestation. The eggs may not be physiologically "normal" possibly due to a delayed toxic effect of TP. Kramen and Johnson (1971) induced ovulation in androgenized rats with LH (10 μg), allowed the females to mate, collected the fertilized ova and transferred these to pseudopregnant recipients. When the dose was 50 μg of TP the fetuses (day 18 of gestation) developed normally. The percentage of surviving embryos significantly decreased when androgenization was induced by 250 μg of TP albeit the eggs appeared morphologically normal (Table 4.18).

Parabiotic Union: Another method of providing an exogenous source of gonadotropins is to join two individuals in a parabiotic union. Several groups have used this technique, found that PE females responded with the formation of corpora lutea but judged that the sensitivity of the ovaries was lower in androgenized females than in controls (Moguilevsky *et al.,* 1967; von Fels *et al.,* 1968; Kurcz and Gerhardt, 1968). Mennin *et al.* (1974) concluded from parabiosis experiments that ovarian sensitivity of normal, pentobarbital blocked females to LH is 5–10 times greater than in androgenized rats.

Hormonal Feedback: The effect of androgenization on hormonal feedback was evaluated in castrated individuals or by modifying the endogenous hormonal milieu by exogenous steroids. In castrated females estradiol will, depending upon conditions, either suppress or stimulate the release of LH (Chapter 2). In andro-

Table 4.18 Mating Response and Egg Viability of Androgenized Rats After LH Treatment.

TP dose, mg	Mating response, %		Ova viability, percent
	before LH	after LH	
0	84	NS	50
50	61	52	53
250	36	35	24
1250	11	34*	NS

* Statistically significant; NS, not studied. From Kramer and Johnson (1971).

genized females castration alone will result in the release of both pituitary hormones (Flerkó and Bardos, 1961; Swanson and van der Werff ten Bosch, 1964a;
van Rees and Gans, 1966; Schiavi, 1969; Turgeon and Barraclough, 1974). Exogenous estrogens fail to induce LH release (Hayashi, 1967b; Petrusz and Flerkó,
1968; Barraclough and Haller, 1970; Gerall and Kenney, 1970; Mennin and Gorski, 1975). The lack of pituitary response to exogenous estradiol suggests that the
cause of sterility in the rat may be due, among other factors, to the loss of sensitivity to gonadal hormones on LH secretion (Harlan and Gorski, 1978; Gogan
et al., 1980). The loss could be mediated via hypothalamic centers or be caused
by "... altered secretion, blood transport or metabolism of endogenous hormones ..." (Joseph and Kincl, 1974; Chiappa and Fink, 1977).

Takasugi (1954a; 1956) gained some success by providing implants of progesterone (8–10 mg), by daily progesterone injections (5 mg), or by implants of cortexone acetate (10 mg) into rats made PE by 5 doses of estrone. Luteinization
was induced in 3/6, and 2/5 adult females, respectively. Mennin and Gorski (1975) induced ovulation with progesterone alone.

Deficiency in the secretion of gonadal hormones is not the only contributing
influence. Neonatal androgenization (or estrogenization) induces both thyroid and
adrenal hypofunction (Chapter 5). In 1952 Eayrs concluded "... absence of thyroid hormone in the developing organism has significant consequences for physiological and behavioral functions than does the absence of androgens ..." (see also
Eayrs, 1964). Takasugi (vide supra) induced corpora lutea formation by prolonged
treatment with cortexone acetate. Machida (1971) reported that constant estrus
rats (induced by the destruction of the medial preoptic area by electrolytic lesions)
formed corpora lutea if exposed to stress. Fujii (1973) induced corpora lutea formation in TP rats (1 mg dose) by thyroidectomy performed at the age of 10 to
12 weeks. Irregular cycles continued for as long as one month.

Transplantation:

The damage to germinal epithelium was repaired by transplanting portions of
gonadal tissue under the skin, under the kidney or spleen capsules, or into variously prepared recipients. When ovaries are homotransplanted under the kidney
capsule the concentrations of LH and FSH remain within normal limits, and more
estrogens circulate in the blood. Transplantation under the spleen capsule results
in increased LH release, decreased FSH secretion and no change in estradiol concentration in the blood (Uilenbroek et al., 1978). The results suggest that transplants in the spleen (in this case the blood drains first through the liver) produce
more inhibin-like activity than those placed under the kidney capsule.

Interpretation of some results is difficult; the number of animals used was
small and statistical analysis absent. The production of gonadal hormones is not
the only parameter which influences graft acceptance:
(i)   The degree of stress to which the recipients are exposed will influence graft
      acceptance. Takasugi reported that changes in electrolyte balance (feeding of
      sodium rich diet) or an excessive continuous supply (implanted tablets) of
      cortexone or cortisone acetate inhibited luteinization (Takasugi, 1959).
(ii) The age of the donors is of importance. Ovaries transplanted from young un-

treated (3 month old) mice will maintain cyclic estrus in 71% of castrated recipients. When the donors are 12 months old only 15% of recipients will respond, and none will if the donors are 20 months old (Mori, 1979).

(iii) The response also depends whether androgens, or estrogens, were used to produce sterility. Luteinization takes place readily in androgenized females when the tissue is autotransplanted under the spleen capsule; autotransplanted ovaries in estrogenized rats will not form CL (Takasugi 1953; 1954b).

(iv) Excessive estrogens suppress luteinization. Hayashi transplanted ovaries under the spleen capsule of PE and control rats. Daily injections (for 40 days) of estrone (2 µg), or no injections, allowed CL to develop; when the amount of injected estrone was increased to 8 µg luteal tissue in the graft was reduced, while the number of follicles increased.

Autotransplantation: Transplantation of the tissue under the skin in the ear affords a visible evidence of the condition of the graft. Transillumination shows the gross morphological condition, i. e. the presence of Graafian follicles, cystic follicles, or corpora lutea. Bradbury (1941) used this method in females androgenized by daily injections of 250 µg of TP from day 5 to day 23 of life. Ovariectomy was done on day 32. In the control group CL formed in 3/4 recipients whereas in the androgenized females CL were present in 2/4 grafts. Kikuyama and Kawashima (1966) androgenized pups by ten injections of TP (10 µg dose) given for the first 10 days of life. The females were ovariectomized when 40 days old; ovaries from control donors were transplanted under the skin and left *in situ* for 40 days. Luteinized tissue formed in 1/5 androgenized and in 5/5 control recipients.

Cystic ovaries obtained from PE rats readily luteinize when transplanted under the spleen capsule since the tissue is exposed to a different hormonal milieu. The main venous drainage from the spleen is through the portal vein leading into the liver. Hormones (androgens and estrogens) produced by the graft transport first through the liver where the bulk is metabolized during the first pass. This results in the inactivation of estrogens, atrophy of the uterus and vagina and the development of castration cells in the pituitary (Biskind and Biskind, 1949; Achilles and Sturgis, 1951; Flerkó and Illei, 1957).

Kovacz (1966) used 5–10 animals per group. A single injection of 2.5 mg of TP on day 1 or 2 of life produced androgenization. Corpora lutea formed in one androgenized ovary provided the second was removed. If the second ovary was left in place, CL were not formed. Corpora lutea also developed in androgenized tissue homotransplanted under the spleen capsule of castrated animals. It did not matter whether the recipient was male or female, or whether it was androgenized or not. The author did not state when the recipients were castrated. Kincl and Maqueo (1967) androgenized females by injecting 250 µg of TP on day 5, performed transplantation on day 45, and left the tissue *in situ* for 30 days. Luteinization took place in 3/8 autotransplanted tissue when one ovary was left intact; the same rate occurred in bilaterally castrated animals. Others (Roy *et al.*, 1964; Zeilmaker, 1964; Hayashi, 1967a, Ladosky *et al.*, 1969) also obtained CL in autotransplanted ovaries in PE rats.

Prolonged transplantation (6–8 months) may lead to the formation of luteomas and granulosa cell tumors (Kovacs, 1965).

Heterotransplantation: The ideas of Pfeiffer and Harris (Chapter 2) influenced for a long time the view of the differences between sexes. Both believed that the hypothalamus-pituitary axis is either masculine or feminine and that the differentiation takes place before birth. The critical evidence cited for the argument was:

(i)   Ovarian tissue transplanted into recipients will become luteinized only if the recipients are females, or males castrated shortly after birth;
(ii) If the recipients are adult males castrated before transplantation, or females androgenized by neonatal TP, luteinization will not take place.

The results of Kawashima (1960), Tramezzani and Poumeau (1967), Hart (1968), and Wagner (1968) support the conclusion. In contrast Oberling *et al.* (1936) obtained only partial success under similar conditions. He had seen luteinization in only 3 of 7 neonatally castrated male recipients. Turner (1938) found no obvious difference of maintaining testicular grafts placed into the ocular cavity regardless of whether males or females were the recipients. Other groups reported the formation of corpora lutea in grafts transplanted into intact, adult males (Sand, 1918; Wang *et al.*, 1925; Hill, 1937; Deanesly, 1938). Harris and Campbell (1966) observed the formation of CL in male rats castrated as adults and stated that since these were few in number and small, such tissue "... almost certainly represents granulosa or thecal luteinization and does not represent a cyclic release of luteinizing hormone, with consequent ovulation and luteinization ..." Swanson (1970) found 6/8 luteinized ovaries in castrated adult male hamsters and 7/8 in males castrated as infants. She believed that the CL indicated ovulation "... since the luteinization was rather diffused and more suggestive of theca luteinization ..."

Kikuyama and Kawashima (1966) obtained cyclic ovarian function in androgenized rats (10 µg TP per g of BW injected either on day 7, 10 or 13) provided the females were ovariectomized at the age of 40 days and used as recipients at the age of 140 days. When the androgenized females were left intact until the age of 140 days, at which time they were ovariectomized and the ovaries were grafted, only 1/5 formed CL.

Henzl *et al.* (1971) studied the response to transplantation under the kidney capsule of male rats and hamsters castrated on day 1, 21, or 50 after birth. Grafting was done when the animals were 50 days old (80 days for those castrated at this age). To judge the viability of the CL formed the authors have included measurement of the progesterone concentration in peripheral plasma. The group had seen luteinization in grafts transplanted into males, regardless of when they were castrated. Formation of CL was highest in groups castrated neonatally, or as adults, provided a 30 day period was allowed between castration and transplantation. The concentration of progesterone in peripheral plasma showed good correlation with the number of CL formed, indicating that the tissue was functional (Table 4.19). Neonatal estradiol (200 µg on day 5) decreased significantly graft acceptance in adult males. Only 30% were viable, but graft luteinization was high in the surviving tissue. A dose of 100 µg had no effect on graft maintenance. The authors concluded "... the failure (of grafts) in males castrated as adults and grafted

Table 4.19 Corpora Lutea Formation and Progesterone Plasma Levels in Male Rats Castrated at Different Ages (ovarian tissue grafted under the kidney capsule).

Age (days at which castrated)	Transplanted	No. of CL	Progesterone mg/ml ± SE	Graft acceptance %
1	50	12.6	14.1 ± 2.5	90
21	50	3.5	5.0 ± 0.9	77
50	50	1.7	3.2 ± 0.7	100
50	80	3.3	6.5 ± 1.0	100
1 (EB, 100 μg)*	50	2.7	NS	100
50 (EB, 200 μg)*	80	3.3	NS	90
1 (EB, 200 μg)*	50	1.0	ND	29
60 (EB, 200 μg)*[a]	90	8.0	NS	29

* EB injected on day 5 of life; [a] female rats were the recipients; NS = not studied; ND = non-detectable; data from Henzl et al. (1971).

at the time of castration may be due to the presence of residual androgen steroids, different patterns of gonadotrophin secretion and/or release, and also possibly immunological incompetence . . ."

Van der Schoot and Zeilmaker (1972; 1973) observed cyclicity of ovarian and vaginal grafts in neonatally castrated males in one strain whereas in another strain, persistent corpora lutea had formed, and the vaginal epithelium exhibited prolonged (23–32 days) diestrus. Even in neonatally castrated males the nature of the grafts changed after a limited number of cycles (9–22). Transplanted vaginal epithelium became permanently cornified, new corpora lutea did not form, and the ovarian tissue assumed the aspect of ". . . large, vesicular follicles . . ." Tissue grafted into female recipients of either strain exhibited cyclic behavior during the whole period of observation.

Vreeburg et al. (1977) used a PDS implant filled with an inhibitor of aromatization (1,4,6-androstatriene-3,17-dione) in neonatal males and castrated the animals at the age of 55 days. The treatment did not interfere with sex function (the males behaved as males), yet even so in 5/9 animals corpora lutea developed in transplanted ovaries. Quinn (1966) obtained luteinization in intact, adult male recipients provided the preoptic area was stimulated with a low intensity DC current (10 μA/20 sec).

4.4.2.2 Estrogenized Animals

Gonads from estrogenized animals of either sex respond poorly to attempts to restore reproductive function. This indicates a direct deleterious effect such as seen in the peripheral sex organs (Section 4.4.3).

Females
Arai and Masuda (1970a) induced luteinization by electrical stimulation of various centers in the hypothalamus in "low dose" estrogenized rats. Females sterilized with repeated doses of estrone did not respond to the treatment. Duluc and Mayer (1967) induced luteinization in estrogenized PE rats by exogenous gonadotropins.

Transplantation: Takasugi (1954a) found different responses between tissue obtained from PD and PE rats grafted into the spleen 80–110 days after the last injection. In PD females only 1/11 grafts was active while in PE females 5/10 grafts were active. In the experiment reported by Kincl and Maqueo (1967) only 1/8 splenic autograft in females injected with 250 µg of EB on day 5 showed signs of luteinization. Hayashi (1967a) and Hayashi and Kawashima (1971) confirmed the poor ability of "estrogenized" ovaries to form CL when autotransplanted under the spleen capsule. Ovaries obtained from estrone injected (32 or 128 µg daily) females formed CL more readily, possibly due to a different feedback to the hypothalamus (Hayashi, 1969).

In mice ovaries transplanted from estrogenized females (20 µg of estradiol daily for three days after birth) will maintain estrus cyclicity in about 50% of castrates which accepted the graft (Mori, 1979).

Males

Males sterilized by neonatal estradiol benzoate responded to prolonged PMSG (injected every third day), or testosterone (injected daily), if the treatment began on day 21 of life. Figure 4.7 illustrates testis morphology of an animal injected neonatally with EB and then treated with PMSG. After the treatment the sperma-

Table 4.20 Restoration of Gondal Function in 50 Day Old Male Rats Treated Neonatally with Estradiol Benzoate.

Treatment		Body weight g±SE	Organ weights, mg±SE				Spermato-genetic index (range)
Gonado-tropin	EB, µg		Testes	Ventral prostate	Seminal vesicles	Levator ani	
0	0	226± 2	2738± 42	187 ±13	172 ±10	174 ± 5	94.5 (76–100)
0	120	192± 1	575±112	48.4± 7.3	27.8±26	72.9± 5.9	49.3 (5.5–94.5)
0	240	195± 2	1323± 77	86.8± 5.2	28.0± 3.4	68.0± 6.6	2.8 (0–17)
PMSG, 30 IU[a]	120	222±14	1619± 53	464 ±21	224 ±15	217 ±12	92.5 (69–100)
TP, 1 mg[b]	120	234± 6	1372± 48	512 ±18	288 ±12	223 ± 6	94.9 (89–100)
PMSG, 10 IU[a]	240	225± 6	1730± 68	378 ±52	291 ±42	202 ± 9	80.3 (62–94)
PMSG, 30 IU[a]	240	246± 2	1673± 56	452 ±32	278 ±10	257 ± 7	93.5 (92–94)
TP, 1 mg[b]	240	224± 5	1434± 45	510 ±26	320 ± 8	244 ± 7	83.7 (66–96)

[a] injected every 3rd day beginning day 21 of life; [b] daily injections beginning on day 21 of life; spermatogenetic index = percent tubules showing normal maturation wave; data from Kincl *et al.* (1964).

Figure 4.7. Partial Restoration of Gonadal Function in Adult Rat Injected with Estradiol Benzoate at the Age of Five Days and PMSG at Puberty. Testis of 70 day old rat injected with 120 µg of estradiol benzoate when 5 days old; lack of spermatozoa in all tubuli (**A** ×64) with marked damage to germinal epithelium and decreased number of Leydig cells. **B** Testis of 70 day old rat injected with 120 µg of estradiol benzoate when 5 days old and PMSG, 30 IU every 3rd day beginning at the age of 21 days; 31% of tubules without spermatozoa; ×164; **C** Same tissue ×260.

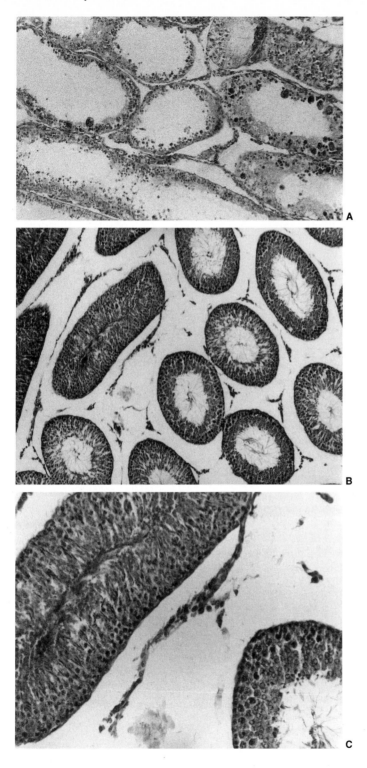

togenic index was 69 in the group exposed to 120 µg and only 33 in those medicated with 240 µg (Table 4.20). Joining the males in a parabiotic union leads also to a restoration of spermatogenesis albeit partially (Kincl et al., 1964). Takasugi and Mitsuhashi (1972) made similar observations in mice.

4.4.3 Peripheral Organs – The Carcinogenic Effect

Peripheral organs exposed neonatally to gonadal hormones become sensitized in the adult to the same hormones. The observation suggests a direct effect on the tissue not mediated through the hypothalamus-pituitary axis.

4.4.3.1 Androgenized Animals

Females
The traumatized uterus of androgenized rats is less sensitive to estradiol and progesterone stimulation (deciduomata formation) than the tissue of untreated castrated females (Zeilmaker, 1964). In females exposed to 10–30 days of TP (0.5 mg daily for 10 days, 1 mg from day 11 to 20, and 2 mg from day 21 to 30) the uterine stroma did not respond to daily injections of estradiol (Takewaki, 1968). Van Rees and Gans (1966) confirm decreased sensitivity of the uterus to estrogens in androgenized rats.

Males
Neonatal androgenization increases the sensitivity of accessory sex organs (the prostate and seminal vesicles) to androgenic stimulation in adult rats (Morrison and Johnson, 1966; Chung and Ferland-Raymond, 1975) and mice (Bronson and Desjardins, 1969). Dixit and Niemi (1973) androgenized rats with 2.5 mg TP at birth, castrated then when 80 days old and stimulated the castrates with 1 mg of TP daily for 10 days. The treatment produced an eight fold increase in the weight, and a threefold increase in RNA and DNA content, in the perineal complex (penis and levator ani muscle). Bronson et al. (1972) found that in androgenized mice the rate of RNA synthesis increased in these organs.

The increased sensitivity may be independent of the presence of testes. Morrison and Johnson (1966) castrated male rats on the day of birth and then gave a single injection of TP (2.5 mg) 4 days later. The response to injected testosterone propionate (50 µg) was tested when the animals were 30 days old. Androgenization produced a somewhat greater growth of the prostate and seminal vesicles (80.0 ± 4.0 and 27.9 ± 1.4) than in controls (72.7 ± 5.4 and 25.2 ± 1.4, respectively).

Neonatal exposure to androgens does not change testosterone plasma levels (see Table 4.11) (Frick et al., 1969; Moguilevski et al., 1977). The increased sensitivity is a reflection of the increased biological half-life of testosterone (Section 4.4.1.5).

4.4.3.2 Estrogenized Animals

The vaginal and uterine epithelium of mice is more sensitive to estrogen treatment than that of rats. During the neonatal period a specific cell line in the endometrium becomes permanently influenced. The cell population slowly replaces the normal epithelium and later gives rise to dyscrasia associated with the development of vaginal cancer. The tumors that develop in response to exposure of neonates can be transplanted (Jones and Pacillas-Verjan, 1979). DES is more toxic in this respect (Chapter 5).

Uterus and Vagina

In estrogenized rats squamous uterine metaplasia is estrogen dependent. Metaplasia develops in ovariectomized females if estrogen therapy begins on the day of castration and persists as long as the exogenous estrogens are supplied (Hayashi, 1975). In PE mice the epithelium remains cornified; later administration of estrogens causes a reduction in the number of mitoses (Takasugi, 1976), columnar cells (Mori, 1976) and nuclear estrogen receptors (Shyamala *et al.*, 1974). The hyperplastic epithelium persists even when the ovaries are removed (Kimura and Takasugi, 1964; Mori and Nishizuka, 1978; 1982).

The development of the PE syndrome can be blocked by concomitant androgen therapy. Arai (1970) inhibited estrogen induced uterine metaplasia by injecting from birth 50 µg of TP for 15 days followed by 100 µg for an additional 15 days.

The Carcinogenic Effect

Epithelial hyperplastic changes induced in BALB/cfC3H mice by estradiol (5 µg or 20 µg daily for the first five days of life) become progressively more severe, and eventually vaginal and cervical cancer develops (Bern *et al.*, 1975) especially after exposure to exogenous estrogens (Kawashima *et al.*, 1980a). The development of basal cell carcinoma is independent of ovarian function; cancer will form even in animals ovariectomized a few days following neonatal estrogens exposure. The transformation is prevented if the ovaries are removed at the age of 30 days but not if left *in situ* for 270 days (Iguchi *et al.*, 1985b).

Bern *et al.* (1975) reported that progesterone alone (100 µg daily for the first five days of life) increased tumor incidence in mice (BALB/cfC3H/Crgl). The strain develops vaginal tumors spontaneously.

4.4.4 Uptake in Various Tissues

Hormones must be present in sufficient concentration in the target tissues to express their physiological effects. The biosynthesis and production rates, metabolic patterns, biological half-lives and clearance rates all contribute to the amounts that reach the target. Alteration in the blood/brain barrier may influence the binding of steroid hormones in the central nervous system.

The neuroendocrine tissue of neonatal rodents does bind androgens and estrogens, but not preferentially. The uptake of testosterone and estradiol in the

Table 4.21 Uptake of Radioactivity in Various Organs of 5 Day Old Male (^{14}C-Estradiol) and Female (^{14}C-testosterone propionate) Pups 6 hours After the Injection (data are expressed in percent radioactivity found).

Tissue studied	Males	Females
Pituitary	0.5	0.2
Cerebrum	2.0	1.8
Hypothalamus	1.1	1.9
Gonads	80.0	66.7
Adrenals	16.4	29.5

hypothalamus and the pituitary of rats (2 days before birth, at birth, and 2 days old) is sex independent, not different from the concentration found in the cortex (Tuohimaa and Johansson, 1971). The uptake of either testosterone or estradiol (after 5, 15, and 60 min) in the hypothalamus, or the pituitary gland, of 2 to 10 day old female rats was about the same as in the cerebral cortex and spleen. Estradiol uptake was higher in the uterus, and both hormones were found in high concentrations in the liver (Alvarez and Ramirez, 1970). Kincl (1970) confirmed that hormone receptors are poor in the CNS at an early age. Radioactivity derived from ^{14}C-testosterone propionate (females), and from ^{14}C-estradiol (males), injected into 5 day old pups (6 h pulse) was found mainly in the gonads, adrenals, and uterus, while the hypothalamic-pituitary area, and the rest of the brain, concentrated insignificant amounts (Table 4.21).

Testosterone, but not dihydrotestosterone, prevents the nuclear uptake of estradiol in the CNS of 2 day old female rats (Sheridan *et al.*, 1974). The observation suggests that the brain of neonatal females may already have the capacity to aromatize androgens.

4.4.4.1 Androgenized Animals

Females
Early reports indicated that the hypothalamus and the pituitary of males and androgenized female rats are similar in that both bind less estradiol than the tissues of normal females (Flerkó and Mess, 1968; McGuire and Lisk, 1969; Vertes and King, 1971). Others reported reduced estradiol binding only in the hypothalamus (Anderson and Grennwald, 1969; McEwen and Pfaff, 1970; Tuohimaa and Johansson, 1971; Maurer and Wooley, 1971; Clark *et al.*, 1972) while still others did not find any difference in the uptake in either tissue between males, androgenized females and normal females (Eisenfeld and Axelrod, 1966; Grenn *et al.*, 1969; Attramadal and Aakvaag, 1970; Anderson *et al.*, 1973; Scott and Traurig, 1973; 1977; Maurer and Woolley, 1974a). Differences in the protocol used may explain some of the variations. Maurer and Wooley (1975) demonstrate that in androgenized females the concentration of hypothalamic estradiol receptors does not differ quantitatively, or qualitatively, from that found in normal females. The animals

used in the study were recently castrated (2 days) adults and 100 and 200 day old females injected neonatally with graded doses of TP (30, 120 and 1250 µg). Concentration of total radioactivity derived from ^{3}H-estradiol after 1 hr pulse was studied in the preoptic area-anterior hypothalamus, median eminence-basal hypothalamus, dorsal hypothalamus, septum, amygdala, cortex, anterior pituitary, uterus, liver, kidney, seminal vesicles and plasma. The group found decreases in radioactivity concentration uptake only in the preoptic-anterior hypothalamic area, only in the 200 day old females, and only in those treated with 1.25 mg TP. However, the investigators used only a single pulse; the analysis does not provide information pertaining to the dynamics of the uptake.

Heffner and van Tienhoven (1979) measured the uptake of estradiol after 2, 4, and 6 h pulse. Neonatal androgenization (30 µg of TP on day 5) decreased the accumulation in the anterior and middle hypothalamus and the uterus of adults.

Flerkó and Mess (1968), Flerkó et al. (1969), Green et al. (1969), Tuohimaa and Johansson (1971), Lobl (1975) and Maurer and Wooley (1975) all confirm decreased uptake of estradiol in the uterus of androgenized rats.

Males
Kincl and Chang (1970) measured testosterone uptake in 200 day old neonatally (day 5) androgenized (TP 1000 µg) male rats. The uptake was higher only in the ventral prostate and not in the CNS, testes, seminal vesicles, epididymis, or liver.

4.4.4.2 Estrogenized Animals

Females
Vaginae of neonatally estrogen treated mice bind labelled estradiol more rapidly than controls (Kohrman and Greenberg, 1968).

Males
Testosterone uptake in males estrogenized with 250 µg of EB on day 5 was higher in the ventral prostate but not in other tissues (Kincl and Chang, 1970).

4.4.5 Hormone Receptors

Receptors for various hormones begin to develop during fetal life, and the formation continues after birth. Csaba and Nagy (1976) speculated that the ontogeny can be deformed during the neonatal period by gonadal hormones.

4.4.5.1 Brain Receptors

Androgen, estrogen, and progesterone receptors are present in the brain after birth in different, sex related amounts. The degree the variation contributes to sex differences of reproductive functions is not known. The testicular feminized mouse (tmf/y), a strain insensitive to androgen stimulation, is an example. The behavior of the males is feminine. It could be expected that testosterone binding would be lower and estradiol binding similar to that of females. In contrast, estradiol binding is different between the sexes. Compared with other strains only the binding of dihydrotestosterone in the brain cytosol is lower, 70–80% (Attardi *et al.*, 1976).

McGuire and Lisk (1969) established that estrogen receptors were less numerous in the pituitary and hypothalamus of estrogen sterilized rats.

4.4.5.2 Receptors in Other Tissues

Females
Terenius *et al.* (1969) found decreased binding of estradiol in androgenized and estrogenized mouse uterus and vagina. Gellert *et al.* (1977) observed decreased synthesis or replenishment of uterine estradiol receptors in estrogenized rats, prior to the development of the anovulatory syndrome. Shyamala *et al.* (1974) found the presence of vaginal nuclear estrogen receptors in PE animals with ovary-dependent cornified epithelium (low dose estrogenization); nuclear estrogen receptors were not detectable in vaginae of mice which exhibited ovary-independent cornification.

Kolena *et al.* (1977) reported decreased hCG binding by ovarian homogenate in 25 day old androgenized (50 μg TP) or estrogenized (100 μg of ED) females. The decrease was not permanent. After day 40 of life the uptake of hCG rose.

Males
Testicular gonadotropin receptors decreased both in males androgenized with 250 μg of testosterone propionate or estrogenized with 400 μg of estradiol dipropionate (Kolena, 1977). The hypothalamus does not preferentially concentrate gonadal hormones. Kincl (1970) found in 5 day old male pups most of the radioactivity derived from ^{14}C-estradiol benzoate (6 h pulse) is in the gonads and adrenals. The hypothalamus-pituitary area, and the rest of the brain, concentrate relatively insignificant amounts. This indicates that at this age the gonads and adrenals are richer in hormone receptors than the central nervous system. In 200 day old estrogenized (250 μg EB) males the uptake of testosterone in the central nervous system, testes, and accessory sex tissues is significant (Kincl and Chang, 1970).

4.4.6 Steroid Biosynthesis and Metabolism

Hormones circulate in blood free and bound to carrier proteins. In guinea pigs, castration of females increases the binding activity of testosterone and progesterone; in males castration has the opposite effect. Neonatal treatment with testosterone (5 mg TP/day for several days) decreased only slightly the binding in females (Diamond *et al.*, 1969). In untreated adult males estrogens alter testosterone metabolism (Lee *et al.*, 1974).

4.4.6.1 Androgenized Animals

Females

Ovaries of cycling rats produce insignificant amounts of androgens. The production of estrone is in the range of 15 to 28 nM and of estradiol 29 to 85 nM. Neonatal androgenization altered biosynthetic patterns (Weisz and Lloyd, 1965). The ovaries (*in vitro* incubation) produced more androgens and estrogens than controls. The default resembled that seen in old rats in constant estrus. Progesterone was transformed in significant amounts to testosterone (35 to 460 nM), 4-androstene-3,17-dione (117 to 1258 nM), estrone (60 to 360 nM) and estradiol (192 to 1340 nM). Goldzieher and Axelrod (1963) and Rosner *et al.* (1965) confirmed increased conversion of progesterone (and of 3β-hydroxy-5-pregnen-20-one) into testosterone and 4-androstene-3,17-dione. Sawada and Ichikawa (1978) androgenized females by injecting 1.25 mg of TP on day 5 and studied the secretion of pregnanes and estrogens, and the response to hCG stimulation, in venous ovarian blood collected *in situ*. The polycystic ovaries exhibited low 5α-reductase activity. Pregnane compounds were not produced by unstimulated PE rats.

Stimulation with hCG increased the secretion of progesterone both in PE rats and in controls. The secretion of 5α-pregnane-3,20-dione and 3α-hydroxy-5α-pregnan-20-one was lower from polycystic ovaries whereas the output of estrone, estradiol and estriol was significantly higher.

Rubinstein and Ahren (1969) noted that ovaries of immature controls (20–26 days old) stimulated with FSH incorporated α-amino isobutyric acid; the ovaries of PE rats (1 mg of TP on day 2) did not.

Males

Kincl and Zbuzkova (1971) employed 200 day old male rats injected at the age of 5 days with TP (1000 μg) and studied the fate of paired ^{3}H-pregnenolone and ^{14}C-DHA, or ^{3}H-androstenedione and ^{14}C-testosterone after a 20 min pulse. Androgenized males exhibited an increase in 17β-hydroxy oxidase activity and a decrease in the activity of 17-oxo reductase. In the brain of treated rats the concentration of testosterone and androstenedione was lower, and the concentration of polar metabolites was higher. The findings suggested either a different metabolism in this tissue or an alteration in the permeability of the blood-brain barrier.

Neonatal sterilization also influences testosterone biological half-life and clearance rate (Kincl and Henderson, 1978). Significant differences exist in the initial part of the elimination curve composed of A_e-$^\alpha$ (Table 4.22). Metabolite formation is dependent on the clearance rate from the central compartment (K_{10}) composed of two rate constants: K_e which is the elimination (renal and hepatic) rate constant and K_f which is the rate of metabolite formation. K_{10} for the TP group (0.408 min^{-1}) was more than twice the clearance rate for control (0.186 min^{-1}). Two explanations are possible: either testosterone tissue levels are high and fewer metabolites are released (i. e. the ventral prostate possesses more receptor sites for DHT), or the liver may metabolize testosterone more slowly. The difference in the volume of the central compartment (Vc) at the same body volume (Vb) may indicate liver and kidney involvement. As a result of the above, the difference in the total metabolic clearance rate (TMCR) in the TP group (27.9 ml/

Table 4.22 The Effect of Neonatal Steroids on Testosterone Disposition Rate Constants (plasma values) in Adult Male Rats Injected at the Age of 5 Days Either with Testosterone Propionate (TP) or Estradiol Benzoate (EB).

Treatment	C	V_b	V_c	TMCR	$T_{1/2\varrho}$	$T_{1/2\alpha}$	$T_{1/2\beta}$	r^2
Control	2	884	185	34.4		0.84	17.8	0.9988
TP	2	812	68	27.9		0.43	20.2	0.9967
EB	2	814	189	32.4		0.94	17.4	0.9984
	3*	1030	155	29.4	0.55	2.37	24.4	0.9999

C = Compartments; V_b body volume (liters); V_c volume of the central compartments (liters); TMCR total metabolic clearance rate (ml/min); $T_{1/2\varrho}$, $T_{1/2\alpha}$, $T_{1/2\beta}$, testosterone half-lives (min^{-1}).
* Data recalculated for a three compartment model. From Kincl and Henderson (1978).

min) was thought to be significant when compared with controls (34.4 ml/min). Decreased rate of metabolite formation and lower TMCR may explain the increased sensitivity of TP treated males to androgens when tested as adults (Section 4.4.1.2); injected testosterone "lasts longer" in androgenized animals.

Females
No information is available on the dynamics of testosterone or estradiol circulation in androgen sterilized females.

4.4.6.2 Estrogenized Animals

Females
No information is available on the dynamics of testosterone or estradiol circulation in estrogen sterilized females.

Males
In males decreased testosterone production is the main injury caused by estrogenization (see Table 4.11). Kincl and Zbuzkova (1971) used 200 day old male rats injected at the age of 5 days with EB (250 µg) and studied the fate of paired ^{3}H-pregnenolone and ^{14}C-DHA, or ^{3}H-androstene dione and ^{14}C-testosterone (20 min pulse). Estrogenization decreased the formation of testosterone by depressing reduction of the 17-ketone and increasing the oxidation of the 17β-hydroxyl. Neonatal estrogen exposure also reduced the cleavage of the pregnane side chain. This results in reduced formation of C-19 steroids (Joseph and Kincl, 1974).

The metabolic clearance rate from the central compartment (K_{10}) in EB animals (0.171 min^{-1}) is not different from controls (0.186 min^{-1}). The results show that testosterone TMCR is not influenced by neonatal estradiol exposure (Kincl and Henderson, 1978).

Table 4.23 Conversion *In Vitro* of Progesterone and Dehydroepiandrosterone into Various Metabolites in the Feminized Male Rat.

	Healthy rats (mean ± SD)	Feminized rats (mean ± SD)	Student's *t*-test probability (p)
Substrate: progesterone			
Untransformed progesterone	2.8 ± 1.3	30.3 ± 8.3	<0.001
17α-Hydroxyprogesterone	8.5 ± 3.7	14.2 ± 6.9	0.01–0.001
Androstenedione	4.8 ± 2.4	0.8 ± 0.2	<0.001
Testosterone	40.7 ± 13.6	3.8 ± 1.1	
Estrone	3.1 ± 0.4	<0.1	
Aqueous fraction	5.0 ± 1.0	4.4 ± 0.3	0.01–0.001
% Radioactivity accounted for	65.0	53.4	
Substrate: dehydroepiandrosterone			
Untransformed dehydroepiandrosterone	1.9 ± 1.0	2.5 ± 1.1	<0.001
Androst-5-ene-3β, 17β-diol	9.8 ± 1.7	0.8 ± 0.6	
Androstenedione	9.2 ± 2.9	22.8 ± 5.8	0.01–0.001
Testosterone	10.2 ± 5.5	29.3 ± 3.9	<0.001
Estrone	3.7 ± 1.2	0.1 ± 0.01	
Estradiol-17β	3.1 ± 1.2	0.1	
Aqueous fraction	9.8 ± 2.5	4.2 ± 1.3	
% Radioactivity accounted for	47.7	59.7	

Data from Bottiglioni *et al.* (1971).

4.4.6.3 Feminized Males

Androgen biosynthesis was modified in male rats feminized by neonatal anti-androgen treatment, cyproterone acetate (Table 4.23). The testes produced testosterone from pregnenolone, but the amounts were decreased at the expense of increased progesterone and 17-hydroxyprogesterone production. Estrogen synthesis was normal (Bottiglioni *et al.,* 1971).

4.5 Other than Reproductive Effects

Steroid hormones influence, during the neonatal period, body growth, thyroid and adrenal functions, and sensitivity to carcinogenic agents.

4.5.1 Somatic Growth

Androgenization produces weight increases in females and may decrease body growth in males. Estrogenized females are lighter while males injected with a low estradiol dose may grow heavier (Barraclough, 1961; Khan, 1963; Swanson and van der Werff ten Bosch, 1963; 1964a, b; Kincl *et al.,* 1963; 1965a; Jacobson, 1964). The increases in body weight are dose dependent (Tarttelin *et al.,* 1975). In females, body growth increase is not due to change in fat content and is augmented by castration (Dubuc, 1976). Lipid metabolism is disturbed (Hácik and Palkovic, 1973).

Ostadalová and Parizek (1968) charted body growth in animals injected with 1250 µg estradiol dipropionate on day 5 and used animals castrated on the same day asa positive control. Retardation of growth, more severe than in castrated animals, became evident from the 25th day of life (Figure 4.8). The authors speculated that neonatal estrogen treatment may have affected the production of growth hormone but provided no evidence or measurements of hormone concentrations. Hughes and Tanner (1974) used 100 µg of hexoestrol in males and 1.25 mg of tes-

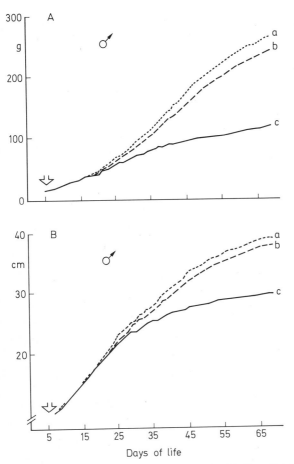

Figure 4.8. Growth of Male Rats Injected with Estradiol Benzoate at the Age of Five Days. Somatic growth (**A** , body weight in grams; **B** , body length in cm) of control rats and rats injected with 1.25 mg of estradiol dipropionate on the 5th day of life. The curve represents average values of 8 animals. Redrawn from Ostadalova and Parizek (1968).

tosterone propionate in females on day five and measured pelvis growth from radiographs taken from day 23 to day 140 of life. Neonatal steroid injections diminished, but did not abolish, sex differences in pelvic size (Tanner and Hughes, 1970).

4.5.1.1 Growth Hormone

In sexually mature rats the concentration of GH is sex dependent. The concentration is higher in the pituitary of males than of females, the pituitaries of males contain more GH, and individual male somatotrophs secrete more hormone than do those of female rats (Birge *et al.,* 1967; Hoeffler and Frawley, 1986).

The hormone is secreted in an episodic, sex-dependent pattern. In males basal levels are below detectable concentrations, and the hormone pulses in 3–3½ h intervals with high amplitudes. Neonatal orchidectomy decreases the amplitude (Jansson *et al.,* 1984). The decrease is restored by neonatal testosterone or by continuous testosterone exposure of adults (Jansson and Frohman, 1987a). Stress during neonatal life depresses growth hormone concentration in blood (Section 4.5.4.1).

In females the pulses are irregular, more frequent, of lower amplitude, but the basal levels are higher (Edén, 1979; Jansson *et al.,* 1984). Androgenization of females results in increases of growth hormone concentration in the pituitary in about 3 months (Table 4.24). Ovarian hormones are necessary to maintain female patterns. Jansson and Frohman (1987b) androgenized female pups with 250 µg of TP on the day of birth and removed the ovaries in some females. In the adult, the GH pulses were greater, and the GH concentration increased to levels comparable with intact males.

Differences in growth hormone patterns influence the formation of prolactin receptors (Saunders *et al.,* 1976) and steroid hepatic metabolism (Section 4.5.4.2).

Table 4.24 Effect of Castration (Male Rats) and Androgenization (Female Rats) on Pituitary Weights and Growth Hormone Concentration in the Anterior Pituitary.

Group	Body weight g ± SE	Pituitary weight, mg ± SE	GH concentration µg/mg ± SE
2 Months Old			
Control males	197 ± 9	5.3 ± 0.3	58.7 ± 8.6
Castrated males	157 ± 7	7.6 ± 0.6	29.4 ± 3.3
Control females	146 ± 4	7.5 ± 0.7	27.1 ± 2.5
Androgenized females	176 ± 6	9.0 ± 0.7	26.4 ± 2.4
4 Months Old			
Control males	266 ± 11	6.0 ± 0.4	79.9 ± 11
Castrated males	201 ± 6	7.1 ± 0.4	43.7 ± 3.7
Control females	185 ± 8	8.1 ± 0.5	28.2 ± 3.1
Androgenized females	251 ± 10	7.7 ± 0.7	62.0 ± 8.7

Adapted from Somana *et al.* (1978).

4.5.2 Brain Development

Sex-linked morphological dissimilarities exist in the brain of rats. In males the neuropil, the region of synapses between axons and dendrites (Raisman and Field, 1973), and the volume of the preoptic area are larger than in females (Gorski *et al.,* 1978; 1980). The pattern of synaptic organization is different in the arcuate nucleus (Matsumoto and Arai, 1980), the medial amygdala (Nishizuka and Arai, 1981), the preoptic area (Raisman and Field, 1973) and dendrite branching (Greenough *et al.,* 1977). Androgenization of females (TP 1.25 mg on day 5) decreased the number of spine synapses, but not of shaft synapses, in the adult (Matsumoto and Arai, 1980).

Steroid hormones (testosterone and estradiol) enhance neuronal growth of cells isolated from the preoptic area of mice (Toran-Allerand, 1976; 1980; Ohtani and Kawashima, 1983). In immature animals estrogens promote neural maturation (Nishizuka, 1978) and synapse formation (Matsumoto and Arai, 1976; Arai and Matsumoto, 1978) in the arcuate nucleus and synaptogenesis in the amygdala (Nishizuka and Arai, 1981).

Testosterone enhances neuritic growth of cells isolated from the neonatal rat hypothalamus-preoptic area but not from the cerebral cortex. Only cells derived from fetuses before birth (day 21 of gestation) respond to estradiol; cells derived from younger fetuses (day 17) are not sensitive to the hormone. 5α-Dihydrotestosterone is not active in this respect (Uchibori and Kawashima, 1985).

In males copulatory function is mediated by the spinal nucleus of the bulbocavernosus; the neurons are larger and more numerous. The growth is testosterone dependent. The number of motoneurons and their size are determined during the neonatal period by androgens. Injection of TP (1 mg) the female pups on the second day of life caused multiplication in the number of neurons. Estradiol benzoate (100 µg) was not effective in causing an increase (Bredlowe *et al.,* 1982).

Neurons in other regions are also sensitive to the influence of gonadal hormones. TP (Dibner and Black, 1978) and estradiol (Wright and Smolen, 1983) stimulated the concentration of nerve growth factor(s) and the growth of adrenergic neuron innervation in the superior cervical ganglion. Bredlowe *et al.* (1982) confirmed the masculinizing effect of TP but not of estradiol in this respect.

Both results show that sex dimorphism in this area is independent of the aromatization of testosterone.

4.5.3 Mammary Glands

Neonatal estrogens influence the differentiation of the glands in the adolescent (Mori, 1968a; Bern *et al.,* 1975; Mori *et al.,* 1976). The lobules of large alveoli become prematurely developed and the ducts dilated suggestive of increased PRL secretion. The effects are not permanent (Mori *et al.,* 1979).

Neonatal exposure to various hormones increases the development of neoplastic lesions in mammary tumor bearing mice (BALB/cfC3H/Crgl strain). The exposure leads to an earlier onset and a higher incidence of mammary tumors (Table 4.25). Tumors develop less readily when the animals are ovariectomized at the age of 40 days. The changes induced by DES are more severe (see Chapter 5).

Table 4.25 Effects of Neonatal Hormone Treatment on Normal, Preneoplastic, and Neoplastic Mammary Growth in Intact Mice.

Strain	Hormones[a]	Treatments		HAN and other dysplasias (%)	Mammary tumor incidence (%)	References
		Daily dose	Injection period (days in post-natal life)			
BALB/ cfC3H[b]	E₂	5	1–5	67(35/52)[g]	63(34/54)[h]	Jones and Bern
	E₂	20	1–5	44(11/25)	42(11/26)	(1977) and Mori
	T	5	1–5	80(12/15)	67(10/15)	et al. (1976)
	T	20	1–5	82(14/17)	93(13/14)	
	P	100	1–5	94(30/32)	72(23/32)	
	PRL	5	1–5	26 (6/23)	17 (4/23)	
	PRL	20	1–5	17 (3/18)	6 (1/18)	
	Vehicle	–	1–5	25(11/44)	20 (9/44)	
BALB/c[c]	E₂	5	1–5	0 (0/15)	0 (0/15)	Mori et al. (1976)
	E₂	20	1–5	0 (0/9)	0 (0/9)	
	PRL	5	1–5	0 (0/13)	0 (0/13)	
	PRL	20	1–5	0 (0/11)	0 (0/11)	
	Vehicle	–	1–5	0 (0/18)	0 (0/18)	
BALB/c[d]	E₂	5	1–5	50 (7/14)	0 (0/14)	Jones and Bern
	E₂	20	1–5	44 (4/9)	0 (0/9)	(1978)
	P	100	1–5	71(17/24)	0 (0/24)	
	Vehicle	–	1–5	0 (0/10)	0 (0/10)	
C3H[e]	E₂	0.5	1–7	–	33 (7/21)	Mori (1968 a, b)
	E₂	20	1–10	–	25 (4/16)	
	Vehicle	–		–	9 (2/22)	
C57BL[f]	E₂	20	1–5	0 (0/15)	0 (0/15)	Mori et al. (1976)
	PRL	20	1–5	0 (0/6)	0 (0/6)	
	Vehicle	–	1–5	0 (0/7)	0 (0/7)	

[a] E₂, 17β-estradiol; T, testosterone; P, progesterone; PRL, prolactin; [b] Mice were killed by one year of age; [c] Mice were killed at one year of age; [d] Mice were killed at 18.5–27 months of age; [e] Mice were killed by 14 months of age; [f] Mice were killed at one year of age; [g] Number of mice with HAN or dysplasias in the gland/total number of mice examined; [h] Number of mice with mammary tumors/total number of mice examined; reprinted from Mori et al. (1980).

4.5.4 Other Endocrine Functions

Neonatal steroids alter adrenal and thyroid functions. Exposure to adrenal and thyroid hormones during embryonic life influences their own and other functions (Chapter 5).

4.5.4.1 The Adrenals

Androgenized females may have hypertrophic adrenals (Barraclough, 1961; Khan, 1963; Harris and Levine, 1965; Levine and Mullins, 1967) or be within the normal range (Brown-Grant, 1964; Hácik, 1966; Frick et al., 1969). If increases occur, it is the zona fasciculata that enlarges (Khan, 1963; Kawashima, 1965; Kawashima et al., 1980b). In younger animals (2–8 months) the zona glomerulosa may show partial lipid depletion while complete lipid depletion may take place in older (1 year or more) animals (Khan, 1963). Mäussle and Fickinger (1976) performed detailed evaluation of the outer zona fasciculata of androgenized (1.25 mg of TP) or estrogenized (300 µg EB) male and female rats. In females both hormones caused cell enlargement with increases in the volumes of the nucleus, mitochondria and liposomes. In males neonatal treatment resulted in a numerical increase of liposomes and mitochondria and an increase in lipoid content. The authors speculated that the adrenal may be capable of increased functional capacity.

Hácik (1966) found in females (1 mg TP on days 2 to 4) increased corticosterone concentration in the adrenals (3.38 ± 0.5 ($\pm$ SE) µg/100 mg adrenal weight vs 1.53 ± 0.14 in controls) and in peripheral plasma (390 ± 33 ng/ml and 159 ± 19, respectively). There were no changes in androgenized males. Despite the differences in adrenal and plasma corticosterone levels, both sexes responded well to stress induced by ether in a manner not different from controls (Schapiro, 1965a). There is no difference in ACTH pituitary content between androgenized and control females (Chiappa and Fink, 1977).

Some strains of PE mice (CBH and BALB/c) show adrenal hypertrophy while in others (C57BL) the adrenals are smaller (Takasugi and Bern, 1964).

4.5.4.2 Thyroid

Androgens
Brown-Grant (1964, 1968) had seen no changes in thyroid weights. The values for ^{131}I uptake and release in males androgenized with 1.25 mg of TP (day 5 of life) were within normal values. In females (same dose) the mean level of thyroid activity was more depressed than occurs during estrus.

Kawashima and Takewaki (1966) found severe toxic effects in female rats injected with 0.5 mg of TP during the first 10 days, 1 mg during the next 10 days and 2 mg during the last 10 days. The glands were atrophied, and the epithelial height and the number of colloid vacuoles were reduced. The number of thyrotrophs in the pituitary gland was decreased.

Fujii (1973) noted an influence of the thyroid on ovulation. Females were injected with 1 mg of TP on day 5 of life. Animals which showed persistent cornification of the vaginal epithelium were thyroidectomized. Within 1–2 days after the operation the animals began to cycle. The cycles were irregular, 2 to 6 days of vaginal diestrus and 3 to 10 days of estrus. Corpora lutea formed in the ovaries of thyroidectomized rats.

Estrogens
The effect of neonatal estrogens on thyroid function was evaluated by Kawashima (1965). Kawashima injected estrone daily, from the day of birth. The dose was 50 µg for the first 10 days, 100 µg for the next 10 days, and 200 µg between days 21 and 30. The treatment did not produce any differences in thyroid weights, but colloid accumulation in the gland of males was higher.

4.5.5 Effects on Metabolizing Enzymes

The development of many enzymes (cf. Knox *et al.*, 1956) including those metabolizing steroid hormones (Hubener and Amelung, 1953; Forchielli and Dorfman, 1956; Yates *et al.*, 1958; Leybold and Staudinger, 1959) is sex dependent. The sex differences appear at puberty. Prepubertal exposure alters sex-specific metabolic patterns (Tabei and Heinrichs, 1974; De Moor *et al.*, 1977). Prepubertal imprinting influences the activity of steroid metabolizing enzymes in the kidneys. Estrogenized males exhibit increases in the activity of kidney 3α-, 17β-, 20α-, and 20β-hydroxysteroid dehydrogenase (Ghraf *et al.*, 1975).

The metabolism of testosterone in the pituitary and brain is sex linked. Pituitaries from intact rats form 2½ times more DHT and 1½ times more $3\alpha,17\beta$-dihydroxy-5α-androstane than do the pituitaries of female rats (Denef *et al.*, 1973).

4.5.5.1 Adrenal Enzymes

The metabolism of adrenocortical hormones is sex-linked (Yates *et al.*, 1958; Deckx *et al.*, 1965; Denef and DeMoor, 1965; 1968; 1972; DeMoor and Denef, 1968; Gustafsson *et al.*, 1974; Ghraf *et al.*, 1975). In female rats the reduction of the 3-ketone to the 3α-hydroxy-5α-dihydro metabolite of cortisol is favored while males are characterized by high levels of the 3β-isomers.

Corticoid metabolic clearance rates are likewise sex-dependent. The secretion of corticosterone is more rapid in female rats and hamsters, while in males the cortisol metabolic clearance rate is higher (Colby and Kitay, 1972). Transcortin concentrations in plasma are higher in females (150 mg/l) than in males (60–70 mg/l). Neonatal gonadectomy increases the concentration in males and decreases the amounts in females (van Baelen *et al.*, 1977).

4.5.5.2 Liver Enzymes

The activity of several hydroxylases and hydrogenases acting on both C-19 (androstane) and C-21 (pregnane) substrates is higher in adult male rodents. The rate is dependent on the presence of the testes. Castrated females treated with testosterone propionate metabolize steroids in a male-like pattern (Jacobson and Kuntzman, 1969). The reduction of ring A is the most prominent feature of sex dimorphism (Table 4.26). 5β-Androstane diols are present in males and absent in females (Schriefer *et al.*, 1972a). The 5α-isomers are present in both sexes, but the female produces about twice as much as the male (Ghraf *et al.*, 1973). Male patterns can be induced in females by prepubertal testosterone treatment (Tabei and Heinrichs, 1975; Tabei *et al.*, 1975). Hoff *et al.* (1977) report that prepubertal testosterone will increase the activity of 3α, 4-hydroxysteroid dehydrogenase but not the activity of 20-keto reductase. Gustafsson and Stenberg (1974c; Gustafsson *et al.*, 1974) found in castrated rats increased 6β-hydroxylation of 4-androstene-3,17-dione and 19-hydroxylation of 5α-androstane-$3\alpha,17\beta$-diol. Estradiol stimulates and testosterone inhibits the activity $4,5\alpha$-steroid dehydrogenase (Yates *et al.*, 1958).

Table 4.26 Sex-Linked Difference in Metabolizing Liver Enzymes of Adult Male and Female Rats.

Substrate used	Reductases				Hydroxylases					Reference
	3α	3β	5α	5β	6β	7α	7β	16α	20β	
Testosterone			♀↓	♂ only						Schriefers et al. (1972)
					♂↑	ND		♂↑		Ghraf et al. (1972)
DHA						♂↑	ND			Tabei and Heinrichs (1974)
								♂↑		Tabei and Heinrichs (1972)
A⁵-diol								♂↑		Tabei and Heinrichs (1972)
Cortisol	♀↑	♂↑	♀↑						♂↑	Denef and DeMoor (1972)

ND = no difference between sexes; ↑ indicates an increase, ↓ indicates a decrease of metabolite formed; DHA = 3β-hydroxy-5-androsten-17-one; A⁵-diol = 5-androstene-3β, 17β-diol.

Gustafsson and co-workers (1977) proposed that the programming and differentiation of rat liver enzymes at puberty are directed by a pituitary hormone "feminotropin" that imprints a basal feminine type upon hepatic enzymes. In males they visualized the presence of an inhibitory hypothalamic "feminostatin-secreting center", programmed at birth and activated at puberty by testicular androgens.

A strong point in the argument of a central control of hepatic enzyme activity was the observation of an obligatory role of the pituitary gland:

(i)   Electrolytic lesions in the hypothalamus changed the metabolic pattern (Gustafsson and Stemberg, 1974a; Gustafsson et al., 1976);
(ii) In hypophysectomized rats transplantation of the pituitary under the renal capsule induced a feminine-like pattern (Gustafsson and Stemberg, 1974b);
(iii) In castrated females FSH induces a male-like liver metabolism (Gustafsson and Stemberg, 1975) most likely the result of increases in the number of receptor sites on the hepatocyte surface (Posner et al., 1974; Kleezik et al., 1976);
(iv) Testosterone inhibited the activity of hepatic 4-steroid dehydrogenase in orchidectomized rats, but only if the pituitary gland was present;
(v)   The stimulation of A ring reduction and increases in metabolic clearance rate by estradiol also required the presence of the hypophysis.

In hamsters the results are opposite to those seen in rats. Testosterone stimulated ring A reduction but again only in the presence of the hypophysis (Colby et al., 1973). Others have confirmed the obligatory role of the pituitary gland (Denef, 1974; Gustafsson and Stemberg, 1974b; Gustafsson et al., 1974).

Mode et al. (1983) identified growth hormone as the causative agent which directs the metabolic activity (Section 4.5.1.1).

Neonatal Animals

A brief exposure to androgens or estrogens alters the programming of liver enzymes which becomes apparent at the age of about 30–40 days (Schriefer et al.,

1972b). In the adult, the testosterone effect is seen within a few days (Berg and Gustafsson, 1973).

In males EB (300 µg on day 2 of life) causes metabolic feminization: the formation of 5β-androstane diols is decreased to insignificant levels at the expense of the formation of 5α-isomers (Schriefer *et al.*, 1972a); TP (1250 µg) causes no alterations in the patterns of hepatic enzymes (Schriefer *et al.*, 1972b). The formation of 6β hydroxy testosterone is decreased to castrated female levels, and 16α-hydroxylation is absent (Ghraf *et al.*, 1973; 1975).

Female rats exposed to testosterone display a male pattern; there is decreased activity of 5α-reductase and increased activity of 5β-reductase, 16α-hydroxylase, and 17α- and 3β-hydroxy steroid reductase (Einarsson *et al.*, 1973; Gustafsson and Stenberg, 1974a).

Other Hepatic Enzymes
The amount of low affinity estrogen binding proteins depends on the presence of androgens during the neonatal period. This class of proteins is 10 times larger in adult males than females (Sloop *et al.*, 1983). Monoamine oxidase (MAO) activity is lower in males; neonatal castration increases the activity. The activity is independent of concomitant administration of TP or DES (Illsley and Lamartiniere, 1980).

Neonatal estrogens decreased permanently and irreversibly the activity of histidase whereas TP had no effect on the activity of the enzyme (Lamartiniere, 1979).

4.5.6 Increased Sensitivity to Carcinogens

Neonatal estrogens enhance the induction of hepatic cancer. Weisburger *et al.* (1966) injected 100 µg of EB at birth and fed the offspring after weaning a diet containing a carcinogenic aromatic amine (N-hydroxy-N-2-fluorenylacetamine). The diet increased the incidence of liver tumors in both sexes. A dose of 500 µg of TP rendered only the females more sensitive to the carcinogen (Weisburger *et al.*, 1966).

B. Pregnanes

Exposure during the neonatal period to adrenocortical hormones provokes decreases in body growth, suppression of circadian corticosteroid periodicity, disturbance in the pituitary-gonadal axis, and suppression of immunological competence. Neonatal animals exposed to stress may show similar injury.

4.6 Adrenocortical Hormones

Neonatal exposure to corticosteroid or ACTH delays the development of evoked potentials to auditory, visual, and sciatic nerve stimulation (Salas and Shapiro, 1970), of adult-like seizure thresholds (Vernadakis and Woodbury, 1971), of swimming ability (Kincl *et al.*, 1970; Schapiro *et al.*, 1970). The growth of dendritic spines is delayed, the development of the blood-brain barrier is retarded (Schapiro, 1968), and the immunological response is impaired (Section 4.11).

4.6.1 Influences on Reproduction

Selye and Friedman (1940) found the development of persistent estrus after giving glucocorticoid for prolonged periods. Takasugi (1953) injected mice with desoxycorticosterone acetate (5 µg) for 20 to 40 day and found cycle irregularity and atrophic vaginal epithelium at the age of 180 days. The changes induced by long treatment were present in animals ovariectomized 5 days after the last injection. If the pups were injected only 5 times they cycled regularly. Turner and Taylor (1977) implanted corticosterone-cholesterol pellets (25 mg/100 g body weight) on day 3, 6, 12, or 28 of life. Persistent vaginal cornification, irregular cycles and absence of corpora lutea took place in 55% of females treated on day 3, in 37% treated on day 6, in 25% treated on day 12, and in 31% treated on day 18. Males were not affected.

Takasugi and Tomooka (1976) noted that a daily dose of 20 µg of cortisone acetate for 15 days increased the sensitivity of neonatal mice (C3H/Tw) to estradiol.

Shipley and Meyer (1965) tested the influence of corticosterone (1 mg), cortexone acetate (1 mg), cortisol (1 mg) and prednisone (0.5 mg) injected on day 5 of life. The presence, or absence, of corpora lutea was the end point. All the compounds were inactive, but the short period of testing (45 days) leaves a doubt as to the validity of the results. Estradiol and testosterone were active in the test.

Injections of cortisone acetate (CA, 20 µg/day for 15 days) into 1 day old C3H/Tw mice potentiated the effect of injected estradiol, 1 µg/day for 5 days beginning after CA (Takasugi and Tomooka, 1976). The authors speculated that the effect of CA on the thymus increased the sensitivity to estradiol (see also Section D).

4.6.2 Other than Reproductive Effects

Neonatal cortisol does not have any permanent effects on the activity of various enzymes in the liver, kidney, heart and brain (Greengard, 1975). Hypothalamic content of GH-releasing factor and pituitary GH and TSH contents are lower in adult animals injected neonatally with either 1 mg of cortisol acetate or 1 mg of cortisone acetate (Sawano *et al.*, 1969).

Ways and Bern (1979) studied the long term effects of cortisol in mice.

In newborn dogs a single injection of cortisol (10 mg/100 g body weight) induced hypothyroidism within 6 weeks. The treatment decreased thyroid ^{131}I uptake and resulted in enlarged acini, filled with colloid, and flattened epithelial layer (Fachet *et al.*, 1968).

4.6.2.1 Toxic Effects

Neonatal adrenocortical steroids bring on cachexia in mice, rats and dogs, decrease in body growth, atrophy of the adrenals and spleen, depletion of small lymphocytes and decrease in immunological competence and greater susceptibility to infection.

In rats mortality induced by cortisol (Table 4.27) is dose dependent (Kincl and Folch Pi, 1963; 1970). The toxic effects become visible within 7 to 10 days after the last injection. The animals fail to grow, their coats are sparse and ruffled, sometimes the skin becomes covered with eruptions, and the animals are prone to infections. Autopsy of dead animals (age 13 to 15 days) revealed pathological changes in the spleen, kidneys, and lungs. The major pathology was bronchopneumonia with massive hemorrhage into the alveoli. The kidneys were pale and appeared enlarged. The tubular epithelial lining showed vacuolization and occasionally a considerable enlargement of endothelial cells with vesicular and polygonal nuclei. In the spleen the most marked change was the scarcity of germinal lymphatic follicles. Granulocytes were present which suggested the occurrence of ex-

Table 4.27 Mortality Induced in Infant Mice by Injection of Adrenocortical Hormones.

Steroid used	Total dose, µg	Mortality	Mean survival days ± SE	Reference
Cortisol acetate				
(day 1)	100	2/10	12.4 ± 0.4	Schlesinger and Mark (1964)
	250	19/19	10.6 ± 0.7	
(day 5)	100	5/8	13.0 ± 1.4	
	250	21/32	14.0 ± 0.9	
(day 14)	500	0/7		
(day 21)	500	1/13	16	
Cortisol acetate[a]	250	4/12	9	Reed and Jutila (1965)
	500	58/64	9.9 ± 0.2	
(germ free animals)	250	3/24	15	
	500	5/23	110 ± 1.4	
Cortisol (day 1)	200	0/15		Fachet et al. (1966)
	1000	5/20	15–25[b]	
	1250	10/18	10–20	
	2000	15/25	8–15	
	2500	20/20	7–11	
Prednisolone sodium succinate (day 1)	250	6/16	7–12	
	2500	20/20	7–12	
Dexamethasone sodium phosphate (day 1)	250	8/16	8–15	
Cortisol (day 3)	50	1/40		Kincl et al. (1970)
(days 4, 5, 6)	1000	27/60		
(day 5)	1000	5/30		
Corticosterone	1000	1/20		
(day 5)	5000	5/20		

[a] Injected into non-germ free animals within 24 h of birth; [b] day of first and last death.

tramedullary hematopoiesis. Macroscopically inguinal and mesenteric lymph nodes were either non-apparent or were very small.

Schlesinger and Mark (1963) and Fachet *et al.* (1966; 1968) confirmed the toxicity of cortisol in newborn rats (Table 4.27). The group of Fachet also reported that corticoid treatment produced loss of weight, atrophy of the adrenals, the thymus and the spleen and depletion of small lymphocytes in the thymus, the spleen and the lymph nodes. Homologous skin drafts survived longer than in untreated controls. Immunological competence, the formation of immunoglobulins and antibodies, was depressed (Fachet *et al.*, 1968; Ulrich *et al.*, 1977). Enzine and Papiernik (1979) injected 250 μg of cortisol hemisuccinate (per gram of body weight) on the first day of life. The activity of the immune response was depressed 15 days post-treatment.

In mice cortisol (0.25 mg on day 1) induces a wasting disease similar to that seen in rats. Animals injected neonatally with cortisol survive significantly longer if raised in a germ-free environment. The observation suggests that neonatal cortisol impairs the maturation of the immune mechanism (Reed and Jutila, 1965).

4.6.2.2 Body Growth

Kincl *et al.* (1967; 1970) gave three injections of cortisol (1 mg) on days 4, 5 and 6. The treatment was toxic. Injection of TP (300 μg per day) or of bovine growth hormone (200 μg per day) for 4½ months did not protect the animals (Table 4.28). Those animals that survived until the age of 60 days had lower body and kidney and liver weights. The weights of the adrenals, spleen, thymus, pituitary, pineal gland, ventral prostate, seminal vesicles, levator ani muscle, and heart were comparable with controls. There was no apparent effect on the morphology of the gonads.

Ways and Bern (1979) used 50 μg of cortisol for 10 days from birth in BALB/c mice and noted decreased body growth and reduced spleen, thymus, and lymph node weights at 60 days of age; at 6 months these weights were no different from controls. Krieger (1974) and Parmer *et al.* (1951) confirmed impaired body growth in rats given corticoid during neonatal life. Howard (1968) found reduction in the brain size and DNA content.

Table 4.28 Influence of Growth Hormone (GH) and Testosterone Propionate (TP) on the Growth of Male Rats Aged 180 Days Injected in Infancy with Cortisol.

Treatment	Body weight g ± SE	Percent	
		surviving	with lung infection
Controls	430 ± 8	85	15
Cortisol (F)[a]	395 ± 17	41	54
F and GH[b]	415 ± 8	69	27
F and TP[c]	341 ± 11	44	71

[a] Cortisol (1000 μg) injected on days 4, 5 and 6 of life; [b] bovine growth hormone (200 μg) injected daily beginning day 40 of life; [c] testosterone propionate (300 μg) injected daily beginning day 40 of life; data from Kincl *et al.* (1970).

4.6.2.3 Adrenocortical Steroids in Blood

Injection of dexamethasone phosphate or acetate (100 µg) on day 1, 2, or 3 of life had no effect on the concentrations of corticosterone, LH, FSH, PRL, and GH in the adults (65–70 days); circadian variation in the secretion of corticosterone and LH were not influenced (Poland *et al.*, 1981). Kincl *et al.* (1970) found adrenal atrophy in both sexes given a 3 mg dose of cortisol on day 1, 3, or 5 of life. Values for corticosterone in plasma were lower only in animals injected on day 1 of life.

Cortisol (500 µg on day 3) or dexamethasone phosphate (1 µg day 2–4) suppressed circadian corticosteroid concentration in adult rats (30 days of age). The mean variation of circadian plasma corticosterone was 228 ng/ml in control, 82 ng/ml in cortisol injected, and 27 ng/ml in dexamethasone injected animals (Krieger, 1972). Ulrich *et al.* (1976) reported a reduced secretion of corticosterone in 20 day old rats; Krieger *et al.* (1974) found increased concentrations in the adult.

4.6.3 Stress

Stressing neonatal animals induces changes in the adult. Ward (1972) noted decreased male behavior patterns in adult male rats exposed prenatally to stress. In stressed rat pups (10 day old removed from mother) the concentration of growth hormone decreases by about one-half. The drop is only temporary, and the values return to normal within 15 min of returning the pup (Kuhn *et al.*, 1978).

Dörner *et al.* (1980a) proposed a relationship between stress and the development of homosexual tendencies in the human. The group compared the year of birth of 865 homosexual males and found that significantly more were born during the last years of the war (1944–1945). The authors suggest that prenatal or perinatal stress may represent an etiological factor in the induction of the syndrome. The hypothesis was not accepted (see Chapter 6).

4.6.3.1 Response to Stress

The pituitary-adrenocortical responsiveness to stress of the adult is diminished by treating the neonate with cortical steroids. Injection of cortisol acetate (0.5 or 1 mg within 24 h of birth) suppresses ACTH release in response to ether-induced burden in 25 day old rats. The concentration of corticosterone rose in stressed females (but not in males) albeit less than in controls (Erskine *et al.*, 1979). Kincl *et al.* (1970) also found increased corticosterone in the blood of stressed females; in males the levels were decreased. Schapiro (1965a, b; 1968) found a normal response to stress in rats subjected to neonatal cortisol. Turner and Taylor (1976) inserted postnatally (day 3) corticosterone:cholesterol (1:1) implants and noted a decrease in basal concentrations in the 30 day old animals; mean plasma levels were 9.5 ± 0.9 µg/ml whereas in controls the concentration was 14.9 ± 0.9 µg/ml. The differences disappeared in 40 day old animals.

Infant manipulation (removal from mother for a few minutes, transfer to a different cage, or injecting saline) makes the adult more responsive to stress (Levine, 1958; 1962). Meaney *et al.* (1988) handled pups daily from birth till weaning to determine the influence of the procedure on the feedback mechanism. Adrenocortical secretion decreased in the adult the concentration of glucocorticoid receptors in the hippocampus, the region critical in the negative feedback of adrenocortical activity. Basal adrenocortical steroid concentration in blood was high in the non-handled old (24 months) rats. This led to an accelerated neuron loss in the hippocampus and the loss of spatial memory ... "almost absent in handled rats ..."

4.7 Progesterone

The natural hormone, progesterone, in a dose of 5 mg or less produces no visible effects in the adult. Higher doses are toxic, and most pups die (Kincl and Maqueo, 1965; Gorski, 1966).

4.8 Other Pregnane Derivatives

Shipley and Meyer (1965) tested the influence of medroxyprogesterone (1 mg) injected on day 5 of life. The presence, or absence, of corpora lutea was the end point. The compound was inactive, but the short period of testing (45 days) leaves doubt as to the validity of results. Estradiol and testosterone were active in the test. Gorski (1966) found medroxyprogesterone acetate (MA) inactive at a dose of 2.5 mg. Cupceancu *et al.* (1971) report that 3 mg of a closely related steroid, me-

Table 4.29 Pregnane Derivatives Found Inactive When Injected into Infant Female Rats.

Steroid used	Dose, µg	Age at autopsy, days	Reference
Females			
Progesterone	3000	50	1
	4000	45	2
	5000	100	3
	2500	90	4
5 β-Pregnane-3,20-dione	1500	90	4
17-Hydroxy progesterone caproate	6250	100	3
Cortisone acetate	1250	100	3
Males			
19-Nor progesterone	3000	50	5
Chlormadinone acetate	1000	50	5
Cortisol	5000	50	5

Data from Kincl and Maqueo, 1965 (1); Shipley and Meyer, 1965 (2); Gorski, 1966 (3); Arai and Gorski, 1968 c (4); Kincl and Dorfman, 1967 (5).

gestrol acetate, injected on day 1 of life was active when evaluated at the age of 70 days. Holzhausen *et al.* (1984) measured the LH concentrations in the adult injected during the first day of life with MA (5 µg/g of body weight). The treatment reduced the proestrus surge of LH by one-half and the total amount secreted during proestrus by about one-third. Neonatal treatment with norethisterone acetate (4 µg/g BW) had no effect on the LH surge.

Takasugi and Kato (1974) report that in mice 5 daily injections of 17-hydroxy progesterone caproate given from birth induce PE syndrome.

A list of pregnane derivatives found inactive at the doses studied is given in Table 4.29.

4.8.1 Estrane Derivatives with Progestational Activity

Shipley and Meyer (1965) injected 1 mg of norethindrone on day 5 and found no CL in the ovaries of 45 day old rats. Cupceancu *et al.* (1971) injected various synthetic steroids on the first day of life and found several 17α-alkylated derivatives of 19 nor testosterone (estrane derivatives) active (Table 4.30).

C. Protection Against Steroid Hormone Damage

Various agents block the activity of androgens and estrogens in the neonate. The Japanese workers were the first to study the "protective" action (Table 4.31). Takasugi (1954a, b) employed 100 day old PE rats (produced by 5 injections of 12.5 µg of estrone from day 1) to study the effects of progesterone, supplied for 30 days in the form of subcutaneous tablets (8.4–9.9 mg) or injected daily (0.5 mg). The presence of progesterone in either form induced intermittent periods of diestrus, but the ovaries did not become luteinized. Injection of 5 mg of testosterone propionate, in addition to progesterone tablets, induced luteinization in a few animals.

Table 4.30 Progestational Agents Found Active When Injected into Infant Female Rats.

Steroid used	Dose tested, mg	Fertility inhibition, %
Norethisterone (NET) acetate	1000	100
NET enanthate	1000	92
Ethynodiol diacetate (ED)	1000	100
Allylestrenol (AE)	10,000	100
Norethandrolone (NED)	1000	77
d,1-Norgestrel (NEG)	1000	40
Gestonorone caproate (GES)	1000	37

NET = 17α-ethynyl-17β-hydroxy-4-estren-3-one; ED = 3,17β-diacetoxy-17α-ethynyl-4-estren; AE = 17α-allyl-4-estren-17β-ol; NED = 17α-ethyl-17β-hydroxy-4-estren-3-one; NEG = 17α-ethynyl-17β-hydroxy-18-methyl-4-estren-3-one; GES = 17-hexanoyloxy-19nor-4-pregnene-3, 20-dione. Data from Cupceancu *et al.* (1971).

Table 4.31 Protection Against the Effects of Exogeneous Androgens in Infant Female Rats by CNS Acting Drugs.

Androgenization procedure	Protective agent	Dose μg	Age evaluated (days)	Incidence of PE[a]	Reference
TP 30 μg on	reserpine	10	120	11/16	Arai and Gorski
day 5	chlorpromazine	500	120	9/12	(1968 c)
(rats)	pentobarbital	600	90	4/15	
	phenobarbital	500	90	4/17	
	none	0	90	8/8	
TP 50 μg on	reserpine	10	90	3/10	Nishizuka (1976)
day 4	pentobarbital	50	90	3/9	
(mice, C3H/TW)	phenoxybenza-mine	50	90	3/10	
	propranolol	100	90	11/16	
	none	0	90	9/11	
TP 30 μg on day 4[b]	phenobarbital	500	150–160	19/23	Brown-Grant et al. (1971)

[a] Incidence of PE = animals with anovulatory syndrome/total number of animals used; [b] 58% of rats injected with 30 μg of TP alone were anovulatory.

Kikuyama (1961; 1962) described the effectiveness of reserpine (6 μg) injected concomitantly with estrone for five successive days commencing on the fifth day of life. Arai and Gorski (1968 b, c) obtained "protection" by reserpine, chlorpromazine, phenobarbital, and pentobarbital provided the agents were injected within a few hours after TP injection. Brown-Grant et al. (1971) could not confirm the results.

4.9 Females

4.9.1 Progesterone

I have recorded in Chapter 3 the evidence that the treatment of gravid rats with androgens masculinize the external genitalia of female offspring yet rarely the ovarian function, and the animals ovulate. The evidence shows that testosterone or its biologically active metabolites cross the placental barrier yet are in some way prevented from damaging the function of the gonads. I have reasoned that progesterone, an anti-androgen and a CNS active hormone, may be the agent which protects the hypothalamus of the embryos against the deleterious effects of testosterone (Dorfman and Kincl, 1963; Dorfman, 1967). Several investigators tested the hypothesis. Kincl and Maqueo (1965) were able to impede the effects induced by 50 or 100 μg of testosterone propionate in 5 day old female by injecting 3 mg of progesterone on the same day (a different site). Vaginal introitus took place earlier in the progesterone-TP females but all fifty day old animals had luteinized ovaries (Table 4.32).

Table 4.32 Protective Action of Progesterone Against the Effects of Exogenous Androgens or Estrogens in Infant Female Rats.

Sterilization procedure	Protective agent	Dose μg	Age evaluated days	Incidence[a] of PE	Reference
TP (1250 μg) on day 3	progesterone	50	120	luteinized ovaries	Cagnoni et al. (1965)
		500	120	No CL	
		1000	120	No CL	
TP (50 μg) on day 5	progesterone[b]	3000	45	0/13	Kincl and Maque (1965)
TP 100 μg		3000	45	1/12	
TP 500 μg		3000	45	22/23	
TP (30 μg) on day 5	progesterone	2500	90	7/10	Arai and Gorski (1968 c)
	cortexone	250	90	9/9	
	pregnadione[b]	1500	90	3/5	
E₂ (250 μg) on day 5	progesterone	3000	45	1/22	Shipley and Meyer (1965)
	cortisol	1000	45	3/16	
	11-dehydro P	1000	45	2/20	
	MAP	1000	45	2/24	

[a] number of sterilized rats/total number used; [b] injected 12 h prior to TP injection; TP = testosterone propionate; E_2 = estradiol; pregnadione = 5β-pregnane-3,20-dione; 11-dehydro P = 4,9(11)-pregnadiene-3,20-dione; MAP = 6α-methyl-17-acetoxy-4-pregnene-3,20-dione.

Cagnoni et al. (1965) used only one dose of TP (1.25 mg) in 3 day old females and varied the progesterone dose. They were able to block the effect with 50 μg of progesterone but not when the dose was 500 or 1000 μg. Shipley and Meyer (1965) reported a "protective" action of progesterone (3 mg) against 250 μg of estradiol. Arai and Gorski (1968 c) confirmed the antagonistic action of progesterone (2.5 mg) and of cortexone (1 mg) in blocking the effect of 30 μg of TP, tested at the age of 90 and 120 days (Table 4.32).

Lloyd and Weisz (1966) made the important observation that females protected with progesterone resembled "delayed ovulation" animals. In their study progesterone blocked the effect of TP, and the females ovulated when they reached sexual maturity (45 days). Ovarian activity declined, and older animals (120 day old) were in constant estrus, and their ovaries lacked corpora lutea.

Progesterone does not block the uptake of ^{3}H testosterone in the hypothalamus but increases the uptake in the cerebrum (Heffner and van Tienhoven, 1973).

4.9.1.1 Synthetic Progestational Agents

Cyproterone acetate (CA) is an effective blocking agent (Wollman and Hamilton, 1967). Brown-Grant (1974) confirmed the activity of CA against lower doses of TP (50 μg), but not against higher amounts (100 μg), or if challenged against estradiol benzoate. Shipley and Meyer (1965) impeded the action of estradiol (250 μg) by medroxyprogesterone acetate (1 mg), cortisol (1 mg), or 11-dehydroprogesterone (1 mg).

4.9.2 Androstanes

McEwen *et al.* (1977) proposed that in males estrogens, rather than androgens, are the agents that "masculinize" the CNS. The experimental evidence was based heavily on observing behavioral changes (see Chapter 6). The group cited the following experimental testimony to support the hypothesis:

(i) Resemblance in behavioral responses between males and androgenized females;
(ii) Testosterone plasma concentrations are higher in neonatal males than in females; presumably testosterone, transported to the CNS and aromatized to estrogens, "masculinizes" the brain;
(iii) Only "aromatizable" androgens can "masculinize" female brain;
(iv) Antiestrogens (MER-25) and inhibitors of aromatization block in neonatal females the action of testosterone propionate.

Vreeburg *et al.* (1977) used an inhibitor of aromatization (1,4,6-androstatriene-3,17-dione, ATD) to test the hypothesis. The agent was not active in males; repeated injections of 1 mg (days 1, 3, 5, 10, 15 of life) had no effect on the sexual maturation, testosterone production, and fertility. In 5/9 animals transplanted ovarian tissue became luteinized. In female rats the insertion of a PDS implant containing ATD on day 3 of life blocked the effect of 250 μg of TP injected on day 5; 5/6 females showed ovarian cycles. McEwen *et al.* (1977) confirmed the blocking effect of ATD and noted that neonatal administration of ATD to males interfered with ejaculation in the adult.

Hayashi (1979) did not obtain a protective action by injecting 1 mg of ATD and 10 or 50 μg of testosterone propionate. In this study the steroid was active *per se*. Implants of ATD-paraffin micropellets placed in the hypothalamus induced sterility in 4/10 females by 90 days of age. In this study MER-25 did prevent the action of TP.

5α-Dihydrotestosterone does not block the effect of testosterone propionate (Ulrich *et al.*, 1972; van der Schoot *et al.*, 1976).

4.9.3 Diverse Agents

4.9.3.1 Antiestrogens

MER-25 (see Appendix for structure) in doses of 100, 250, or 500 μg blocks the action of 30 μg of TP, or 10 μg of EB, provided the antagonist is given at least 6 h prior to androgen handling. Only 5 day old pups responded; no protection took place in 1 day old females (Doughty and McDonald, 1974). Brown-Grant (1974) injected MER-25 6 h before injecting 10 μg of EB on day 4. The number of females that mated and ovulated (160 days old) remained low (<25%).

4.9.3.2 Retinoids

Vitamin A and its derivatives possess the ability to inhibit and reverse hyperplasia induced by carcinogens (Lasnitzky and Goodman, 1974; Felix *et al.*, 1975; Sporn *et al.*, 1976; Kurata and Micksche, 1977; Lotan, 1980) and steroid hormones (Chopra and Wilkoff, 1977; Iguchi and Takasugi, 1979). Neonatal injections of vitamin A acetate (retinol acetate) blocked precancerous or cancerous changes seen in the vaginae of PE mice of advanced age. The vitamin must be given simultaneously with the sex hormone (Mori, 1968b; 1969b; Yasui and Takasugi, 1977; Yasui *et al.*, 1977; Tachibana and Takasugi, 1980). If the sex hormones are injected first retinoids do not protect the neonate (Mori, 1969b; Iguchi and Takasugi, 1979; Tachibana *et al.*, 1984a). Daily doses of 100 µg of retinal, retinol, or retinol acetate, or 200 µg of retinol palmitate block the effects of 20 µg of estradiol. Retinoic acid (20 or 100 µg) given with estradiol is toxic leading to death of all experimental animals (Tachibana *et al.*, 1984b).

Vitamin A acetate (200 IU) also protects against the changes (the uterine and vaginal epithelium) induced by 5 µg of DES given for 5 days from birth.

4.10 Males

Kincl and Maqueo (1965) tested the action of progesterone dispensed to 5 day old male pups at the same time as mestranol (MEE). Mature (50 day old) males injected with MEE alone had atrophied testes, ventral prostate, seminal vesicles, and were azoospermic. Only 8.5% of tubules had an active epithelium. Concomitant injection of 3 mg of progesterone blocked the effect of MEE (Table 4.33).

4.10.1 Testosterone

Continuous exposure to testosterone propionate (15 daily injections from birth) will negate the effects of estradiol given during the same period to mice. Mature males (60 days old) show active spermatogenesis and secretory activity in the prostate and seminal vesicles (Takasugi and Furukawa, 1972).

Table 4.33 Protective Effect of Progesterone in 5 Day Old Male Rat Treated with Mestranol (MEE).

Treatment daily dose	No. of rats	Body wt. g ± SE	Organ weight, mg ± SE			Spermato-genetic index (range)
			Testes	Ventral prostate	Seminal vesicles	
0	23	200 ± 3	2261 ± 29	129.0 ± 5.7	104.0 ± 5.7	98.8 (96–98)
MEE (1 µg)	12	192 ± 2	942 ± 80	78.0 ± 4.8	18.6 ± 1.0	8.5 (0–22)
MEE (1 µg) and P (3 mg)	7	194 ± 5	1886 ± 91	85.6 ± 6.4	59.1 ± 2.2	91.2 (81–99)

Mestranol = 17α-ethynyl estradiol 3-methyl ether; P = progesterone; data from Kincl and Maqueo (1965).

4.10.2 Gonadotropins

Injections of gonadotropins protect male pups against estradiol. Takasugi and Mitsuhashi (1972) gave simultaneous injections of estradiol (20 μg) and 1 IU of either PMSG or hCG during the first 10 days of life. The dose of both the estrogen and gonadotropin was doubled during the next 5 days. The presence of gonadotropin was enough to offset the toxic manifestations of estradiol. In 60 day old animals spermatogenesis was active, and the weights of gonads and sex organs were comparable with controls.

4.10.3 Thymocytes

Kincl et al. (1965b) reported that males injected with a suspension of thymic cells on day one of life reacted poorly to an injection of estradiol benzoate on day 5 (Table 4.34). Females injected with TP reacted similarly (Table 4.35). The effect of thymocytes was specific. Cell suspension prepared from the spleen, liver, or salivary glands did not protect the pups against the deleterious effect of gonadal hormones. The pups were not protected when injected with thymic cells suspension at the same age.

D. Lymphoid System

4.11 Role of the Thymus

A connection between the function of the thymus and the reproductive system has long been suspected yet the nature of the relationship remains obscure (Chapter 2). In the adult estrogens and adrenocortical steroids influence immune responses. Prolonged treatment with estrogens induces involution of the thymus gland (Bedivan, 1974), inhibits graft rejection (Franks et al., 1975) and mitogenic

Table 4.34 Prevention of Estradiol Benzoate-Induced Sterility in 5 Day Old Male Rats by Injecting Thymocytes into 1 Day Old Pups.

Treatment		No. of rats	Index of spermato-genesis ±SE	Body wt. g±SE	Tissue weights, mg±SE		
Thymocytes	Estradiol benzoate, μg				Testes	Ventral prostate	Seminal vesicles
Autopsy at 50 Days of Age[b]							
−	0	21	87±1.3	197± 5.9	2490± 40	141.8±7.2	127.3±4.8
+	0	16	92±1.8	184± 4.2	2130±130	116.2±8.2	120.0±7.0
+	120	14	77±3.1	184± 6.8	1540± 89	98.2±7.0	46.1±4.4
+	240	15	19±5.3	184± 6.6	1010± 88	81.1±7.8	25.9±2.3
Autopsy at 160 Days of Age[c]							
+	0	8/8[a]	94±2.0	380±14	2890±109	NS	NS
−	120	2/10[a]	51±7.0	367±13	1920± 96	NS	NS
+	120	9/9[a]	99±0.8	398± 7	3330± 97	NS	NS

NS, not determined; [a] number of fertile rats/total number of rats; [b] Sprague-Dawley strain; [c] Wistar strain.

Table 4.35 Prevention of Testosterone Propionate-Induced Sterility in 5 Day Old Female Rats by Injecting Thymocytes into 1 Day Old Pups.

Treatment		Body wt. g ± SE	Ovarian response[a]	Average no. CL ± SE	Vaginal introitus days ± SE	Tissue weight, mg ± SE	
Thymo-cytes	Testosterone propionate, μg					Ovaries	Uterus
−	0	138 ± 3.8	13/17	9.0 ± 1.3	41 ± 1.3	43.9 ± 4.5	161 ± 8.8
+	0	147 ± 5.4	15/16	13.0 ± 1.4	39 ± 0.8	51.9 ± 3.7	203 ± 15.2
−	500	158 ± 4.4	1/8	2.0 ± 0.8	15 ± 0.2	17.1 ± 1.8	125 ± 44.2
+	100	157 ± 4.0	8/8	4.2 ± 1.2	17 ± 2.6	29.2 ± 2.4	218 ± 19.5
+	500	167 ± 3.9	13/16	6.0 ± 1.0	13 ± 1.3	32.5 ± 4.0	233 ± 19.3
+[b]	500[b]	139 ± 13.6	1/8	1.0 ± 0.9	12 ± 0.3	23.3 ± 3.2	129 ± 31.9

[a] Ovarian response = number of rats with corpora lutea/total number of rats (at the age of 50 days); [b] injected thymocytes, day 5 of life.

response (Kalland *et al.*, 1979) and suppresses lymphocyte transformation *in vitro*. Estradiol enhances the differentiation of B cells by inhibiting the activity of suppressor T cells (Paavonen *et al.*, 1981). In the spleen of male mice T lymphocytes are less abundant and B lymphocytes more abundant than in females. Neonatal androgenization of females modifies the T/B ratio towards the male type (Dörner *et al.*, 1980b). The gland becomes sensitive to sex hormones after birth (Barr *et al.*, 1984).

Removal of the thymus during the first few days after birth in mice and rats produces a wasting disease similar to that seen after neonatal cortisol and causes a significant delay in the age at which vaginal opening takes place (Besedovsky and Sorkin, 1974). Grafting thymus tissue will restore the opening to the time expected after puberty (Allen *et al.*, 1984).

Controversy exists pertaining to the effect of thymectomy on gonadal dysgenesis. Some studies reported cessation of sex function in both sexes of mice and rats, while others have found only marginal consequences. Possibly strain differences and/or incomplete removal of the thymus tissue may explain divergent results. Thyroiditis develops often in thymectomized infant rodents (Nishizuka *et al.*, 1973a; Kojima *et al.*, 1976). The complication confuses further the evaluation of results.

4.11.1 Mice

Females
Nishizuka and Sakakura (1969) studied male and female mice thymectomized at the age of 3 days and autopsied at monthly intervals from the age of 30 days to the age of 180 days to illustrate the injury caused to early folliculogenesis. Ovarian dysgenesis developed after the age of 60 days. In 90 day old females, when no ectopic thymus existed, the ovaries became small, primordial, primary, and secondary follicles were few in numbers, and corpora lutea were absent; interstitial cells were hyperplastic and hypertrophic. Particularly striking was the absence of even

small ripening follicles which are present in hypophysectomized animals. Gonadal development was normal in animals with residual thymus tissue (detected by microscopic examination) in the area of the parathyroid. Thymectomy performed at the age of 7, 20, or 40 days also did not affect gonadal development (Nishizuka and Sakakura, 1971).

Nishizuka *et al.* (1973 b) studied steroid biosynthesis in dysgenic ovaries.

Michael *et al.* (1980) confirmed the development of ovarian dysgenesis in mice thymectomized at the age of 3 days and noted a reduced secretion of LH and of GH but not FSH and PRL. Lintern-Moore and Norback-Sorensen (1976) reported reduction in the follicle population and absence of corpora lutea after neonatal thymectomy in the ovaries of 4 week old mice. In this study only 20% of the animals reacted; the authors did not search for residual thymus tissue in those that did not react.

Males
Thymectomy did not affect the reproductive function in males. The weights of the testes were comparable with controls, and mature sperm were present. The intervention did not result in the development of a wasting disease, did not influence body growth or the development of the spleen, liver, kidneys, mesenteric lymph nodes, salivary glands and the adrenal glands (Nishizuka *et al.*, 1973 b).

4.11.2 Rats

Ovarian dysgenesis develops in the gonads of neonatally thymectomized female rats (Hattori and Brandon, 1977; 1979; Lintern-Moore, 1977). The dysgenesis develops in a similar manner as in mice. Primary and secondary follicles are reduced by degeneration, interstitial cells hypertrophy, and corpora lutea, when present, develop cysts. Body weights and the weights of the ovaries, adrenal glands and uterus are within normal limits. Males develop azoospermia, the diameter of the seminiferous tubules is decreased, and sperm and spermatogonia may be absent. The prostate becomes atrophied in about 75% at the age of 170 days (Hattori and

Table 4.36 Potentation of Sterilizing Effects of Estradiol Benzoate Injected on Day 5 by Adjuvant Injected on Day 1 of Life.

Treatment	N[a]	Body wt.[b], g ± SE	Organ weights, mg = SE			Spermatogenetic index[d] ± SE			
			Testes	Ventral prostate	Seminal vesicles	Sperm	Spermatids	Spermatocytes	Spermatogonia
Control	19	203 ± 4	2485 ± 31	123 ± 7	111 ± 5	94.5 ± 0.9	98.5 ± 0.5	100	100
EB 120 µg	18	204 ± 2	1487 ± 43	102 ± 6	35.4 ± 3.0	76.8 ± 3.7	86.0 ± 2.7	80.1 ± 4.3	95.8 ± 1.2
240 µg	10	196 ± 4	846 ± 75	69.3 ± 6.6	30.2 ± 1.8	1.4 ± 0.7	5.5 ± 1.7	68.5 ± 6.0	86.9 ± 4.0
Adjuvant EB 120 µg	17	182 ± 5	832 ± 49	55.4 ± 4.4	25.4 ± 3.2	0	0	63.0 ± 7.0	81.0 ± 5.5
240 µg	9	175 ± 8	580[c]	32.5 ± 6.7	24.3 ± 2.1	0	2.5 ± 1.4	53.5 ± 7.8	67.6 ± 7.9

[a] N number of animals used; [b] autopsied 50 days old; [c] range 88–1036 mg; [d] spermatogenic index, percent tubules containing apparently normal cell elements (20 tubules randomly evaluated in each testis); Kincl (unpublished data).

Brandon, 1979). Rats of the Lewis strain are genetically predisposed to autoimmune orchitis, and neonatal thymectomy leads to an active manifestation of the disease (Lipscomb *et al.*, 1979).

The Wistar strain is possibly less sensitive (Zbuzkova and Kincl, 1970). Thymectomy in rats within 24 h after birth increased the sensitivity to the injury by steroid hormones. Both sexes were sensitive to the intervention. In males given EB (100 µg) on day 5 the weights of the testes (1826 ± 126 mg, $\pm$SE), ventral prostate (52 ± 5.9) and seminal vesicles (66.7 ± 10) were significantly lower when compared with sham operated and EB injected pups (2380 ± 84, 127 ± 15, and 134 ± 12, respectively). In females thymectomy on day 1 potentiated the effect of TP (100 µg), or EB (10 µg), given on day 5.

Injection of adjuvant on day 1 potentiated the sterilizing consequence of estradiol benzoate in males (Table 4.36). The observation suggests an association between the immune system, or some components of the system, and neonatal sterilization. The nature of this connection is not known.

References

Abdel-Aziz MT, Bialy G, Keith WB, Williams KIH (1969) Excretion of radioactive mestranol in the milk of animals. Int J Fertil 14: 39–47

Achilles WE, Sturgis SH (1951) The effect of intrasplenic ovarian graft on pituitary gonadotropins. Endocrinology 49: 720–731

Alklint T, Norgren A (1971) Effects of neonatally injected nonesterified testosterone on reproductive functions in female rats. Acta Endocrinol (Copenh) 66: 720–726

Allen LS, McClure JE, Goldstein AL, Barkley MS, Michael SD (1984) Estrogen and thymic hormone interactions in the female mouse. J Reprod Immunol 6: 25–31

Alvarez EO, Ramirez VD (1970) Distribution curves of ³H-testosterone and ³H-estradiol in neonatal female rats. Neuroendocrinology 6: 349–360

Anderson CH, Greenwald GS (1969) Autoradiographic analysis of estradiol uptake in the brain and pituitary of the female rat. Endocrinology 85: 1160–1165

Anderson CO, Zarrow MX, Denenberg VH (1970) Maternal behavior in the rabbit: effects of androgen treatment during gestation upon the nest-building behavior of the mother and her offspring. Horm Behav 1: 337–345

Anderson JN, Peck EJ Jr, Clark JH (1973) Nuclear receptor estrogen complex: accumulation, retention and localization in the hypothalamus and pituitary. Endocrinology 93: 711–717

Arai Y (1964a) Changes in the hypothalamic-pituitary system after ovariectomy in persistent-estrus and diestrus rats. J Fac Sci (Tokyo) Imp Univ See *IV*, 10: 369–379

Arai Y (1964b) Long-lasting effects of early postnatal treatment with estrogen on pituitary-gonadal system in male rats. Endocrinol Jpn 11: 153–158

Arai Y (1968) Metaplasia in male rat reproductive accessory glands induced by neonatal estrogen treatment. Experientia 24: 180–181

Arai Y (1970) Nature of metaplasia in rat coagulating glands induced by neonatal treatment with estrogen. Endocrinology 86: 918–920

Arai Y (1971) A possible process of the secondary sterilization: delayed ovovuletin syndrome. Experientia 27: 463–465

Arai Y (1972) Effect of 5-dihydrotestosterone on differentiation of masculine pattern of the brain in the rat. Endocrinol Jpn 19: 389–393

Arai Y, Gorski RA (1968a) Critical exposure time for androgenization of the rat hypothalamus determined by antiandrogen injection. Proc Soc Exp Biol Med 127: 590–593

Arai Y, Gorski RA (1968b) Critical exposure time for androgenization of the developing hypothalamus in the female rat. Endocrinology 82: 1010–1014

Arai Y, Gorski RA (1968c) Protection against the neural organizing effect of exogenous androgen in the neonatal rat. Endocrinology 82: 1005–1009

Arai Y, Gorski RA (1968d) Critical exposure time for androgenization of the developing hypothalamus in the female rat. Endocrinology 82: 1010–1014

Arai Y, Kasuma (1968) Effect of neonatal treatment with estrone on hypothalamic neurons and regulation of gonadotrophin secretion. Neuroendocrinology 3: 107–114

Arai Y, Masuda S (1968) Long-lasting effects of prepubertal administration of androgen on male hypothalamic-pituitary-gonadal system. Endocrinol Jpn 15: 375–378

Arai Y, Masuda S (1970a) Effect of electrochemical stimulation on the preoptic area of induction of ovulation in estrogenized rats. Endocrinol Jpn 17: 237–239

Arai Y, Masuda S (1970b) Effect of neonatal treatment with oestrogen on the pituitary reaction to neonatal castration in the male rat. J Endocrinol 46: 279–280

Arai Y, Chen CY, Nishizaka Y (1978) Cancer development in male reproductive tract in rats given diethylstilbestrol at neonatal age. Gann 69: 861–862

Aschheim P, Pasteels JL (1963) Etude histophysiologique de la sécrétion de prolactine chez les rates seniles. C R Acad Sci 257: 1373–1375

Attardi B, Geller LN, Ohno S (1976) Androgen and estrogen receptors in brain cytosol from male, female, and testicular feminized (tfm/y ♀) mice. Endocrinology 98: 864–874

Attramadal A, Aakvaag A (1970) The uptake of ^{3}H-oestradiol by the anterior hypophysis and hypothalamus of male and female rats. Z Zellforsch 104: 582–596

Barraclough CA (1961) Production of anovulatory sterile rats by single injection of testosterone propionate. Endocrinology 68: 62–67

Barraclough CA (1966) Modification in the CNS regulation of reproduction after exposure of prepubertal rats to steroid hormones. Recent Prog Horm Res 22: 503–529

Barraclough CA (1968) Alterations in reproductive function following prenatal and early postnatal exposure to hormones. In: McLaren A (ed) Advances in reproductive physiology. Academic Press, New York, vol 3, pp 81–112

Barraclough CA, Fajer AB (1968) Progestin secretion by gonadotropin-induced corpora lutea in ovaries of androgen sterilized rats. Proc Soc Exp Biol Med 128: 781–785

Barraclough CA, Gorski RA (1961) Evidence that the hypothalamus is responsible for androgen-induced sterility in the female rat. Endocrinology 68: 68–79

Barraclough CA, Haller EW (1970) Positive and negative feedback effects of estrogen on pituitary LH synthesis and release in normal and androgen-sterilized female rats. Endocrinology 86: 542–551

Barraclough CA, Leathem JH (1954) Infertility induced in mice by a single injection of testosterone propionate. Proc Soc Exp Biol Med 85: 673–674

Barraclough CA, Leathem JH (1959) Influence of age on the response of male mice to testosterone propionate. Anat Rec 134: 239–255

Baum MJ, Gallagher CA, Martin JT, Damassa DA (1982) Effects of testosterone, dihydrotestosterone, or estradiol administered neonatally on sexual behavior of female ferrets. Endocrinology 111: 773–780

Beach IA, Noble RG, Orndoff RK (1969) Effects of perinatal androgen treatment on responses of male rats to gonadol hormones in adulthood. J Comp Physiol Psychol 68: 490–497

Bedivan M (1974) Atrophie des formations lymphoides et troubles de development sous l'action prolongée des oestrogéns chez rat. Rev Roum Endocr 11: 97–102

Beeman EA (1947) The effect of male hormone on aggressive behavior in mice. Physiol Zool 20: 373–405

Bern HA, Gorski RA, Kawashima S (1973) Long-term effects of perinatal hormone administration. Science 181: 189–190

Bern HA, Jones LA, Mori T, Young PN (1975) Exposure of neonatal mice to steroids: long-term effects on the mammary gland and other reproductive structures. J Steroid Biochem 6: 673–676

Bern HA, Jones LA, Mills KT, Kohrman A, Mori T (1976) Use of the neonatal mouse in studying long-term effects of early exposure to hormones and other agents. J Toxicol Environ Health, Suppl 1: 103–116

Bern HA, Mills KT, Ostrander PL, Schoenrock B, Graveline B, Plapinger L (1984) Cervicovaginal abnormalities in BALB/c mice treated neonatally with sex hormones. Teratology 30: 267–274

Bernard RM (1977) Studies on lactation and contraception in WHO's research programme. J Biosoc Sci Suppl 4: 113

Besedovsky HO, Sorkin E (1974) Thymus involvement in female sexual maturation. Nature 249: 356-358

Biddulph (1939) The effect of testosterone propionate on gonadal development and gonadotropic hormone secretion in young male rats. Anat Rec 73: 447-463

Birge CA, Peake GT, Mariz IK, Daughaday WH (1967) Radioimmunoassayable growth hormone in the rat pituitary gland: effects of age, sex and hormonal state. Endocrinology 81: 195

Biro J, Endroczi E (1977) Nuclear RNA content and synthesis in anterior pituitary in intact, castrated and androgen sterilized rats. Endocrinol Exp (Bratisl) 11: 163-168

Booth JE (1976) Effects of 19-hydroxylated androgens on sexual differentiation in the neonatal female rat. J Endocrinol 70: 319-320

Borvendég J, Hermann H, Bajusz S (1972) Ovulation induced by synthetic lutenizing hormone releasing factor in androgen sterilized female rats. J Endocrinol 55: 207-208

Bottiglioni F, Collins WP, Flamingni C, Neumann F, Sommerville IF (1971) Studies on androgen metabolism in experimentally feminized rats. Endocrinology 89: 553-559

Bradbury T (1941) Permanent after-effects following masculinization of the infantile rat. Endocrinology 28: 101-106

Branham WS, Sheehan DM, Zehr DR, Ridlon E, Nelson CJ (1985) The postnatal ontogeny of rat uterine glands and age-related effects of 17β-estradiol. Endocrinology 115: 2229-2237

Breedlowe SM, Jacobson CD, Gorski RA, Arnold AP (1982) Masculinization of the female rat's spinal cord following a single neonatal injection of testosterone propionate but not estradiol benzoate. Brain Res 237: 173-181

Bronson FH, Desjardins C (1969) Aggressive behavior and SV function in mice: differential sensitivity to androgens given neonatally. Endocrinology 85: 971-973

Bronson FH, Desjardins C (1970) Neonatal androgen administration and adult aggressiveness in female mice. Gen Comp Endocrinol 15: 320-325

Bronson FH, Whitsett JM, Hamilton TH (1972) Responsiveness of accessory glands of adult mice to testosterone: priming with neonatal injections. Endocrinology 90: 10-16

Brown-Grant K (1964) The effect of testosterone during the neonatal period on the thyroid gland of male and female rats. J Physiol 176: 91-104

Brown-Grant K (1968) The effects of a single injection of estradiol benzoate on thyroid function in intact male rats. Can J Physiol Pharmacol 46: 697-700

Brown-Grant K (1974a) Steroid hormone administration and gonadotropin secretion in the gonadectomized rats. J Endocrinol 62: 319-332

Brown-Grant K (1974b) Failure of ovulation after administration of steroid hormones and hormone antagonists to female rats during the neonatal period. J Endocrinol 62: 683-684

Brown-Grant K (1975) On critical periods during the postnatal development of the rat. INSERM 32: 357-368

Brown-Grant K, Sherwood MR (1971) The early androgen syndrome in the guinea pig. J Endocrinol 49: 277-291

Brown-Grant K, Quinn DL, Zarrow MX (1964) Superovulation in the androgen-treated immature rat. Endocrinology 74: 811-813

Brown-Grant K, Munck A, Naftolin F, Sherwood MR (1971) The effects of the administration of testosterone propionate alone or with phenobarbitone and of testosterone metabolites to neonatal female rats. Horm Behav 2: 173-182

Brown-Grant K, Fink G, Greig F, Murray MAF (1975) Altered sexual development in male rats after oestrogen administration during the neonatal period. J Reprod Fertil 44: 25-42

Bruni JF, Van Vugt D, Mashall S, Meitis J (1977) Effects of naloxane, morphine and methionine enkephalin on serum prolactin, luteinizing hormone, follicle stimulating hormone, thyroid stimulating hormone and growth hormone. Life Sci 21: 461-466

Brunner JA, Witschi E (1946) Testosterone-induced modifications of sex development in female hamsters. Am J Anat 79: 293-320

Buchanan R (1975) Breast-feeding - aid to infant health and infertility control. Population Reports Series J 4: 49

Burin P, Thevenot-Duluc AJ, Mayer G (1963) Exploration des potentialités de l'hypophyse et des effecteurs des hormones genitales chez les Rattes en oestrus permanent provoqué par une injection posnatale de testostérone. C R Soc Biol 157: 1258-1262

Butenandt A, Kudszus H (1935) Über Androstendion, einen hochwirksamen männlichen Prägnungsstoff. Hoppe-Seylers Z Physiol Chem 237: 75-86

Butterstein GM, Freis ES (1978) Effect of parturition time on the response to neonatal androgens in female rats. Proc Soc Exp Biol Med 158: 179-182

Cagnoni M, Fantini F, Morace G, Ghetti A (1965) Failure of testosterone propionate to induce the "early-androgen" syndrome in rats previously injected with progesterone. J Endocrinol 33: 527-528

Campbell HJ (1964) The effects of neonatal hormone treatment on reproduction in the rabbit. J Physiol 176: 29-30

Carlevaro E, Riboni L, D'Albora H, Dominguez R, Buño W (1969) Effects of hemispaying on guinea pigs injected with testosterone on the first day of life. Acta Physiol Lat Am 19: 315-323

Chiappa SA, Fink G (1977) Releasing factor and hormonal changes in the hypothalamic-pituitary-gonadotrophin and adrenocorticotrophin systems before and after birth and puberty in male, female and androgenized female rats. J Endocrinol 72: 211-224

Chopra DP, Wilkoff LJ (1977) β-retinoic acid inhibits and reverses testosterone induced hyperplasia in mouse prostate organ culture. Nature 265: 339-341

Chopra JG (1972) Effect of steroid contraceptives on lactation. Am J Clin Nutr 25: 1202-1214

Chung LWR, Ferland-Raymond G (1975) Differences among rat sex accessory glands in their neonatal androgen dependence. Endocrinology 97: 145-153

Clancy D-AP, Edgren RA (1968) The effects of norgestrel, ethinyl -estradiol, and their combination, ovral, on lactation and the offspring of rats treated during lactation. Int J Fertil 13: 133-141

Clark JH, Zarrow MX (1968) Effect of pretreatment with androgen in the neonatal rat on LH-induced ovarian cholesterol depletion. Proc Soc Exp Biol Med 127: 626-629

Clark JH, Campbell PS, Peck EJ Jr (1972) Receptor estrogen complex in the nuclear fraction of the pituitary and hypothalamus of male and female immature rats. Neuroendocrinology 10: 218-228

Clemens JA, Meites J (1971) Neuroendocrine status of old constant-estrous rats. Neuroendocrinology 7: 249-256

Clemens LG, Hiroi M, Gorski RA (1969) Induction and facilitation of female mating behavior in rats treated neonatally with low doses of testosterone propionate. Endocrinology 84: 473-478

Colby HD, Kitay JI (1972) Sex and substrate effects in hepatic corticosteroid metabolism in the rat. Endocrinology 90: 473-438

Colby HD, Gaskin JH, Kitay JI (1973) Requirements of the pituitary gland for gonadal hormone effects on hepatic corticosteroid metabolism in rats and hamsters. Endocrinology 92: 769-774

Colombo JA (1968) Precocious vaginal opening in neonatally spayed testosterone-treated rats. Lack of luteinization in the ovarian graft. Acta Physiol Lat Am 18: 266-267

Cortes V, McCracken JA, Lloyd CW, Weisz J (1971) Progestin production by the ovary of the testosterone-sterilized rat treated with an ovulatory dose of LH, and the normal, proestrous rat. Endocrinology 89: 878-885

Crossley DA, Swanson HH (1968) Modification of sexual behavior of hamsters by neonatal administration of testosterone propionate. J Endocrinol 41: 8-9

Csaba G, Nagy SU (1976) Plasticity of the hormone receptors and possibility of their deformation in neonatal age. Experientia 32: 651

Cunha GR (1975) Age-dependent loss of sensitivity of female urogenital sinus to androgenic conditions as a function of the epithelial-stromal interaction in mice. Endocrinology 97: 665-673

Cupceancu B, Neumann F, Steinbeck H (1971) Beeinflusung der Ovarialfunktion erwachsener weiblicher Ratten durch neonatale Behandlung mit verschiedenen Steroiden mit bekannter Gestagen-wirkung. Acta Endocrinol (Copenh) 67: 337-344

Curtis EM (1964) Oral-contraceptive feminization of a normal male infant. Report of a case. Obstet Gynecol 23: 295-299

D'Albora H, Carlevaro E, Riboni L, de los Reyes L, Zipitria D, Dominguez R (1974) Advanced puberty in female guinea pigs treated with human chorionic gonadotrophin (HCG) or testosterone enantate (TE) at birth. Horm Res 5: 344-350

Danguy A, Pasteels JL, Ectors F (1977) Sensitivity of anterior hypothalamic areas to gonadal steroid implantation in androgenized female rats. J Endocrinol 74: 315-322

Dantchakoff V (1938a) Sur les effets de l'hormone male dans une jeune cobaye femelle traité depuis un stade embryonnaire (inversions sexuelles). C R Soc Biol 127: 1255-1258

Dantchakoff V (1938b) Effet du traitement hormonal de l'embryon de cobaye par la testostérone sur ses facultés procreatrices et sur sa progeniture. C R Soc Biol 128: 891-895

Dantchakoff V (1938c) Sur les effets de l'hormone male dans un jeune cobaye male traité depuis un stade embryonnaire (production d'hypermales). C R Soc Biol 127: 1259-1262

Deanesly R (1938) The androgenic activity of ovarian grafts in castrated male rats. Proc R Soc Lond (Biol) 126: 122-135

Deckx R, Raus J, Denef C, De Moor P (1965) Sex differences in the metabolism of cortisol by rat liver. Steroids 6: 129-141

De Moor P, Van Baelan H, Verhoeven G, Boeckx W, Adam-Heylen M, Vandoren G (1977) Role of the hypothalamo-hypophyseal axis in neonatal androgenisation and its postpubertal expression. J Steroid Biochem 8: 579-584

Denef C (1973) Differentiation of steroid metabolism in the rat and mechanisms of neonatal androgen action. Enzyme 15: 254-271

Denef C (1974) Effects of hypophysectomy and pituitary implants at puberty on the sexual differentiation of testosterone metabolism in rat liver. Endocrinology 94: 1577-1582

Denef C, De Moor P (1968) The "puberty" of the rat liver. II. Permanent changes in steroid metabolizing enzymes after treatment with a single injection of testosterone propionate at birth. Endocrinology 83: 791-798

Denef C, De Moor P (1969) "Puberty" of the rat liver. IV. Influence of estrogens upon the differentiation of cortisol metabolism induced by neonatal testosterone. Endocrinology 85: 259-269

Denef C, De Moor P (1972) Sexual differentiation of steroid metabolising enzymes in the rat liver. Further studies on predetermination by testosterone at birth. Endocrinology 91: 374-384

Denef C, Magnus C, McEwen BS (1973) Sex differences and hormonal control of testosterone metabolism in rat pituitary and brain. J Endocrinol 59: 605-621

Diamond M, Rust N, Westphal U (1969) High affinity binding of progesterone, testosterone, and cortisol in normal and androgen treated guinea pigs during various reproductive stages. Relationship to masculinization. Endocrinology 84: 1143-1151

Dieterlen F (1959) Das Verhalten des syrischen Goldhamster. Z Tierpsychol 16: 47-103

Dixit VP, Niemi M (1973) Action of testosterone administered neonatally on the rat perineal complex. J Endocrinol 59: 379-380

Dorfman RI (1967) The anti-estrogenic and anti-androgenic activity progesterone in the defence of a normal fetus. Anat Rec 157: 547-558

Dorfman RI, Kincl FA (1963) A new function of progesterone. Fourth Ann Meeting Mex Endocrin Soc Ixtapan de la Sal, Mexico, pp 171-175

Döcke F, Dörner G (1975) Anovulation in adult female rats after neonatal intracerebral implantation of oestrogen. Endokrinologie 65: 375-381

Dörner G, Hinz G, Schlenker G (1977) Demasculinizing effect of prenatal oestrogen on sexual behavior in domestic pigs. Endokrinologie 69: 347-350

Dörner G, Geier Th, Ahrens L, Krell L, Munx G, Sieler H, Kittner E, Muller H (1980a) Prenatal stress as possible aetiogenetic factor of homosexuality in human males. Endokrinologie 75: 365-368

Dörner G, Eckert R, Hinz G (1980b) Androgen-dependent sexual dimorphism of the immune system. Endokrinologie 76: 112-114

Doughty C, McDonald PG (1974) Hormonal control of sexual differentiation of the hypothalamus in the neonatal female rat. Differentiation 2: 275-285

Doughty C, Booth JE, McDonald PG, Parrott RF (1975) Effects of oestradiol-17β oestradiol benzoate and the synthetic oestrogen RU 2858 on sexual differentiation in the neonatal female rat. J Endocrinol 67: 419-424

Dubuc PU (1976) Body weight regulation in female rats following neonatal testosterone. Acta Endocrinol (Copenh) 81: 215-224

Duluc AJ, Mayer G (1967) Apparititon et stimulation experimentales des corps jaunes chez la ratte sterilisée par administration postnatale d'hormones genitales. C R Acad Sci Paris 264: 377-379

Dunlay JL, Preis LK Jr, Geral AA (1972) Compensatory ovarian hypertrophy as a function of age and neonatal androgenization. Endocrinology 90: 1309-1314

Dupont A, Cusan L, Labrie F, Coy DH, Li CH (1977) Stimulation of prolactin release in the rat by intraventricular injection of β-endorphins and methionine-enkephalin. Biochem Biophys Res Commun 75: 76-82

Eayers JT (1952) Sex differences in the maturation and function of the nervous system in the rat. Ciba Found Coll Endocrinol 3: 18-33

Eayers JT (1964) Effect of neonatal hyperthyroidism on maturation and learning in the rat. Anim Behav 12: 195-199

Eden S (1979) Age- and sex-related differences in episodic growth hormone secretion in the rat. Endocrinology 105: 555-560

Eguchi Y, Sakamoto Y, Arishima K, Morikawa Y, Hashimoto Y (1975) Hypothalamic control of the pituitary-testicular relation in fetal rats: measurement of collective volume of Leydig cells. Endocrinology 96: 504-507

Einarsson K, Gustafsson J-Å, Stenberg Å (1973) Neonatal imprinting of liver microsomal hydroxylation and reduction of steroids. J Biol Chem 248: 4987-4997

Eisenfeld AJ, Axelrod J (1966) Effect of steroid hormones, ovariectomy, estrogen pretreatment, sex and immaturity on the distribution of ^{3}H-estradiol. Endocrinology 79: 38-42

Elsaesser F, Parivizi N (1979) Estrogen feedback in the pig: sexual differentiation and the effect of prenatal testosterone treatment. Biol Reprod 20: 1187-1193

Enzine S, Papiernik M (1979) Effect de l'hemisuccinate d'hydrocortisone injecté pendant la gestation et à la naisssance sur le developpement du système immunitaire de la souris. INSERM 89: 105-112

Erskine MS, Geller E, Yuwiler A (1979) Effects of neonatal hydrocortisone treatment on pituitary and adrenocortical response to stress in young rats. Neuroendocrinology 29: 191-199

Everett JW, Holsinger JW, Zeilmaker GH, Redmond WC, Quinn DL (1970) Strain differences for preoptic stimulation of ovulation in cyclic, spontaneously persistent-estrous, and androgen sterilized rats. Neuroendocrinology 6: 98-108

Fachet J, Palkovits M, Vallent K, Stark E (1966) Effect of a single glycocorticoid injection on the first day of life in rats. Acta Endocrinol (Copenh) 51: 71-76

Fachet J, Stark E, Palkovits M (1968) Effect of a single neonatal glycocorticoid injection on the thymus-lymphatic and endocrine system and on the growth of the rat and the dog. Acta Med Acad Sci Hung 25: 395-407

Falvo RE, Buhl A, Nalbandov AV (1974) Testosterone concentrations in the peripheral plasma of androgenized female rats and in the estrous cycle of normal female rats. Endocrinology 95: 26-29

Felix EL, Loyd BC, Cohen MH (1975) Inhibition of the growth and development of a transplantable murine melanoma by vitamin A. Science 189: 886-888

Flerkó B, Bardos V (1961) Luteinization induced in "constant oestrus rats" by lowering estrogen production. Acta Endocrinol (Copenh) 37: 418-422

Flerkó B, Illei Gy (1957) Nach Gonadalhormoneverabreichung eintretende Gewebsreaktionen der intralinienalen Ovarialtransplantate von Ratten. Acta Morphol Hung 7: 377-39

Flerko B, Mess B (1968) Reduced oestradiol-binding capacity of androgen sterilized rats. Acta Physiol Acad Sci Hung 33: lll-113

Flerkó B, Mess B, Illei-Donoffer A (1969) On the mechanism of androgen sterilization. Neuroendocrinology 4: 164-169

Forchielli E, Dorfman RI (1956) Separation of 5-ene-5 and 4-ene 5α-hydrogenases from rat liver homogenates. J Biol Chem 222: 443-448

Forsberg J-G (1969) The development of atypical epithelium in the mouse uterine cervix and vaginal fornix after neonatal oestradiol treatment. Br J Exp Pathol 50: 187-195

Forsberg J-G (1979) Developmental mechanism of estrogen-induced irreversible changes in the mouse cervicovaginal epithelium. Natl Cancer Inst Monogr 51: 41-56

Forsberg J-G, Kalland T (1981) Neonatal estrogen treatment and epithelial abnormalities in the cervicovaginal epithelium of adult mice. Cancer Res 41: 721-734

Franks CR, Perkins FT, Bishop D (1975) The effect of sex hormones on the growth of HeLa tumor nodules in male and female mice. Br J Cancer 31: 100-110

Frick J, Chang CC, Kincl FA (1969) Testosterone plasma in adult male rats injected neonatally with estradiol benzoate or testosterone propionate. Steroids 13: 21-27

Friend JP (1977) Persistence of maternally derived ³H-estradiol in fetal and neonatal rats. Experientia 33: 1235-1236

Fujii T (1973) Reappearance of cyclicity of vaginal smears in androgen-sterilized rats following thyroidectomy. Endocrinol Jpn 20: 425-428

Fuxe K, Hokfelt T, Nilsson O (1972) Effect of constant light and androgen-sterilization on the amine turnover of the tuberoinfundibular dopamine neurons: blockade of cyclic activity and induction of a persistent high dopamine turnover in the median eminence. Acta Endocrinol (Copenh) 69: 625-639

Gala RR, Clarke WP, Haisenleder DJ, Pan JT, Pieper DR (1984) The influence of blinding, olfactory bulbectomy, and pinealectomy on twenty four-hour plasma prolactin levels in normal and neonatally androgenized female rats. Endocrinology 115: 1256-1261

Gellert RJ, Lewis J, Pétra PH (1977) Neonatal treatment with sex steroids: relationship between the uterotropic response and the estrogen "receptor" in prepubertal rats. Endocrinology 100: 520-528

Gerall AA, Kenney A McM (1970) Neonatally androgenized females' responsiveness to estrogen and progesterone. Endocrinology 87: 560-566

Ghraf R, Lax ER, Hoff HG, Schriefers H (1973) Sex dependency of steroid retention in rat-liver slices incubated with testosterone. Eur J Biochem 35: 57-61

Ghraf R, Hoff HG, Lax ER, Schriefers H (1975) Enzyme activity in kidney, adrenal and gonadal tissue rats treated neonatally with androgen or oestrogen. J Endocrinol 67: 317-326

Gogan F, Beattie IA, Hery M, Laplante E, Kordon C (1980) Effect of neonatal administration of steroids or gonadectomy upon oestradiol-induced luteinizing hormone release in rats of both sexes. J Endocrinol 85: 69-74

Goldman AS (1970) Virilization of the external genitalia of the female rat fetus by dehydroepiandrosterone. Endocrinology 87: 432-435

Goldman BD, Gorski RA (1971) Effects of gonadal steroids on the secretion of LH and FSH in neonatal rats. Endocrinology 89: 112-115

Goldzieher JW, Axelrod LR (1963) Clinical and biochemical features of polycystic ovarian disease. Fertil Steril 16: 631-635

Gorski RA (1963a) Effects of low dosages of androgen on the differentiation of hypothalmic regulatory control of ovulation in the rat. Endocrinology 73: 210-216

Gorski RA (1963b) Modification of ovulatory mechanisms by postnatal administration of estrogen to the rat. Am J Physiol 205: 842-844

Gorski RA (1966) Localization and sexual differentiation of the various structures which regulate ovulation. J Reprod Fertil Suppl 1: 67-88

Gorski RA (1968) Influence of age on the response to perinatal administration of a low dose of androgen. Endocrinology 82: 1001-1004

Gorski RA, Barraclough CA (1963) Effects of low dosage of androgens on the differentiation of hypothalamic regulatory control of ovulation in the rat. Endocrinology 73: 210-216

Gorski RA, Gordon JH, Shryne JE, Southam AM (1978) Evidence for a morphological sex difference within the medial preoptic area of the rat brain. Brain Res 148: 333-346

Gorski RA, Harlan RE, Jacobson CD, Shryne JE, Southam AM (1980) Evidence for the existence of a sexually dimorphic nucleus in the preoptic area of the rat. J Comp Neurol 193: 529-539

Gottlieb H, Gerall AA, Theil A (1974) Receptivity in female hamsters following neonatal testosterone, testosterone propionate, and MER-25. Physiol Behav 12: 61-68

Green R, Luttge WG, Whalen RE (1969) Uptake and retention of tritiated estradiol in brain and peripheral tissue of male, female, and neonatally androgenized female rats. Endocrinology 85: 373-378

Greene RD, Burrill MW (1940) The recovery of testes after androgen induced inhibition. Endocrinology 26: 516-522

Greene RD, Burril MW (1941) Postnatal treatment of rats with sex hormones. The permanent effect on the ovary. Am J Physiol 133: 302-309

Greengard O (1975) Cortisol treatment of neonatal rats: effects on enzymes in kidney, liver and heart. Biol Neonate 27: 352-360

Greenough WT, Carter CS, Steerman C, DeVoogd TJ (1977) Sex differences in dendritic patterns in hamster preoptic area. Brain Res 126: 63-72

Griffiths EC, Hooper KC (1973a) Peptidase activity in the hypothalami of rats treated neonatally with oestrogen. Acta Endocrinol (Copenh) 72: 9-17

Griffiths EC, Hooper KC (1973b) The effects of orchidectomy and testosterone propionate injection on peptidase activity in the male rat hypothalamus. Acta Endocrinol (Copenh) 72: 1-8

Grossman CJ (1984) Regulation of the immune system by sex steroids. Endocrinol Rev 5: 435-455

Grossman CJ, Nathan P, Taylor BB, Sholiton LJ (1979) Rat thymic dihydrotestosterone receptors: preparation, location and physicochemical properties. Steroids 34: 539-553

Gustafsson J-Å, Stenberg Å (1974a) Neonatal programming of androgen responsiveness of liver of adult rats. J Biol Chem 249: 719-723

Gustafsson J-Å, Stenberg Å (1974b) Masculinization of rat liver enzyme activities following hypophysectomy. Endocrinology 95: 891-896

Gustafsson J-Å, Stenberg Å (1974c) Irreversible androgenic programming at birth of microsomal and soluble rat liver enzymes active on 4-androstene-3,17-dione and 5α-androstane-3β,17α-diol. J Biol Chem 249: 711-718

Gustafsson J-Å, Stenberg Å (1975) Partial masculinization of rat liver enzyme activities following treatment with FSH. Endocrinology 96: 545-553

Gustafsson J-Å, Gustafsson SA, Ingelman-Sundberg M, Pousette Å, Stenberg A, Wrange O (1974) Sexual differentiation of hepatic steroid metabolism in the rat. J Steroid Biochem 5: 855-859

Gustafsson J-Å, Ingelman-Sundberg M, Stenberg Å, Hökfelt T (1976) Feminization of hepatic steroid metabolism in male rats following electrothermic lesion of the hypothalmus. Endocrinology 98: 922-926

Gustafsson J-A, Eneroth P, Pousette Å, Skett P, Sonnenschein C, Stenberg Å, Ahlen A (1977) Programming and differentiation of rat liver enzymes. J Steroid Biochem 8: 429-443

Gustafsson PO, Beling CG (1969) Estradiol-induced changes in beagle pups: effect of prenatal and postnatal administration. Endocrinology 85: 481-491

Háčik T (1966) Effects of a single dose of testosterone administered to rats in early postnatal period on the adrenal function in adult life. Arch Int Physiol Biochem LXXIV:1-8

Háčik T, Palkovic M (1973) Effect of perinatal administration of sex hormones on lipid metabolism in adult rats. Endocrinol Exp 7: 53-61

Hale HB (1944) Functional and morphological alterations of the reproductive system of the female rat following prepubertal treatment with estrogens. Endocrinology 35: 499-506

Harlan RE, Gorski RA (1977) Steroid regulation of luteinizing hormone secretion in normal and androgenized rats at different ages. Endocrinology 101: 741-749

Harlan RE, Gorski RA (1978) Dissociation between release of luteinizing hormone and prolactin and induction of female sexual behavior in normal and androgenized rats. Biol Reprod 19: 439-446

Harris GW, Campbell HJ (1966) The pituitary gland, vol 2, p 99. California Press, Berkeley

Harris GW, Levine S (1965) Sexual differentiation of the brain and its experimental control. J Physiol 181: 379-400

Hart BL (1968) Neonatal castration: influence on neural organization of sexual reflexes in male rats. Science 160: 1135-1136

Hattori M, Brandon MR (1977) Infertility in rats induced by neonatal thymectomy. In: Boettcher B (ed), Immunological influence on human fertility. Academic Press, New York, pp 311-322

Hattori M, Brandon MR (1979) Thymus and the endocrine system: ovarian dysgenesis in neonatally thymectomized rats. J Endocrinol 83: 101-lll

Hayashi S (1967a) Difference in luteinization in intrasplenic ovarian grafts between normal and persistent-estrus rats following ovariectomy. J Fac Sci (Tokyo) Imp Univ Sec IV 2: 235-242

Hayashi S (1967b) Difference in the responsiveness to estrogen of the anterior hypophysis between normal and neonatally estrogenized rats. J Fac Sci (Tokyo) Imp Univ Sec IV 2: 227-234

Hayashi S (1968) Hyperplasia and metaplasia of the uterine epithelium following estrone injections in ovariectomized adult rats given neonatal injections of sex steroids. Endocrinol Jpn 15: 229-234

Hayashi S (1969) Suppression by estrogen injections of luteinization in intrasplenic ovarian grafts in ovariectomized "normal" and neonatally estrogenized rats. Ann Zool Jpn 42: 13–20

Hayashi S (1974) Failure of intrahypothalmic implants of antiestrogen, MER-25, to inhibit androgen sterilization in female rats. Endocrinol Jpn 21: 453–457

Hayashi S (1975) Estrogen dependent metaplasia in uterine epithelium of adult rats given neonatal injections of estradiol benzoate. Ann Zool Jpn 48: 155–160

Hayashi S (1976a) Sterilization of female rats by neonatal placement of estradiol micropellets in the anterior hypothalamus. Endocrinol Jpn 23: 55–60

Hayashi S (1976b) Failure of intrahypothalamic implants of an estrogen antagonist, ethamoxytriphetol (MER-25), to block neonatal androgen-sterilization. Proc Soc Exp Biol Med 152: 389–392

Hayashi S (1977) Intrahypothalamic implants of testosterone or dihydrotestosterone in neonatal female rats with reference in induction of sterility. Endocrinol Jpn 24: 595–599

Hayashi S (1979) Influence of intrahypothalamic implants of antioestrogen or aromatase inhibitor on development of sterility following neonatal androgenization in female rats. J Steroid Biochem ll: 537–541

Hayashi S, Gorski RA (1974) Critical exposure time for androgenization by intracranial crystals of testosterone propionate in neonatal female rats. Endocrinology 94: 1161–1167

Hayashi S, Kawashima S (1971) Effect of estrogen on the development of corpora lutea in subcutaneous ovarian transplants in neonatally estrogenized rats. J Fac Sci Tokyo Sec IV 12: 359–363

Heffner LJ, van Tienhoven A (1973) Effects of progesterone on uptake and retention of ^{3}H-testosterone in neonatal female rat. Neuroendocrinology 12: 129–141

Heffner LJ, van Tienhoven A (1979) Effects of neonatal ovariectomy upon ^{3}H-estradiol uptake by target tissues of androgen sterilized female rats. Neuroendocrinology 29: 237–246

Heinrich J, Delille GP, Tramezzazni JH (1964) Action sur le tractus genital de la rate d'une dose unique de benzoate d'oestradiol administrée avant le 5 jour de la vie. Acta Anat 59: 141–162

Hendricks SE, Gerall AA (1970) Effect of neonatally administered estrogen on development of male and female rats. Endocrinology 87: 435–439

Henzl MR, Chang CC, Sundaram K, Kincl FA (1971) The influence of castration and neonatally administered steroids on the development of transplanted ovarian tissue under the kidney capsule in male rats and hamsters. Acta Endocrinol (Copenh) 66: 547–557

Hoeffler JP, Frawley LS (1986) Capacity of individual somatotropes to release growth hormone varies according to sex: analysis by reverse hemolytic plaque assay. Endocrinology 119: 1037–1041

Hoff HG, Ghraf R, Lax ER, Schriefers H (1977) Androgen dependency of hepatic hydroxysteroid dehydrogenases in the rat: prepubertal responsiveness and unresponsiveness towards exogenous testosterone. Experientia 33: 540–541

Holzhausen C, Murphy S, Birke LIA (1984) Neonatal exposure to a progestin via milk alters subsequent LH cyclicity in the female rat. J Endocrinol 100: 149–154

Howard E (1968) Reductions in size and total DNA of cerebrum and cerebellum in adult mice after corticosterone treatment in infancy. Exp Neurol 22: 191–208

Hubener HJ, Amelung D (1953) Enzymatische Umwandlungen von Steroiden. Hoppe-Seylers Z Physiol Chem 293: 137–141

Huffman JW (1941) Effect of testosterone propionate upon reproduction in the female. Endocrinology 29: 77–79

Hughes PCR, Tanner JM (1974) The effect of a single neonatal sex hormone injection on the growth of the rat pelvis. Acta Endocrinol (Copenh) 77: 612–624

Iguchi T (1984) Effects of sex hormones on neonatal mouse vaginal epithelium in vitro. Proc Jpn Acad 60 (Ser B): 414–417

Iguchi T (1985) Occurrence of polyovular follicles in ovaries of mice treated neonatally with diethylstilbestrol. Proc Jpn Acad 61 (Ser B): 288–291

Iguchi T, Takasugi N (1979) Blockade by vitamin A of the occurrence of permanent vaginal changes in mice treated neonatally with 5α-dihydrotestosterone. Anat Embryol 155: 127–134

Iguchi T, Takasugi N (1983) Comparison of the effect of 5α- and 5β-dihydrotestosterone on mouse vaginal and uterine epithelia administered neonatally or postpubertally. IRCS Med Sci 11: 696–697

Iguchi T, Ohta Y, Takasugi N (1976) Mitotic acitivity of vaginal epithelial cells following neonatal injections of different doses of estrogen in mice. Dev Growth Differ 18: 69–78

Iguchi T, Tachibana H, Takasugi N (1979) Inhibitory effect of anti-estrogen (MER-25) on the occurence of permanent vaginal changes in neonatally estrogen-treated mice. IRCS Med Sci 7: 579–583

Iguchi T, Ostrander PL, Mills KT, Bern HA (1985a) Induction of abnormal epithelial changes by estrogen in neonatal mouse vaginal transplants. Cancer Res 45: 5688–5692

Iguchi T, Iwase Y, Kato H, Takasugi N (1985b) Prevention by vitamin A of the occurrence of permanent vaginal and uterine changes in ovariectomized adult mice treated neonatally with diethylstilbestrol and its nullification in the presence of ovaries. Exp Clin Endocrinol 85: 129–137

Illei-Donhoffer A, Tima L, Flerkó B (1970) Ovulation induced by hypothalamic lesions in persistent oestrus rats. Acta Biol Acad Sci Hung 21: 197–206

Illsley NP, Lamartiniere CA (1980) The imprinting of adult hepatic monoamine oxidase levels and androgen responsiveness by neonatal androgen. Endocrinology 107: 551–556

Jacobsohn D (1964) Development of female rats injected shortly after birth with testosterone or an "anabolic steroid". Acta Endocrinol (Copenh) 45: 402–414

Jacobsohn D, Kuntzman R (1969) Studies on testosterone hydoxylation and pentobarbital oxidation by rat liver microsomes: selective effects of age, sex, castration and testosterone propionate treatments. Steroids 13: 327–341

Jacobsohn D, Norgren A (1965) Early effects of testosterone propionate injected into 5 day old rats. Acta Endocrinol (Copenh) 49: 453–465

Jansson J-O, Frohman LA (1987a) Differential effects of neonatal and adult androgen exposure on the growth hormone secretory pattern in male rats. Endocrinology 120: 1551–1557

Jansson J-O, Frohman LA (1987b) Inhibitory effect of the ovaries on neonatal androgen imprinting of growth hormone secretion in female rats. Endocrinology 121: 1417–1423

Jansson J-O, Ekberg S, Isaksson OGP, Eden S (1984) Influence of gonadal steroids on age- and sex-related patterns of growth hormone in the rat. Endocrinology 114: 1287–1294

Jimenez J, Ochoa M, Paz Soler M, Portales P (1984) Long-term follow-up of children breast-fed by mothers receiving depot medroxyprogesterone acetate. Contraception 30: 523–533

Johnson DC (1963) Hypophysial LH release in androgenized female rats after administration of sheep brain extracts. Endocrinology 72: 832–836

Johnson DC (1967) Gonadotropin patterns in male and female rats. Plasma LH and FSH at various ages evaluated by the method of parabiosis. Acta Endocrinol (Copenh) 56: 165–176

Johnson DC (1971) Serum follicle-stimulating hormone (FSH) in normal and androgenized male and female rats. Proc Soc Exp Biol Med 138: 140–144

Johnson DC (1979) Maintenance of functional corpora lutea in androgenized female rats treated with PMSG. J Reprod Fertil 56: 263–269

Johnson DC, Witschi E (1963) Hypophyseal gonadotropins following gonadectomy in male and female androgenized rats. Acta Endocrinol (Copenh) 44: 119–127

Johnson DC, Yasuda M, Sridharan BN (1964) Prepuberal development of the androgenized male rat. J Endocrinol 29: 95–96

Jones LA, Bern HA (1979) Cervicovaginal and mammary gland abnormalities in BALB/cCrgl mice treated neonatally with progesterone and estrogen, alone or in combination. Cancer Res 39: 2560–2567

Jones LA, Pacillas-Verjan R (1979) Transplantability and sex steroid hormone responsiveness of cervicovaginal tumors derived from female BALB/cCrgl mice neonatally treated with ovarian steroids. Cancer Res 39: 2591–2594

Joseph AA, Kincl FA (1974) Neonatal sterilization of rodents with steroid hormones. 5. A note on the influence of neonatal treatment with estradiol benzoate or testosterone propionate on steroid metabolism in the brain and testes of adult male rats. J Steroid Biochem 5: 227–231

Josso N (1970) Action of testosterone on the Wolffian duct of the rat fetus in organ culture. Arch Anat Microsc Morphol Exp 59: 37–50

Jost A (1972) Use of androgen antagonist and anti-androgens in studies on sex differentiation. Gynecol Invest 2: 180–202

Kalland T, Strand O, Forsberg J-G (1980) Long-term effects of neonatal estrogen treatment on mitogen responsiveness of mouse spleen lymphocytes. J Natl Cancer Inst 63: 413–421

Kalter H (1978) Elimination of fetal mice with sporadic malformations by spontaneous resorption in pregnancies of older females. J Reprod Fertil 53: 407–410

Kamal I, Hefnawi F, Ghoneim M, Talaat M, Younis N, Tagui A, Abdalla M (1969) Clinical, biochemical, and experimental studies on lactation. 1. Lactation pattern in Egyptian women. Am J Obstet Gynecol 105: 314–323

Kawashima S (1960) Secretion of androgen by ovarian grafts in male rats castrated at birth. J Fac Sci (Tokyo) Imp Univ Sec IV 9: 117–125

Kawashima S (1965) The effect of early postnatal injections of estrogen on the hypophysial-gonadal, adrenal and thyroid systems in the rat, with special reference to neurosecretion. J Fac Sci (Tokyo) Imp Univ Sec IV 10: 497–523

Kawashima S, Shinoda A (1968) Spontaneous activity of neonatally estrogenized female rats. Endocrinol Jpn 15: 305–312

Kawashima S, Takewaki K (1966) Basophils and erythrosinophils in the anterior hypophysis of steroid-sterilized female rats. Ann Zool Jpn 39: 23–29

Kawashima S, Asai T, Wakabayashi K (1974) Prolactin secretion in normal and neonatally estrogenized persistent-diestrous rats at advanced ages. In: Hatotami N (ed) Proc workshop conf int soc psychoneuroendocrinology. S Karger, Basel, pp 128–135

Kawashima S, Mori T, Kimura T, Arai Y, Nishizuka Y (1980a) Effects of estrogen treatment on persistent hyperplastic lesions of the vagina in neonatally estrogenized mice. Endocrinol Jpn 27: 533–539

Kawashima S, Wakahayashi K, Nishizuka Y (1980b) Lower incidence of nodular hyperplasia of the adrenal cortex after ovariectomy in neonatally estrogenized mice than in the controls. Proc Jap Acad Sci 56 Ser B:350–355

Khan MM (1963) Studies on the effects of a single injection of testosterone propionate into the female rat. Ph D Thesis, Indiana Univ, pp 1-127

Kikuyama S (1961) Inhibitory effect of reserpine on the induction of persistent estrus by the sex steroids in the rat. Ann Zool Jpn 34: 111-116

Kikuyama S (1962) Inhibition of induction of persistent estrus by chlorpromazine in the rat. Ann Zool Jpn 35: 6-11

Kikuyama S (1963) Doses of estrone in securing persistent-estrus rats. Ann Zool Jpn 36: 145–148

Kikuyama S (1965) Response of vaginal epithelium to estrogen in rats given injections of androgen in early postnatal life. Ann Zool Jpn 38: 64–69

Kikuyama S, Kawashima S (1966) Formation of corpora in ovarian grafts in ovariectomized adult rats subjected to early postnatal treatment with androgen. Sci Pap Coll Gen Ed Univ Tokyo 16: 69–74

Kimura T, Nandi S (1967) Nature of induced persistent vaginal cornification in mice. IV. Changes in the vaginal epithelium of old mice treated neonatally with estradiol or testosterone. J Natl Cancer Inst 39: 75–93

Kimura T, Takasugi N (1964) Persistent hyperplasia of vaginal epithelium following ovariectomy in rats given injections of androgen in early postnatal life. J Fac Sci (Tokyo) Imp Univ Sec IV 10: 391–396

Kimura T, Basu SL, Nandi S (1967) Nature of induced persistent vaginal cornification in mice. I. Effects of neonatal treatment with various doses of steroids. Exp Zool 165: 71–88

Kincl FA (1970) Neonatal sterilization of rodents with steroid hormones. 1. The uptake of radioactivity from injected estradiol or testosterone propionate in various organs of 5-day-old male and female rats. Endocrinol Exp (Bratisl) 4: 139–141

Kincl FA (1980) Debate on the use of hormonal contraceptives during lactation. Res Reprod 12(No 2): 1

Kincl FA, Chang CC (1970) Neonatal sterilization of rodents with steroid hormones. 2. Distribution of intravenously injected testosterone in adult male rats treated at the age of 5 days with testosterone propionate or estradiol benazoate. Endocrinol Exp (Bratisl) 4: 207–213

Kincl FA, Dorfman RI (1967) Activity of various steroids in inhibiting sexual development in rats when injected at the age of 5 days. Acta Endocrinol (Copenh) 55: 78–82

Kincl FA, Folch Pi A (1963) Efecto del tratamiento con esteroides en el cuyo. Ciencia (Mex) 22: 209–211

Kincl FA, Henderson SB (1978) The influence of neonatal steroid exposure on testosterone metabolism in adult rats. In: Dorner G, Kawakami M (eds) Hormones and brain development. Elsevier/North-Holland, New York, pp 147–152

Kincl FA, Maqueo M (1965) Prevention by progesterone of steroid induced sterility in neonatal male and female rats. Endocrinology 77: 859–862

Kincl FA, Maqueo M (1967) The effect of intrasplenic ovarian autografts on luteinization in rats treated neonatally with steroid hormones. Steroids 9: 583–589

Kincl FA, Maqueo M (1972) Neonatal sterilization of rodents with steroid hormones. The influence of estradiol benzoate injected into five day old rats on the weight of testes and ventral prostate and histology of the seminiferous epithelium and the pituitary gland during sexual maturation. Endocrinol Exp (Bratisl) 6: ll–15

Kincl FA, Zbuzková V (1971) Neonatal sterilization of rodents with steroid hormones. 3. Influence of neonatal treatment with testosterone propionate or estradiol benzoate on steroids in plasma of adult male rats. J Steroid Biochem 2: 215–221

Kincl FA, Folch Pi A, Herrera Lasso L (1962) Efecto de las hormonas sexuales en la rata prepuber., 3 rd Meeting Soc Méx Nutric Endocrinol San Jose Purua, Mich, Mexico, pp 23–28

Kincl FA, Folch Pi A, Herrera Lasso L (1963) Effect of estradiol benzoate treatment in the newborn male rat. Endocrinology 72: 966–968

Kincl FA, Maqueo M, Folch Pi A (1964) Recovery of gonadal function in male rats treated neonatally with 17β-oestradiol benzoate. Acta Endocrinol (Copenh) 47: 200–208

Kincl FA, Folch Pi A, Maqueo M, Herrera Lasso L Oriol A, Dorfman RJ (1965 a) Inhibition of sexual development in male and female rats treated with various steroids at the age of five days. Acta Endocrinol (Copenh) 49: 198–205

Kincl FA, Oriol A, Folch Pi A, Maqueo M (1965 b) Prevention of steroid-induced sterility in neonatal rats with thymic cell suspension. Proc Soc Exp Biol Med 120: 252–255

Kincl FA, Sickles JS, Henzl M (1967) Inhibition of sexual development in birds with steroid hormones. Gen Comp Endocrinol 9: 401–405

Kincl FA, Henzl M, Rudel HW (1970) The influence of neonatal injection of cortisol in rats. In: Kazda S, Denenberg VH (eds) The postnatal development of phenotype. Academia, Prague, pp 307–318

Kleezik GS, Marshal S, Campbell GA, Gelato M, Meites J (1976) Effects of castration, testosterone, estradiol, and prolactin on specific prolactin binding activity in ventral prostate of male rats. Endocrinology 98: 373–379

Knox WE, Auerbach VH, Lin ECC (1956) Enzymatic and metabolic adaptions in animals. Physiol Rev 36: 164–179

Kohrman AF, Greenberg RE (1968) Permanent effects of estradiol on cellular metabolism of the developing mouse vagina. Dev Biol 18: 632–650

Kojima A, Tanaka-Kojima Y, Sakakura T, Nishizuka Y (1976) Spontaneous development of autoimmune thyroiditis in neonatally thymectomized mice. Lab Invest 34: 550–557

Kolena J (1977) Testicular binding of ^{125}I-HGG in developing estrogenized and androgenized rats. Endokrinologie 69: 266–268

Kolena J, Háčik T, Šebökova E (1977) Postnatal development of gonadotropin binding sites and c-AMP synthesis in ovaries and estradiol plasma levels in estrogenized or androgenized female rats. Endocrinol Exp (Bratisl) ll: 219–225

Korenbrot CC, Paup DC, Gorski RA (1975) Effects of testosterone propionate or dihydrotestosterone propionate on plasma FSH and LH levels in neonatal rats and on sexual differentiation of the brain. Endocrinology 97: 709–717

Kovács K (1966) Production of tumors in the ovary of androgenized rats after autotransplantation into the spleen. Endocrinology 77: 654–658

Kovács K (1967) The development of corpora lutea in androgenized female rats with the ovary transplanted into the spleen. Acta Anat (Basel) 63: 167–178

Kramen MA, Johnson DC (1971) Mating, fertilization, and ovum viability in the avovulatory, persistent-estrous rat. Fertil Steril 22: 745–754

Krieger DT (1972) Circadian corticosteroid periodicity: critical period for abolition by neonatal injection of corticosteroid. Science : 1205–1207

Krieger DT (1974) Effect of neonatal hydrocortisone on corticosteroid ciradian periodicity, responsiveness to ACTH and stress in prepuberal and adult rats. Neuroendocrinology 16: 355–363

Kubo K, Mennin SP, Gorski RA (1975) Similarity of plasma LH release in androgenized and normal rats following electrochemical stimulation of the basal forebrain. Endocrinology 96: 492–500

Kuhn CM, Butler SR, Schanberg SM (1978) Selective depression of serum growth hormone during maternal deprivation in rat pups. Science 201: 1034–1036

Kurata T, Micksche M (1977) Suppressed tumor growth and metastasis by vitamin A+BCG in Lewis lung tumor bearing mice. Oncology 34: 212–215

Kurcz M, Gerhardt VJ (1968) Gonadotrophic activity in androgenized female rats studied by the method of parabiosis. Endocrinol Exp (Bratisl) 2: 29–38

Kurcz M, Kovalz K, Tiboldi T, Orosz A (1967) Effect of androgenization on adenohypophysial prolactin content in rats. Acta Endocrinol (Copenh) 54: 663–667

Ladosky W, Kesikowski WM (1969) Testicular development in rats treated with several steroids shortly after birth. J Reprod Fertil 19: 247–254

Ladosky W, Noronha JGL, Gaziri IB (1969) Luteinization of ovarian grafts in rats related to treatment with testosterone in the neonatal period. J Endocrinol 43: 253–258

Lamartiniere CA (1979) Neonatal estrogen treatment alters sexual differentiation of hepatic histidase. Endocrinology 105: 1031–1035

Lasnitzki I, Goodman DS (1974) Inhibition of the effects of methylcholanthrene on mouse prostate in organ culture by vitamin A and its analogs. Cancer Res 34: 1564–1571

Laumas KR, Malkani PK, Bhatnagar S, Laumas V (1967) Radioactivity in the breast milk of lactating women after oral administration of ^{3}H-norethynodrel. Am J Obst Gynecol 98: 411–413

Laumas V, Malkani PK, Laumas KR (1970) The possibility of estrogenic activity in the human and goat milk after administration of oral gestagens. Contraception 2: 331–338

Leathem JH (1956) The influence of steroids on prepuberal animals. Proc III Congress Anim Reprod, Cambridge, England

Lee DKH, Janikowsky A, Bird CE, Clark AF (1974) Kinetics of [1,2-3H]-testosterone metabolism in normal adult male rats: effects of estrogen administration. J Steroid Biochem 5: 27–32

Levine S (1958) Differential maturation of an adrenal response to cold stress in rats manipulated in infancy. J Comp Physiol Psychol 51: 774–778

Levine S (1962) Plasma corticosteroid response to electric shock in rats stimulated in infancy. Science 135: 795–796

Levine S, Mullins RF Jr (1967) Neonatal androgen or estrogen treatment and the adrenal cortical response to stress in adult rats. Endocrinology 80: 1177–1179

Leybold K, Staudinger H (1959) Geschlechts unterschiede im Steroid-stoffwechsel von Ratten Leber mikrosomen. Biochem Z 331: 389–398

Lieberman MW (1963) Early developmental stress and later behavior. Science 141: 824–825

Liggins GC, Howie RN (1972) A controlled trial of antepartum glucocorticoid treatment for prevention of the respiratory distress syndrome in premature infants. Pediatrics 50: 515–525

Lintern-Moore S (1977) Effect of athymia on the initiation of follicular growth in the rat ovary. Biol Reprod 17: 155–161

Lintern-Moore S, Norbaek-Sørensen I (1976) Effect of neonatal thymectomy on follicle numbers in the post-natal mouse ovary. Mech Ageing Dev 5: 235–239

Lipscomb HL, Gardner PJ, Sharp JG (1979) The effect of neonatal thymectomy on the induction of autoimmune orchitis in rats. J Reprod Immunol 1: 209–217

Lloyd CN, Weisz J (1966) Interrelationship of steroids and neural mechanism. Excer Medica Int Congr Series lll: 86

Lobl RT (1975) Androgenization: alterations in the mechanism of oestrogen action. J Endocrinol 66: 79–84

Lobl RT, Gorski RA (1974) Neonatal intrahypothalomic androgen administration: the influences of dose and age on androgenization of female rats. Endocrinology 94: 1325–1330

Lunaas T (1963) Transfer of 17 β-estradiol to milk in cattle. Nature 198: 288–289

Lynch KM Jr (1955) Recovery of the rat testis following estrogen therapy. Ann NY Acad Sci 55: 734

Machida T (1971) Luteinization of ovaries under stressful conditions in persistent estrous rats bearing hypothalamic lesions. Endocrinol Jpn 18: 427–431

Mallampati RS, Johnson DC (1974) Gonadotropins in female rats androgenized by various treatments: prolactin as an index to hypothalamic damage. Neuroendocrinology 15: 255-266

Maqueo M, Kincl FA (1964) Testicular histo-morphology of young rats treated with oestradiol-17β benzoate. Acta Endocrinol (Copenh) 46: 25-30

Maqueo M, Burt A, Kincl FA (1966) Pituitary histology of rats treated neonatally with steroid hormones. Acta Endocrinol (Copenh) 53: 438-442

Marriq P, Oddo G (1974) La gynecomastie induite chez le nouveaune par le lait maternel? Nouv Press Med 3: 2579

Martin T, Ladosky W, Veloso L (1961) Diferenciacõĭo sexual do tipo regulacão da actividade das gônadas feminina: ciclica, masculina: constante ou aciclica. Inversões experimentais em mamiferos. Ann Acad Bras Sci 33:XLIII

Martinez C, Bittner JJ (1956) A non-hypophyseal sex difference in estrous behavior of mice bearing pituitary grafts. Proc Soc Exp Biol Med 36: 825-828

Matsumoto A, Arai Y (1980) Sexual dimorphism in wiring pattern in the hypothalamic arcuate nucleus and its modification by neonatal hormonal environment. Brain Res 190: 238-242

Matsumoto A, Arai Y (1981) Effect of androgen on sexual differentiation of synaptic organization in the hypothalamus arcuate nucleus: an ontogenic study. Neuroendocrinology 33: 166-169

Maurer RA, Woolley DE (1971) Distribution of ³H-estradiol in clomiphene treated and neonatally androgenized rats. Endocrinology 88: 1281-1287

Maurer RA, Woolley DE (1974) Demonstration of nuclear ³H-estradiol finding in hypothalmus and amygdala of female, androgenized female, and male rats. Neuroendocrinology 16: 137-147

Maurer RA, Woolley DE (1975) ³H-estradiol distribution in female, androgenized female, and male rats at 100 and 200 days of age. Endocrinology 96: 755-765

Mausle E, Fickinger G (1976) Ultramorphometric studies of the adrenal cortex of adult rats after neonatal administration of sex steroids. Acta Endocrinol (Copenh) 81: 537-547

Mayer G, Duluc AJ (1970) Post-natal administration of sexual steroids: a method of exploration of the reproductive functions in the rat. In: Kazda S, Denenberg VH (eds) The postnatal development of phenotype. Academia, Prague, pp 253-262

Mazer H, Mazer C (1939) The effect of prolonged testosterone propionate administration on the immature and adult female rat. Endocrinology 24: 175-181

McDonald PG, Doughty C (1974) Effect of neonatal administration of different androgens in the female rat: correlation between aromatization and the induction of sterilization. J Endocrinol 61: 95-103

McEwen BS, Pfaff DW (1970) Factors influencing sex hormone uptake by rat brain regions. l. Effects of neonatal treatment, hypophysectomy, and competing steroid on estradiol uptake. Brain Res 21: l-16

McEwen BS, Liebergurg I, Chaptal C, Krey LC (1977) Aromatization: important for sexual differentiation of the neonatal rat brain. Horm Behav 9: 249-263

McGuire JL, Lisk RD (1969) Oestrogen receptors in androgen or oestrogen sterilized female rats. Nature 22: 1068-1069

McKenzie SA, Selley JA, Agnew JE (1970) Secretion of prednisolone into breast milk. Arch Dis Child 50: 894-896

Meaney MJ, Aitken DH, van Berkel C, Bhatnagar S, Sapolsky RM (1988) Effect of neonatal handling on age-related impairments associated with the hippocampus. Science 239: 766-768

Mennin SP, Gorski RA (1975) Effects of ovarian steroids on plasma LH in normal and persistent estrous adult female rats. Endocrinology 96: 486-491

Mennin SP, Kubo K, Gorski RA (1974) Pituitary responsiveness to luteinizing hormone-releasing factor in normal and androgenized female rats. Endocrinology 96: 412-416

Merklin RJ (1953) Reproductive performance of female mice treated prepubertally with a single injection of estradiol dipropionate. Endocrinology 53: 342-343

Michael SD, Taguchi O, Nishizuka Y (1980) Effect of neonatal thymectomy on ovarian development and plasma LH, FSH, GH and PRL in the mouse. Biol Reprod 22: 343-350

Mode A, Norstedt G, Eneroth P, Gustafsson J-A (1983) Purification of liver feminizing factor from rat pituitaries and demonstration of its identity with growth hormone. Endocrinology 113: 1250-1260

Moguilevsky JA, Fels E, Rubinstein L, Bur GE, Libertun C (1967) Effect of gonadectomy on gonadotrophin secretion in androgenized rats studied in parabiosis. Acta Physiol Lat Am 17: 200–203

Moguilevsky JA, Scacchi P, Rubinstein L (1977) Effect of neonatal androgenization on luteinizing hormone, follicle-stimulating hormone, prolactin and testosterone levels in male rats. J Endocrinol 74: 143–144

Moore CR, Price D (1930) The question of sex hormone antagonism. Proc Soc Exp Biol Med 28: 38–40

Moore CR, Price D (1932) Gonadal hormone formations and the reciprocal influence between gonads and hypophysis with its bearing on the problem of sex hormone antagonism. Am J Anat 50: 13–71

Moore CR, Price D (1938) Some effects of testosterone and testosterone propionate in the rat. Anat Rec 71: 59–78

Mori T (1967) Changes in alkaline phosphatase activity and mitotic rate in vaginal epithelium following estrogen injections in neonatally estrogenized mice. Ann Zool Jpn 40: 82–90

Mori T (1968 a) Changes in the reproductive and some other organs in old C3H/Ms mice given high dose estrogen injections during neonatal life. Ann Zool Jpn 41: 85–94

Mori T (1968 b) Effects of neonatal injections of estrogen in combination with vitamin A on the vaginal epithelium of adult mice. Ann Zool Jpn 41: 113–118

Mori T (1969 a) Mitotic activity in uterine and vaginal epithelium following postpuberal estrogen injections in neonatally estrogenized mice. Proc Jpn Acad 45: 931–936

Mori T (1969 b) Further studies on the inhibitory effect of vitamin A on the development of ovary-independent vaginal cornification in neonatally estrogenized mice. Proc Jpn Acad 45: 115–120

Mori T (1975) Effect of postpubertal oestrogen injections on mitotic activity of vaginal and uterine epithelial cells in mice treated neonatally with oestrogen. J Endocrinol 64: 133–140

Mori T (1976) Ultrastructural characteristics of the vaginal epithelium of neonatally estrogenized mice in response to subsequent estrogen treatment. Endocrinol Jpn 23: 341–345

Mori T (1977) Ultrastructure of the uterine epithelium of mice treated neonatally with estrogen. Acta Anat (Basel) 99: 462–468

Mori T (1979) Age-related changes in ovarian responsiveness to gonadotropins in normal and neonatally estrogenized mice. J Exp Zool 207: 451–458

Mori T, Nishizuka M (1978) Additional effects of postpuberal estrogen injections on the vaginal epithelium in neonatally estrogenized mice. Acta Anat (Basel) 100: 369–374

Mori T, Nishizuka Y (1982) Morphological alterations of basal cells of vaginal epithelium in neonatally oestrogenized mice. Experientia 38: 389–390

Mori T, Bern HA, Mills KT, Young PN (1976) Long term effects of neonatal steroid exposure on mammary gland development and tumorigenesis in mice. J Natl Cancer Inst 57: 1057–1062

Mori T, Mills KT, Bern HA (1979) Lack of evidence for a direct permanent effect of neonatal sex steroid exposure on mammary gland. IRCS Med Sci 7: 201–205

Mori T, Nagasawa H, Bern HA (1980) Long-terms effects of perinatal exposure to hormones on normal and neoplastic mammary growth in rodents: a review. J Environ Pathol Toxicol 3: 191–205

Mori T, Iguchi T, Takasugi N (1983) Origin of permanently altered epithelial cells of the vagina in neonatally estrogen-treated mice. J Exp Zool 225: 99–105

Morishita H, Naftolin F, Todd RB, Wilen R, Davies IJ, Ryan KJ (1975) Lack of an effect of dihydrotestosterone on serum luteinizing hormone in neonatal female rats. J Endocrinol 67: 139–140

Morrison RL, Johnson DC (1966) The effects of androgenization in male rats castrated at birth. J Endocrinol 34: 117–123

Moulton BC, Koenig BB (1984) Uterine deoxyribonucleic acid synthesis during preimplantation in precursors of stromal cell differentiation during decidualization. Endocrinology 115: 1302–1307

Nadler RD (1972) Intrahypothalamic locus for induction of androgen sterilisation in neonatal female rats. Neuroendocrinology 9: 349–353

Nagasawa H, Yanai R, Kikuyama S, Mori J (1973) Pituitary secretion of prolactin, luteinizing hormone and follicle-stimulating hormone in adult female rats treated neonatally with oestrogens. J Endocrinol 59: 599–604

Nagasawa H, Yanai R, Jones LA, Bern HA, Mills KT (1978) Ovarian dependence of the stimulatory effect of neonatal hormone treatment on plasma levels of prolactin in female mice. J Endocrinol 79: 391–392

Napoli AM, Gerall AA (1970) Effect of estrogen and anti-estrogen on reproductive function in neonatally androgenised female rats. Endocrinology 87: 1330–1337

Neill JD (1972) Sexual differences in the hypothalmic regulation of prolactin secretion. Endocrinology 90: 1154–1159

Neumann F, Steinbeck H (1974) Antiandrogens. In: Eichler O, Farah A, Herken H, Welch AD (eds) Handbook of experimental pharmacology. Springer, Berlin Heidelberg New York, pp 1–484

Neumann F, Elger W, von Berswordt-Wallrabe R (1966a) The structure of the mammary glands and lactogenesis in the feminized male rats. J Endocrinol 36: 353–356

Neumann R, Elger W, Kramer M (1966b) Development of a vagina in male rats by inhibiting androgen receptors with an anti-androgen during the critical phase of organogenesis. Endocrinology 78: 628–632

Neumann F, Elger W, von Berswordt-Wallrabe R, Kramer M (1966c) Beeinflussung der Regelmechanismen des Hypophysenzwischens von Ratten durch einen Testosteron-Antagonisten, Cyproteron (1,2-methylen-6-chlor-$\Delta^{4,6}$-pregnadien-17α-ol-3,20-dion). Arch Pharmak Exp Path 255: 221–235

Neumann F, Hahn JD, Kramer M (1967) Hemmung von Testosteronabhangigen Differenzierungsvorgangen der mannlichen Ratte nach der Geburt. Acta Endocrinol (Copenh) 54: 227–240

Nilsson S, Nygren K-G (1979) Transfer of contraceptive steroids to human milk. Res Reprod ll, No l: l-2

Nilsson S, Nygren K-G (1980) Debate on the use of hormonal contraceptives during lactation. Res Reprod 12, No 2: l-2

Nilsson S, Mellbin T, Hofvander Y, Sundelin C, Valentin J, Nygren K-G (1986) Long-term follow-up of children breast-fed by mothers using oral contraceptives. Contraception 34: 443–457

Nishizuka M (1976) Neuropharmacological study on the induction of hypothalamic masculinization in female mice. Neuroendocrinology 20: 157–165

Nishizuka Y, Sakakura T (1969) Thymus and reproduction: sex linked dysgenesia of the gonad after neonatal thymectomy in mice. Science 166: 753–755

Nishizuka Y, Sakakura T (1971) Ovarian dysgenesis induced by neonatal thymectomy in the mouse. Endocrinology 89: 886–893

Nishizuka Y, Tanaka Y, Sakakura T, Kojima A (1973a) Murine thyroiditis induced by neonatal thymectomy. Experientia 29: 1396–1398

Nishizuka Y, Sakakura T, Tsujimura T, Matsumoto K (1973b) Steroid biosynthesis in vitro by dysgenetic ovaries induced by neonatal thymectomy in mice. Endocrinology 93: 786–792

Nishizuka Y, Sakakura T, Taguchi O (1979) Mechanism of ovarian tumorigenesis in mice after neonatal thymectomy. NCI Monograph 51. Perinatal carcinogenesis, pp 89–96

Oberling C, Guérin M, Guérin P (1936) La production experimentale de tumeurs hypophysaires chez le rat. C R Soc Biol (Paris) 123: 1152

Ohta Y, Takasugi N (1974) Ultrastructural changes in the testis of mice given neonatal injection of estrogen. Endocrinol Jpn 21: 183–199

Ohtani R, Kawashima S (1983) The effects of estradiol-17β on the neuritic growth of neonatal mouse hypothalamus and preoptic area in primary culture. Ann Zool Jpn 56: 275–281

Orth JM, Gunsalus GL, Lamperti AA (1988) Evidence from Sertoli cell-depleted rats indicates that spermatid number in adults depends on numbers of Sertoli cells produced during perinatal development. Endocrinology 122: 787–794

Ošťádalová I, Pařizek J (1968) Delayed retardation of somatic growth of rats injected with oestrogens during the first days of postnatal life. Physiol Bohenoslov 17: 217–228

Ošťádalová I, Babický A, Lojda Z, Kolář M, Deyl Z, Pařizek J (1970) Some delayed effects of the administration of steroids to the newborn rats. In: Kazda S, Denenberg VH (eds) The postnatal development of phenotype. Academia, Prague, pp 267–283

Paavonen T, Andersson LC, Adlercreutz H (1981) Sex hormone regulation of in vitro immune response. Estradiol enhances human B cell maturation via inhibition of suppressor T cells in pokeweed mitogen-stimulated cultures. J Exp Med 154: 1935–1941

Packman PM, Boshans RL, Bragdon MJ (1977) Quantitative histochemical studies of the hypothalamus: dehydrogenase enzymes following androgen sterilization. Neuroendocrinology 23: 330–340

Pantic V, Genbacev O (1972) Pituitaries of rats neonatally treated with oestrogen. I. Luteotropic and somatotropic cells and hormones. Z Zellforsch 126: 41–52

Pařizek J, Figarová V, Lišková I (1963) Pozdní důsledky dodaní nekterych steriodu v ranem postnatalním období. Čs Fysiol 12: 347

Parmer LG, Katonah F, Angrist AA (1951) Comparative effects of ACTH, cortisone, corticosterone, desoxycorticosterone, pregnenolone on growth and development of infant rats. Proc Soc Exp Biol Med 77: 215–218

Petrusz P, Flerkó B (1968) Effect of ovariectomy and oestrogen administration on the pituitary and uterine weight in androgen sterilized rats. Acta Biol Acad Sci Hung 19: 159–162

Petrusz P, Nagy E (1967) On the mechanism of sexual differentiation of the hypothalamus; decreased hypothalamic oestrogen sensitivity in androgen-sterilized female rats. Acta Biol Acad Sci Hung 18: 21–26

Pfeiffer CA (1936) Sexual differences of the hypophyses and their determination by the gonads. Am J Anat 58: 195–225

Pincus G, Bialy G, Layne DS, Paniagna N, Williams KH (1966) Radioactivity in the milk of subjects receiving radioactive 19 steroids-nor. Nature 212: 924–925

Poland RE, Rubin RT, Weichsel ME Jr (1981) Neonatal dexamethasone administration. II. Persistent anteration of circadian serum anterior pituitary hormone rhythms in rats. Endocrinology 108: 1055–1059

Posner BI, Kelly PA, Friesen HG (1974) Induction of a lactogenic receptor in rat liver: influence of estrogen and the pituitary. Proc Natl Acad Sci USA 71: 2407–2410

Presl J, Jirásek J, Horský J, Henzl M (1966) Effect of early postnatal oestrogen administration on pituitary cytology in the adult female rat. J Endocrinol 34: 409–410

Presl J, Jirásek J, Horský J, Henzl M (1967) Early estrogen syndrome in the rat by the non-steroidal estrogen diethylstilbestrol. Experientia 23: 374–375

Quinn DL (1966) Luteinizing hormone release following preoptic stimulation in the male rat. Nature 209: 891–892

Raisman G, Field PM (1971) Sexual dimorphism in the preoptic area of the rat. Science 173: 731–733

Raisman G, Field PM (1973) Sexual dimorphism in the neuropil of the preoptic area of the rat and its dependence on neonatal androgen. Brain Res 54: 1–29

Ratner A, Adamo NJ (1971) Arcuate nucleus region in androgen sterilized female rats: ultrastructural observations. Neuroendocrinology 8: 26–35

Raynaud A (1942) Modification experimentale de la differenciation sexuelles des embryons de souris, par action des hormones androgènes et oestrogènes. Actual Sci Indus Nos 925 et 926, Hermann &Cie, Paris, pp 66–134

Reed ND, Jutila JW (1965) Wasting disease induced with cortisol acetate studies in germ free mice. Science 150: 356–357

Reilly RW, Thompson JS, Bielski RK, Severson CD (1967) Estradiol induced wasting syndrome in neonatal mice. J Immunol 98: 321–329

Rosner J, Delille GP, Tramezzani JH, Cardinalli D (1965) Production in vitro d'androgènes par l'ovaire de la ratte sterile. C R Seances Acad Sci [III] 261: 1113–1115

Roy S, Greenblatt RB, Mahesh VB (1964) Effects of clomiphene and intrasplenic ovarian autotransplantation on the anovulatory cystic ovaries of rats having androgen-induced persistent-estrus. Fertil Steril 15: 310–315

Rubinstein L, Ahrén K (1969) Uptake of α-aminoisobutyric acid (AIB) in isolated ovaries from androgenized rats. Endocrinology 84: 803–807

Rudel HW, Kincl FA (1966) The biology of anti-fertility steroids. Acta Endocrinol (Copenh) 51 (Suppl) 105: 1–45

Rúzsás C, Trentini GP, Mess B (1977) The role of the pineal gland in the regulation of LH-release in rats with different types of the anovulatory syndrome. Endokrinologie 70: 142–149

Salaman DF (1970) RNA synthesis in the rat anterior hypothalamus and pituitary: relation to neonatal androgen and oestrous cycle. J Endocrinol 48: 125–137

Salas M, Schapiro S (1970) Hormonal influences upon the maturation of the rat brain's responsiveness to sensory stimuli. Physiol Behav 5: 7–11

Sand K (1918) Studier over konskarakterer. Copenhagen, Denmark (cited from Pfeiffer)

Sarkar DK, Fink G (1979) Mechanism of the first spontaneous gonadotrophin surge and that induced by pregnant mare serum and effects of neonatal androgen in rats. J Endocrinol 83: 339-354

Saunders A, Terry LC, Audet J, Brazean P, Martin JB (1976) Dynamic studies of growth hormone and prolactin secretion in the female rat. Neuroendocrinology 21: 193-198

Saunders FJ (1967) Effects of norethynodrel combined with mestranol on the offspring when administered during pregnancy and lactation in rats. Endocrinology 80: 447-452

Savu L, Nuñez E, Jayle M-F (1974) Haute affinité du serum d'embryon de souris pour les oestrogenes. Biochem Biophys Acta 359: 273-281

Sawada T, Ichikawa S (1978) Sites of production of sex steroids: secretion of steroids from X-irradiated and polycystic ovaries of rats. Endocrinology 102: 1436-1444

Sawano S, Arimura A, Schally AV, Redding TW, Schapiro S (1969) Neonatal corticoid administration: effects upon adult pituitary growth hormone and hypothalamic growth hormone releasing hormone activity. Acta Endocrinol (Copenh) 61: 57-64

Schapiro S (1965a) Androgen treatment in early infancy: effect upon adult adrenal cortical response to stress and adrenal and ovarian compensatory hypertrophy. Endocrinology 77: 585-587

Schapiro S (1965b) Neonatal cortisol administration: effect of growth, the adrenal gland and pituitary adrenal response to stress. Proc Soc Exp Biol Med 120: 771-774

Schapiro S (1968) Some physiological, biochemical, and behavioral consequences of neonatal hormone administration: cortisol and thyroxine. Gen Comp Endocrinol 10: 214-228

Schapiro S, Salas M, Vukovich K (1970) Hormonal effects on ontogeny of swimming ability in the rat: assessment of central nervous system development. Science 168: 147-151

Schiavi RC (1968) Adenohypophyseal and serum gonadotropins in male rats treated neonatally with estradiol benzoate. Endocrinology 82: 983-988

Schiavi RC (1969) Effect of castration on pituitary and serum LH and FSH in testosterone-sterilized rats. Neuroendocrinology 4: 101-lll

Schlesinger M, Mark R (1964) Wasting disease induced in young mice by administration of cortisol acetate. Science 143: 965-966

Schriefers H, Ghraf R, Lax ED (1972a) Sex-specific aglucone patterns of testosterone metabolism in rat liver and their alteration following interference with sexual differentiation. Z Physiol Chem 353: 371-377

Schriefers H, Hoff HG, Ghraf R, Ockenfels H (1972b) Geschlechts und altersabhängige Entwicklung der Aktivittattsmuster von Enzymen des Steroidhormonstowechsels in der Rattenleber nach neonatalem Eingriff in die sexuelle Differenzierung. Acta Endocrinol (Copenh) 69: 789-800

Schuetz AW, Meyer RK (1963) Effect of early postnatal steroid treatment on ovarian function in prepuberal rats. Proc Soc Exp Biol Med 112: 875-880

Schultz KD, Haarmann H, Harland A (1972a) Female endocrine control mechanism during the neonatal period. I. The effects of oestrogens. Acta Endocrinol (Copenh) 70: 396-408

Schulz KD, Harland A, Haarmann H (1972b) Female endocrine control mechanisms during the neonatal period. Acta Endocrinol (Copenh) 71: 431-442

Schwartz BD, Gerall AA (1979) Influence of photoperiod and neonatally administered androgen on estrous cycles and behavior in hamsters. Biol Reprod 21: 1115-1124

Scott JN, Traurig HH (1973) Radioactivity in reproductive organs and the liver of neonatal and adult rats following [³H] testosterone administration. Biol Neonate 23: 346-358

Scott JN, Traurig HH (1977) Radioactivity in pituitaries and cerebrums of neonatal rats following [³H]-estradiol administration. Endokrinologie 69: 293-298

Segal SJ, Johnson DC (1959) Inductive influence of steroid hormones on the neural system: ovulation controlling mechanism. Arch Anat Microsc Morphol Exp 48: 261-274

Self LW (1966) CNS regulation of reproduction. Recent Prog Horm Res 22: 533-535

Selye H (1940) Production of persistent changes in the genital organs of immature female rats treated with testosterone. Endocrinology 27: 657-660

Selye H, Friedman SM (1940) The action of various steroid hormones on the ovary. Endocrinology 27: 857-866

Shaar CJ, Frederickson RCA, Dininger NB, Jackson L (1977) Enkephalin analogues and naloxane modulate the release of growth hormone and prolactin. Evidence for regulation by an endogenous opioid peptide in the brain. Life Sci 21: 853-860

Shapiro BH, Goldman AS, Steinbeck HF, Neumann F (1976) Is feminine differentiation of the brain hormonally determined? Experientia 32: 650-651

Shay H, Gerson-Cohen J, Paschkis KE, Fels SS (1939) The effect of a large dose of testosterone propionate (Oreton) on the female genital tract of the very young rats. Production of ovarian cysts. Endocrinology 25: 933-943

Sheehan DM, Branham WS, Medlock KL, Olson ME, Zehr DR (1981) Uterine responses to estradiol in the neonatal rat. Endocrinology 109: 76-82

Sheridan PJ, Zarrow MX, Denenberg VH (1973) Androgenization of the neonatal female rat with very low doses of androgen. J Endocrinol 57: 33-45

Sheridan PJ, Sar M, Stumpf WE (1974) Autoradiographic localization of ^{3}H-estradiol or its metabolites in the central nervous system of the developing rat. Endocrinology 94: 1386-1390

Shipley EG, Meyer RK (1965) Effect of corticoids and progestine on pituitary gonadotropic functions in immature rats. In: Dorfman RI (ed) Hormonal steroids, biochemistry, pharmacology and therapeutics. Academic Press, New York, vol 2, pp 293-300

Short RV (1974) Sexual differentiation of the brain of sheep. ISERM 32: 121-142

Shyamala G, Mori T, Bern HA (1974) Nuclear and cytoplasmic oestrogen receptors in vaginal and uterine tissue of mice treated neonatally with steroids and prolactin. J Endocrinol 63: 275-284

Sloop TC, Clark JC, Rumbaugh RC, Lucier GW (1983) Imprinting of hepatic estrogen-binding protein neonatal androgens. Endocrinology 112: 1639-1646

Somana R, Visessuwan S, Samridtong A, Holland RC (1978) Effect of neonatal androgen treatment and orchidectomy on pituitary levels of growth hormone in the rat. J Endocrinol 79: 399-400

Sporn MB, Dunlop NM, Newton DL, Smith JM (1976) Prevention of chemical carcinogen by vitamin A and its analogs (retinoids). Fed Proc 35: 1332-1338

Steinberger E, Duckett GE (1965) Effect of estrogens on initiation and maintenance of spermatogenesis in the rat. Endocrinology 76: 1184-1189

Swanson HE (1966) Modification of the reproductive tract of hamsters of both sexes by neonatal administration of androgen or oestrogen. J Endocrinol 36: 327-328

Swanson HE (1970) Effects of castration at birth in hamsters of both sexes on luteinization of ovarian implants, estrus cycles and sexual behavior. J Reprod Fertil 21: 183-186

Swanson HE, van der Werff ten Bosch JJ (1963) Sex differences in growth of rats, and their modification by a single injection of testosterone propionate shortly after birth. J Endocrinol 26: 197-207

Swanson HE, van der Werff ten Bosch JJ (1964a) The "early androgen" syndrome; its development and the response to hemispaying. Acta Endocrinol (Copenh) 45: 1-12

Swanson HE, van der Werff ten Bosch JJ (1964b) The "early androgen" syndrome; differences in response to pre-natal and post-natal administration of various doses of testosterone propionate in female and male rats. Acta Endocrinol (Copenh) 47: 37-50

Swanson HE, van der Werff ten Bosch JJ (1965a) Modification of male rat reproductive tract development by a single injection of testosterone propionate shortly after birth. Acta Endocrinol (Copenh) 50: 310-316

Swanson HE, van der Werff ten Bosch JJ (1965b) The "early androgen" syndrome; effects of pre-natal testosterone propionate. Acta Endocrinol (Copenh) 50: 379-390

Tabei T, Heinrichs WL (1974) Enzymatic oxidation and reduction of C_{19}-Δ^5-3β-hydroxysteroids by hepatic microsomes. III. Critical period for the neonatal differentiation of certain mixed-function oxidases. Endocrinology 94: 97-103

Tabei T, Heinrichs WL (1975) Enzymatic oxidation and reduction of C_{19}-3β-hydroxysteroids by hepatic microsomes. V. Testosterone as a neonatal determinant in rats of the 7α- and 16α-hydroxylation and reduction of 3β-hydroxyandrost-5-en-17-one (DHA). Endocrinology 97: 448-424

Tabei T, Fukushima K, Heinrichs WL (1975) Enzymatic oxidation and reduction of C_{19}-Δ^5-3 β-hydroxysteroids by hepatic microsomes. IV. Induction of DHA hydroxylases and aminopyrine N-demethylase in immature male rats by androgens. Endocrinology 96: 815-819

Tachibana H, Takasugi N (1980) Restoration of normal responsiveness of vaginal and uterine epithelia to estrogen in neonatally estrogenized, A-vitaminized adult mice. Proc Jpn Acad 56 Ser B:162-166

Tachibana H, Iguchi T, Takasugi N (1984a) Different perinatal periods of vitamin A administration for prevention of the occurrence of permanent vaginal changes in mice treated neonatally with estrogen. Zool Sci 1: 777-785

Tachibana H, Iguchi T, Takasugi N (1984b) Comparative study of blocking effects of various retinoids on the occurrence of permanent proliferation of vaginal epithelium in mice treated neonatally with estrogen. Endocrinol Jpn 31: 645-650

Taeusch HW Jr (1975) Glucocorticoid prophylaxis for respiratory distress syndrome. A review of potential toxicity. J Pediatr 87: 617-623

Takahashi S, Kawashima S (1981) Responsiveness to estrogen of pituitary glands and prolactin cells in gonadectomized male and female rats. Ann Zool Jpn 54: 73-84

Takasugi N (1952) Einflüsse von Androgen und Estrogen auf die Ovarien der neugeborenen und reifen, weiblichen Ratten. Ann Zool Jpn 25: 120-131

Takasugi N (1953) Intrasplenische Transplantation von Ovarien auf die weiblichen Ratten, denen von Geburt an Oestrogen, Androgen, Progesteron, Desoxycorticosteronacetat oder Cholesterin injiziert wurden. Ann Zool Jpn 26: 91-98

Takasugi N (1954a) Veranderungen der hypophysaren, gonadotropen Akivität der reifen, weiblichen Ratten, denen von Geburt an zwei gemischte Arten von hormonischen Steroiden injiziert wurden. J Fac Sci (Tokyo) Imp Univ Sec IV 7: 153-159

Takasugi N (1954b) Einflüsse von Androgen und Progestogen auf die Ovarien der Ratten, denen sofort nach der Geburt Oestrogeninjektion durchgefuhrt wurde. J Fac Sci (Tokyo) Imp Univ Sec IV: 299-311

Takasugi N (1956) Untersuchungen über die hypophysare, gonadotrope Aktivität der daueroestrischen und normalen Ratten unter Stress-Situationen. J Fac Sci Univ Tokyo 7: 605-623

Takasugi N (1958) Abnormalitäten bei den langfristig gehaltenen dauerestrischen Ratten. Ann Zool Jpn 31: 74-81

Takasugi N (1959) Veranderungen der gonadotropen Aktivitat des Hypophysenvorderlappens der Wistarratten, die mit natriumreicher oder natriummangelnder Kost gefuttert wurden. J Fac Sci Univ Tokyo 8: 549-561

Takasugi N (1963) Vaginal cornification in persistent-estrus mice. Endocrinology 72: 607-619

Takasugi N (1964) Hyperplasia of vaginal epithelium unaffected by ovariectomy in rats receiving early postnatal injections of estrogen. J Fac Sci (Tokyo) Imp Univ Sec IV 10: 403-407

Takasugi N (1966) Persistent changes in vaginal epithelium in mice induced by short-term treatment with estrogen beginning at different early postnatal ages. Proc Jpn Acad 42: 151-155

Takasugi N (1970) Testicular damages in neonatally estrogenized adult mice. Endocrinol Jpn 17: 277-281

Takasugi N (1971) Morphogenesis of estrogen-independent proliferation and cornification of the vaginal epithelium in neonatally estrogenized mice. Proc Jpn Acad 47: 193-198

Takasugi N (1972) Carcinogenesis by vaginal transplants from ovariectomized, neonatally estrogenized mice into ovariectomized normal hosts. Gann (Jap J Cancer Res) 63: 73-77

Takasugi N (1976) Cytological basis for permanent vaginal changes in mice treated neonatally with steroid hormones. Int Rev Cytol 44: 193-224

Takasugi N (1979) Development of permanently proliferated and cornified vaginal epithelium in mice treated neonatally with steroid hormones and the implication in tumorigenesis. Natl Cancer Inst Monogr 51: 57-66

Takasugi N, Bern HA (1962) Crystals and concretions in the vaginae of persistent-estrus mice. Proc Soc Exp Biol Med 109: 622-624

Takasugi N, Bern HA (1964) Tissue changes in mice with persistent vaginal cornification induced by early postnatal treatment with estrogen. J Natl Cancer Inst 33: 855-863

Takasugi N, Furukawa M (1972) Inhibitory effect of androgen on induction of permanent changes in the testis by neonatal injections of estrogens in mice. Endocrinol Jpn 19: 417-422

Takasugi N, Kamishima Y (1973) Development of vaginal epithelium showing irreversible proliferation and cornification in neonatally estrogenized mice: an electron microscope study. Devel Grow Diff 15: 127-140

Takasugi N, Kato T (1984) Acceleration by vitamin A of the permanent proliferation of mouse vaginal epithelium induced by neonatal treatment with progestins. Zool Sci 1: 120-125

Takasugi N, Mitsuhashi Y (1972) Effects of gonadotropins on the occurrence of permanent changes in the testis of mice given neonatal estrogen injections. Endocrinol Jpn 19: 423-428

Takasugi N, Tomooka Y (1976) Alterations of the critical period for induction of persistent oestrus by early postnatal treatment with gonadal steroids in neonatally cortisone-primed mice. J Endocrinol 69: 293–294

Takasugi N, Bern HA, DeOme KB (1962) Persistent vaginal cornification in mice. Science 138: 438–439

Takasugi N, Kimura T, Mori T (1970) Irreversible changes in mouse vaginal epithelium induced by early post-natal treatment with steroid hormones. In: Kazda S, Denenberg VH (eds) The postnatal development of phenotype. Academia, Prague, pp 229–251

Takewaki K (1968) Reproductive organs and anterior hypophysis of neonatally androgenized female rats. Science Rep Tokyo Woman's Christ Col 1-6: 31–47

Takewaki K, Kawashima S (1967) Some effects of neonatal administration of androgen or estrogen in female rat. Gunma Symp Endocr 4: 195–211

Takewaki K, Ohta Y (1976) Deciduoma formation in rats with cornified vagina. Experientia 32: 224–225

Tanner JM, Hughes PCR (1970) A system of assessing skeletal maturity in the growing rat: with notes on the effects of a single neonatal sex hormone injection on bone development, nose-to-rump and tail lengths. In: Kazda S, Dennenberg VH (eds) The postnatal development of phenotype. Academia, Prague, pp 285–294

Tarttelin MF, Shryne JE, Gorski RH (1975) Patterns of body weight change in rats following neonatal hormone manipulation: a "critical period" for androgen induced growth increases. Acta Endocrinol (Copenh) 79: 177–191

Terenius L, Meyerson J, Palis A (1969) The effect of neonatal treatment with 17β-estradiol or testosterone on the binding of 17β-estradiol by mouse uterus and vagina. Acta Endocrinol (Copenh) 62: 671–678

Teresava E, Kawakami M, Sawyer CH (1969) Induction of ovulation by electrochemical stimulation in androgenized and spontaneously constant-estrous rats. Proc Soc Exp Biol Med 132: 497–501

Tima L, Flerkó B (1967) Ovulation induced by autologous pituitary extracts in persistent estrous rats. Endocrinol Exp 1: 193–199

Tima L, Flerkó B (1968) Ovulation induced by autologous pituitary extracts in androgen and light-sterilized rats. Arch Anat Hist Embr Norm Expt 51: 699–705

Tima L, Flerkó B (1974) Ovulation induced by norepinephrine in rats made anovulatory by various experimental procedures. Neuroendocrinology 15: 346–354

Tima L, Flerkó B (1975) Ovulation induced by the intraventricular infusion of norepinephrine in rats made anovulatory by neonatal administration of various doses of testosterone. Endokrinologie 66: 218–220

Toran-Allerand CD (1976) Sex steroids and the development of the newborn mouse hypothalamus and preoptic area in vitro: implications for sexual differentiation. Brain Res 106: 407–412

Toran-Allerand CD (1980) Sex steroids and the development of the newborn mouse hypothalamus and preoptic area in vitro. II. Morphological correlates and hormonal specificity. Brain Res 189: 413–427

Tramezzani JH, Poumeau DG (1967) Polycystic ovaries in the rat induced by intraocular ovarian graft. Life Sci 6: 1507–1511

Tramezzani JH, Voloschin LM, Nallar R (1963) Effect of a single dose of testosterone propionate on vaginal opening in the rat. Acta Anat 52: 244–251

Treloar OL, Wolf RC, Meyer RK (1972) Failure of a single neonatal dose of testosterone to alter ovarian function in the rhesus monkey. Endocrinology 90: 281–284

Tuohimaa P, Johansson R (1971) Decreased estradiol binding in the uterus and anterior hypothalamus of androgenized female rats. Endocrinology 88: 1159–1164

Turgeon JL, Barraclough CA (1974) Pulsatile LH rhythm in normal and androgen-sterilized ovariectomized rats: effects of estrogen treatment. Proc Soc Exp Biol Med 145: 821–824

Turner BB, Taylor AN (1976) Persistent alterations of pituitary adrenal function in the rat by prepuberal corticosterone treatment. Endocrinology 98: 1–9

Turner BB, Taylor AN (1977) Effects of postnatal corticosterone treatment on reproductive development in the rat. J Reprod Fertil 51: 309–314

Turner CD (1938) Intraocular homotransplantation of prepuberal testes in the rat. Am J Anat 63: 101–159

Turner CD (1941) Permanent genital impairment in the adult rat resulting from the administration of estrogen during early life. Am J Physiol 133: 471–476

Uchibori M, Kawashima S (1985) Effects of sex steroids on the growth of neuronal processes in neonatal rat hypothalamus preoptic area and cerebral cortex in primary culture. Int J Dev Neurosci 3: 169–176

Uilenbroek JThJ, Gribling-Hegge LA (1977) Pituitary responsiveness to LH-RH in intact and ovariectomized androgen-sterilized rats. Neuroendocrinology 23: 43–51

Uilenbroek JThJ, van der Werff ten Bosch JJ (1972) Ovulation induced by pregnant mare serum gonadotrophin in the immature rat treated neonatally with a low or a high dose of androgen. J Endocrinol 65: 533–541

Uilenbroek JThJ, Arendsen de Wolff-Exalto E, Blankenstein MA (1976) Serum gonadotrophins and follicular development in immature rats after early androgen administration. J Endocrinol 68: 461–468

Uilenbroek JThJ, Tiller R, deJong FH, Vels F (1978) Specific suppression of follicle-stimulating hormone secretion in gonadectomized male and female rats with intrasplenic ovarian transplants. J Endocrinol 78: 399–406

Ulrich R, Yuwiller A, Geller E (1972) Failure of 5 α-dihydrotestosterone to block androgen sterilization in female rat. Proc Soc Exp Biol Med 139: 411–413

Ulrich R, Yuwiler A, Geller E (1976) Neonatal hydrocortisone: effect on the development of the stress response and the diurnal rhythm of corticosterone. Neuroendocrinology 21: 49–57

Ulrich R, Yuwiler A, Geller E, Archer C (1977) Neonatal hydrocortisone treatment: effects on adrenocortical response to stress and ACTH and the central threshold to dexamethasone feedback. 59 Ann Meeting Endocr Soc, Chicago, p 143 (abstract)

van Baelen H, Adam-Heylen M, Vandoren G, DeMoor P (1977) Neonatal imprinting of serum transcortin levels in the rat. J Steroid Biochem 8: 735–736

van der Molen HJ, Hart PG, Wijmenga HG (1969) Studies with 4-^{14}C lynestrenol in normal and lactating women. Acta Endocrinol (Copenh) 61: 225–274

van Rees GP, Gans E (1966) Effect of gonadectomy and oestrogen on pituitary LH-content and organ weights in androgen-sterilized female rats. Acta Endocrinol (Copenh) 52: 471–477

van der Schoot P, Zeilmaker GH (1972) Aspects of the function of ovarian grafts in neonatally castrated male rats. Endocrinology 91: 389–395

van der Schoot P, Zeilmaker GH (1973) Regulation of luteal activity in neonatally castrated male rats bearing ovarian grafts. Endocrinology 92: 674–678

van der Schoot P, van der Vaart PDM, Vreeburg JTM (1976) Masculinization in male rats is inhibited by neonatal injections of dihydrotestosterone. J Reprod Fertil 48: 385–387

Vernadakis A, Woodbury DM (1971) Effects of cortisol on maturation of the central nervous system. In: Influence of hormones on the nervous system. Karger, Basel, pp 85–97

Vértes M, King RJB (1971) The mechanism of oestradiol binding in rat hypothalamus; the effect of androgenization. J Endocrinol 51: 271–282

von Fels E, Moguilevsky J, Rubinstein L, Libertun C (1968) Die Reaktion des androgenisierten Ovars auf Gonadotrophine (Parabioseuntersuchungen). Endokrinologie 52: 352–355

von Fels E, Bosch LR, Libertun C (1972) Die Reaktion des androgenisierten Rattenovars auf die Verabfolgung von Gonadotrophinen. Endokrinologie 59: 197–202

Vorherr H (1973) Contraception after abortion and post partum. Am J Obstet Gynecol 117: 1002–1025

Vreeburg JTM, van der Vaart PDM, van der Schoot P (1977) Prevention of central defeminization but not masculinization in male rats by inhibition neonatally of oestrogen biosynthesis. J Endocrinol 74: 375–382

Wagner JW (1968) Luteinization of ovarian transplants in gonadectomized, pregnant mares' serum-primed immature male rats. Endocrinology 83: 479–484

Wagner JW, Erwin W, Critchlow V (1966) Androgen sterilization produced by intracerebral implants of testosterone in neonatal female rats. Endocrinology 79: 1135–1142

Ward IL (1972) Prenatal stress feminized and demasculinizes the behavior of males. Science 175: 82–84

Watts AG, Fink G (1984) Pulsatile luteinizing hormone release, and the inhibitory effect of estradiol-17β in gonadectomized male and female rats: effects of neonatal androgen or exposure to constant light. Endocrinology 115: 2251–2259

Ways SC, Bern HA (1979) Long-term effects of neonatal treatment with cortisol and/or estrogens in the female BALB/c mouse. Proc Soc Exp Biol Med 160: 94–98

Weichert CK, Kerrigan S (1942) Effects of estrogens upon the young of injected lactating rats. Endocrinology 30: 741–752

Weisburger JH, Yamomoto RS, Korzia J, Weisburger EK (1966) Liver cancer: neonatal estrogen enhances induction by a carcinogen. Science 154: 673–674

Weisz J, Ferin M (1970) Pituitary gonadotrophins and circulating LH in immature rats – a comparison between normal females and males and females treated with testosterone in neonatal life. In: Butt WR, Crooke AC, Ryle M (eds) Gonadotrophins and ovarian development. Livingstone, Edinburgh, pp 339–350

Weisz J, Lloyd C (1965) Estrogen and androgen production *in vitro* from 7-^{3}H-progesterone by normal and polycystic rat ovaries. Endocrinology 77: 735–744

Whalen RE, Etgren AM (1978) Masculinization and defeminization induced in female hamsters by neonatal treatment with estradiol benzoate and RU-2858. Horm Behav 10: 170–177

Whalen RE, Luttge WG (1971) Perinatal administration of dihydrotestosterone to female rats and the development of reproductive function. Endocrinology 89: 1320–1322

Whalen RE, Edwards DA, Luttge WG, Robertson TR (1969) Early androgen treatment and male sexual behavior in female rats. Physiol Behav 4: 33–39

Wijmenga HG, van der Molen HJ (1969) Studies with 4-^{14}C mestranol in lactating women. Acta Endocrinol (Copenh) 61: 665–677

Wilson JD (1973) Testosterone uptake by the urogenital tract of the rabbit embryo. Endocrinology 92: 1192–1199

Wilson JG (1943) Reproductive capacity of adult female rats treated prepuberally with estrogenic hormone. Anat Rec 86: 341–359.

Wilson JG, Wilson HC (1943) Reproductive capacity in adult male rats treated prepuberally with androgenic hormone. Endocrinology 33: 353–360

Wilson JG, Young WC, Hamilton JB (1940) A technic suppressing development of reproductive function and sensitivity to estrogen in the female rat. Yale J Biol Med 13: 189–202

Wolfe JM, Wilson JG, Hamilton JB (1944) The effect of early postnatal injection of testosterone propionate on the structure of the anterior hypophysis of male and female rats. Yale J Biol Med 17: 341–349

Wollman AL, Hamilton JB (1967) Prevention by cyproterone acetate of androgenic, but not of gonadotrophic, elicitation of persistent estrus in rats. Endocrinology 81: 350–356

Wright LL, Smolen AJ (1983) Effects of 17 α-estradiol on developing superior cervical ganglion neurons and synapses. Dev Brain Res 6: 299–303

Yanai R, Mori J, Nagasawa H (1977) Long-term effects of prenatal and neonatal administration of 5 -dihydrotestosterone on normal and neoplastic mammary development in mice. Cancer Res 37: 4456–4459

Yasui T, Takasugi N (1977) Prevention by vitamin A of the occurrence of permanent vaginal changes in neonatally estrogen-treated mice. Cell Tissue Res 179: 475–482

Yasui T, Iguchi T, Takasugi N (1977) Blockage of the occurrence of permanent vaginal changes in neonatally estrogen-treated mice by vitamin A: parabiosis and transplantation studies. Endocrinol Jpn 24: 393–398

Yates FE, Herbst AL, Urquhart J (1958) Sex difference in rate of ring A reduction of 4-ene-3-keto-steroids *in vitro* by rat liver. Endocrinology 64: 887–902

Zbuzková V, Kincl FA (1970) The influence of thymectomy and steroid hormones in neonatal rats. Proc Soc Exp Biol Med 135: 874–877

Zeilmaker GH (1964) Aspects of the regulation of corpus luteum function in androgen sterilized female rats. Acta Endocrinol (Copenh) 46: 571–579

Zimbelman RG, Lauderdale JW (1973) Failure of prepartum or neonatal steroid injections to cause infertility in heifer, gilts and bitches. Biol Reprod 8: 388–391

5 Nonsteroidal Agents

A variety of inorganic and organic chemical compounds and heavy metals are toxic to the fetus and the neonate. Of the total number of chemicals available in the US (perhaps 55,000) only about 3,000 have been tested for teratogenic effects. Perhaps about 1,000 are animal teratogens, and some 14 chemicals (or classes of chemicals) have been definitely shown to be human teratogens (Baum, 1987). One of the problems associated with determining which chemicals are human teratogens, or affect the reproductive function, is the difficulty of measuring that function in humans. Of necessity all parameters used, whether they be libido, menstruation patterns, sperm count, mutation or postnatal development, are very variable; only the sperm count can be expressed in meaningful numbers.

Most damage is done to the fetus (Chapter 3), but the sensitivity may continue during the neonatal period. For example Diaz and Samson (1980) obtained poor brain development (20% reduction in brain weight) and poor motor coordination and other behavioral abnormalities by feeding ethanol to baby rats from day 4 through day 7 after birth.

Cholesterol metabolism, a step essential in the biosynthesis of steroid hormones, is sensitive to neonatal influences. Baby rats raised on milk high in cholesterol were able to withstand cholesterol challenge in adult life without developing hypercholesteremia (Reiser and Sidelman, 1972). Cholestyramine resin was used to study the effect of hypocholesteremia. The resin is an agent which decreases cholesterol concentrations in blood by stimulating cholesterol catabolism, by increasing 7α-hydroxylation, and by sequestering bile acids in the intestines. Li *et al.* (1980) fed infant guinea pigs the resin and noted in the adults lower cholesterol concentration in blood, higher activity of 7α-hydroxylase, and increased secretion of bile acids.

Similar observations were made in children who consumed a low cholesterol diet; their plasma cholesterol concentration was lower in adolescence (Hodgson *et al.*, 1976).

Manipulation of diet during weaning can induce metabolic disturbances manifested in later life (Hahn and Kirby, 1973). Nutrition status is equally important. Ficková and Macho (1981) adjusted nutrition by reducing the number of pups in the litter (leaving 14, 8 or 4 pups to nurse). They evaluated the outcome by measuring insulin binding to small and large adipocytes obtained from 30, 60, 120 and 500 day old males. Smaller adipocytes from underfed rats (14 per litter) showed higher binding affinity to insulin; the total number of insulin receptors was the same in both types of cells regardless of neonatal nutritional intake.

5.1 Hormones

5.1.1 Gonadotropins

Neonatal testes and ovaries respond to stimulatory effects of hCG or LH by increased steroidogenesis; the treatment has no permanent effects in later life except for an increased sensitivity to hormones. Bradbury and Gaensbauer (1939) described masculinization of female rats by gonadotrophic extracts. Von Fels *et al.* (1968) provided increased gonadotropin source by joining rats in parabiosis. In a later study (von Fels *et al.*, 1972) the group used hCG.

5.1.1.1 Rats

Neonatal testes (Arai and Gorski, 1974) and ovaries (Neumann and Buchholz, 1971) respond to stimulation by hCG or LH. The data of Neuman and Buchholz (1971) are typical. Daily injections of hCG (1 IU from birth for 14 days) increased the sensitivity to a subsequent treatment with a low dose of TP (50 μg). All females exposed to both agents were, as adults, in constant estrus, and the ovaries contained only vesicular follicles; no CL were present. Females subjected to hCG or to TP alone developed normally.

I have used 200 μg of hCG (15 IU) in 5 day old rat pups and found stimulated growth which persisted into adulthood; body weights in males were increased by 15% and the levator ani muscle by 24%. The weights of testes were augmented by 17%, the ventral prostate by 16% and the seminal vesicles by 10%. Increases in organ weights in females were not significant.

5.1.1.2 Guinea Pigs

D'Albora *et al.* (1974) injected newborn pups with a high dose of hCG (1500 RU, RU = rat unit). The ovaries were smaller but contained corpora lutea. However, the day of vaginal opening occurred earlier (by 15–20 days) than in controls.

Schulz *et al.* (1972) used 5 IU of LH in newborn females and noted that the ovary was not refractory to the stimulation. Incorporation (*in vitro*) of ³H-leucine into ovarian and uterine proteins increased within 1 h after stimulation.

5.1.1.3 Mice

Gonads of both sexes react to stimulation by gonadotropin. Takasugi and Mitsuhashi (1972) injected hCG, PMSG, or a mixture of both to male pups (C3H/Tw strain). The dose was 1 IU daily for the first 10 days of life and double the dose for the next 5 days. Autopsy was at the age of 60 days. The treatment blocked the effects of simultaneous injection of 10 μg of estradiol. Iguchi *et al.* (1986a) induced persistent estrus in 30–50% of mice (ICR/JCL strain) by injecting 1 IU of hCG for 5 days if the treatment was begun when the animals were 10 days or older. Handling begun earlier was not active.

Schulz *et al.* (1972) utilized 5 IU of LH/100 g BW within 24-36 h of birth and noted increased incorporation of ^{3}H uridine and ^{3}H leucine in the ovaries and uterus but not in the cerebral cortex, hypothalamus, liver and adrenal glands. Long term effects, if any, were not reported.

5.1.2 Prolactin

Dopamine (DA) neurons in the tuberoinfundibular nucleus (TIN) stimulate the release of prolactin (PRL); TRH stimulates the system and bromocriptine (a DA suppressor) suppresses PRL release (Chapter 2). During lactation PRL accumulates in the milk, is ingested by the neonate, and passes in significant quantities into the bloodstream of the nursing individual (Whitworth and Grosvenor, 1978). The ingested amounts are deemed sufficient to influence the development of TIN (Shyr *et al.*, 1986). A brief period of neonatal PRL deficiency, induced by treating nursing mothers with bromocriptine, rendered the adult offspring more responsive to the stimulatory action of TRH, and unresponsive to the suppressive action of bromocriptine (Shah *et al.*, 1988).

5.1.3 Thyroid Hormones

The hypothalamus-pituitary axis is developed in newborn rats. The pituitary is sensitive to the stimulation of the hypothalamic hormone (thyrotropin releasing hormone, TRH) (Fisher *et al.*, 1977), and the neurovascular link between the hypothalamus and the pituitary is fully developed in 1 day old pups (Strbak, 1983). Despite the established CNS-pituitary link the neonatal rat is unique in that the release of the pituitary hormone (thyroid stimulating hormone, TSH) is independent of hypothalamic control (Theodoropoulos *et al.*, 1979). The regulation is established between days 5 and 12 of postnatal life (Strbak and Greer, 1979).

Thyroid hormones are present in appreciable amounts in mother's milk of rats, rabbits, pigs and primates but not in cows (Strbak *et al.*, 1983), but it is not certain whether maternal thyroid hormones contribute to the development of the offspring. In rats maternal thyroidectomy leads to a decrease of thyroxine in the plasma of pups 3 days later, but the neonate is able to compensate without any adverse effects. The concentrations return to levels seen in sham-operated pups within 2-3 days (Strbak *et al.*, 1974).

5.1.3.1 TRH and TSH

Repeated injection of TRH (50 μg) to lactating rats leads to an increase of serum TSH in the pups and lower body weights (Strbak *et al.*, 1983). Treatment with TSH results in a decreased TSH response in the adult (Csaba and Nagy, 1976).

5.1.3.2 Induced Hyperthyroidism

A brief increase in thyroid hormones in the neonate induces a permanent dysfunction of the hypothalamus-pituitary-thyroid axis.

Thyroid

Meserve and Leathem (1974) fed to mother rats desiccated thyroid (0.05% of the diet) during gestation and lactation. In 12 day old pups the exposure resulted in an accelerated maturation of the hypothalamus-pituitary-adrenal axis; the offspring responded to ether stress by an abnormally high release of corticoid hormones.

Thyroxine

Kikuyama (1966) employed in females 1 µg of thyroxine sodium salt daily, for seven days, beginning on the day of birth, then doubled the dose for the next 7 days. The treatment caused a premature eye opening (11-12 days) as the eyes opened during days 14-15 in controls. In addition to thyroxine some females were also injected with testosterone propionate on day 10 of life. Thyroxine reduced the incidence of PE. In the group given the combined treatment only 2/7 animals lacked corpora lutea; in the group given TP alone the proportion was 5/7. The author felt "...thyroxine may accelerate the maturation of the hypothalamus function..."

Kincl and co-workers used high doses of L-thyroxine sodium salt (100 µg and 1000 µg) in five day old male and female rats and evaluated the results at the age of 60 days. Both doses were toxic (about 40% of pups died within three weeks). Body and adrenal weights were decreased in males; the testes were atrophied and showed moderate azoospermia. No significant changes were observed in females (Kincl *et al.*, 1970).

Doses between 5 and 30 µg, often given in multiple injections to the neonate, resulted in the adult rats in impaired pituitary and thyroid growth (Bakke *et al.*, 1972; 1976); reduced ^{131}I uptake, protein bound iodine and serum concentration (Bakke *et al.*, 1974); marked reductions in TSH secretory reserves (Azizi *et al.*, 1974; Bakke *et al.*, 1975); greater response to hCG stimulation of androgenized females (Phelps and Leathem, 1976); decreased pituitary growth hormone content (Pascual-Leone *et al.*, 1976); and decreased sebum production (Toh, 1979).

Bakke *et al.* (1974) used only males given a total of 135 µg to 225 µg. In the adult (age 135 days) basal TSH secretion reduced from 30.2 ± 3.3 mU/min (con-

Table 5.1 Growth and Concentrations of Thyroid Hormones in Adult Rats Exposed to Transient Neonatal Hyperthyroidism (NH).[a]

Treatment	Body wt., g ± SE	GH[b]	Serum concentration, (ml) ± SE		
			TSH, ng	T4, ng	T3, pg
Controls	438 ± 11	100	746 ± 45	70 ± 2	949 ± 20
NH	329 ± 12*	54	543 ± 36*	62 ± 2*	751 ± 27*

[a] Neonatal hyperthyroidism was induced by thyroxine 0.4 µg/g BW, injected for 12 days from birth; [b] relative concentration of growth hormone (GH percent; in the pituitary controls 0.3 mg/pituitary/100 g BW); animals 120 days old at autopsy; data from Walker and Courtin (1985).

trols) to 19.7 ± 1.4 (treated group). Walker and Courtin (1985) injected thyroxine 0.4 µg/g BW daily during the first 12 postnatal days. The treatment significantly decreased the concentration of thyroid hormones in 120 day old animals (Table 5.1). Lengvári et al. (1977a,b) injected 10 µg/g BW for 3 days beginning either on day 2 or 8 after birth. Circadian corticosterone fluctuation was higher in 23 day old animals. Higher dose of thyroxine (20 µg/g BW) abolished circadian cycles (Taylor and Lengvári, 1977). The authors suggested that elevated thyroxine influences the maturation of those neural structures which participate in the regulation of the diurnal rhythm of ACTH secretion.

5.1.3.3 Induced Hypothyroidism

Removal of the thyroid shortly after birth results in stunted growth (Scow and Marx, 1945) and provokes delay in sexual maturation (Scow and Simpson, 1945). Hypothyroidism influences proper development of the brain: myelinization (Balazs et al., 1969), amine synthesis (Schwark and Keesey, 1976) and metabolism (Schwark and Keesey, 1975), protein synthesis (Dainat and Rebiere, 1976); the brain excitability is decreased (Maisami et al. (1971).

An agent (thiouracil, TU) which blocks iodine uptake and diminishes the activity of the gland was frequently used to induce hypothyroidism. Hughes (1944) reported retarded body growth, hypertrophy of thyroid glands and cretinism. Goddard (1948) found no permanent consequence on reproduction in rats with propylthiouracil-induced dwarfism. Kikuyama (1969) induced hypothyroidism in female baby rats by feeding propylthiouracil to mothers starting 4 days before expected delivery (day 18) and continuing during lactation for two weeks. The exposure to goitrogen was without effects on gonadal and thyroid functions in 150 day old animals but increased the sensitivity of the pups to androgens; after injection of TP at the age of 10 days only 3/14 females had luteinized ovaries while 8/16 in the group injected with TP alone exhibited regular cycles.

I have studied the effect of injecting 200 µg of thiouracil on days 3,5,7,9 and 12 of life (Group A) or a single injection of 5 mg on day 5 (group B). The treatment had no effect on thyroid and gonadal function in 90 day old male (Table 5.2) and female (Table 5.3) rats. The concentrations of triiodothyronine (T_3) and thyroxine (T_4) in peripheral blood were within normal values. The concentrations were (per 100 ml serum): in group A: (T_3) males 119 ± 13.8 ng ($\pm$SE), females 138 ± 4.2 ng; (T_4) males 4.2 ± 0.7 µg, females 3.7 ± 0.2 µg; Group B: (T_3) males 5.4 ± 0.3 ng, females 3.9 ± 0.3 ng; in controls: (T_3) males 94.7 ± 11.4 ng, females 145 ± 6 ng; (T_4) males 3.0 ± 0.7 µg, females 3.4 ± 0.1 µg. Some animals injected with a 5 mg dose developed patchy fur and had spots of bare skin which persisted during the period of observation.

Crutchfield and Dratman (1980) exposed newborn rats to a combination of stresses by feeding radioactive iodine (^{125}I), or by feeding a low protein and salt (Remington) diet and separating the pups from mother three times weekly. Body growth was temporarily retarded in males at puberty but plasma concentrations of thyroid hormones (thyroxine and triiodothyronine) were not different from controls. Females were not affected.

Table 5.2 The Influence of Neonatal Hypothyroidism on Organ Weights of 90 Day Old Male Rats.

Treatment	Body Wt. g + SD	Organ weight, mg $\pm$ SD				
		Thyroid	Adrenals	Testes	VP	SV
Controls[a]	287 $\pm$ 43	22.5 $\pm$ 1.1	45.4 $\pm$ 2.1	3160 $\pm$ 247	216 $\pm$ 45	186 $\pm$ 21
Thiouracil[a] (200 µg, 5 $\times$)	317 $\pm$ 65	20.5 $\pm$ 4.0	45.4 $\pm$ 4.9	3150 $\pm$ 311	202 $\pm$ 79	253 $\pm$ 67
Thiouracil[b] (5 mg)	352 $\pm$ 6	20.1 $\pm$ 3.4	41.7 $\pm$ 6.3	3342 $\pm$ 188	238 $\pm$ 61	287 $\pm$ 36

[a] 5 animals; [b] 10 animals; VP, ventral prostate; SV, seminal vesicles.

Table 5.3 The Influence of Neonatal Hypothyroidism on Organ Weights of 90 Day Old Female Rats.

Treatment	Body wt. g $\pm$ SD	Organ weight, mg $\pm$ SD			
		Thyroid	Adrenals	Uterus	Ovaries
Control[a]	212 $\pm$ 17	18.7 $\pm$ 1.5	41.6 $\pm$ 7.6	387 $\pm$ 46	86.6 $\pm$ 7.9
Thiouracil[b] (200 µg 5 $\times$)	273 $\pm$ 23	17.7 $\pm$ 1.9	52.4 $\pm$ 6.3	373 $\pm$ 57	88.8 $\pm$ 7.1
Thiouracil[c] (5 mg)	236 $\pm$ 10	17.1 $\pm$ 2.6	52.8 $\pm$ 3.4	355 $\pm$ 56	99.8 $\pm$ 18.6

[a] 5 animals; [b] 10 animals; [c] 9 animals.

5.1.4 Diethylstilbestrol

Triphenylethylenes (DES), chemicals with estrogen-like (uterotrophic) activity, are more toxic in the neonate than estradiol. The high toxicity of synthetic estrogens in neonatal rats is the result of the low binding to α-fetoprotein (AFP), a major plasma α-globulin present during embryonic and neonatal life in many mammalian species (Vannier and Raynaud, 1975; Payne and Katzenellenbogen, 1979; Sheehan and Young, 1979). The binding of estradiol to pregnancy plasma proteins is much greater than of synthetic estrogens while the binding to hormone receptors is similar (Table 5.4). As a result, estradiol is a less active estrogenic hormone than synthetic estrogens. After birth AFP disappears from plasma with a half-life of 3½ to 4½ days (see Pasqualini and Kincl, 1985), and the differences in estrogenic potencies vanish in the adult (Table 5.5).

5.1.4.1 Effects in Females

Ovaries

Polyovular follicles develop spontaneously in rodents and several other mammalian species including humans (see Iguchi *et al.*, 1986). In the Swiss strain mice the incidence is low, 1% or less. Iguchi (1985) injected daily graded doses (10^{-4}, 10^{-3}, 10^{-2}, 10^{-1}, 1 or 5 µg) of DES for five days starting on the day of birth to females

Table 5.4 Relative Binding of Various Estrogens to Fetoprotein (AFP) and Uterine Cytosol Receptors (ER).

Compound	Relative binding	
	AFP	ER
Estradiol	100	100
EE[a]	3.0	191
DES	1.4	248
Hexestrol	0.2	349

[a] EE, 17α-ethinyl estradiol; data from Payne and Katzenellenbogen (1979).

Table 5.5 Relative Uterotrophic Potency of Various Estrogens in Neonatal and Adult Rats.

Estrogen	Dose[a] µg/kg per day	
	Neonatal	Adult
Estradiol	300	3
EE	1.2	1
DES	4	3

[a] amounts needed to provide a half-maximal response; EE, 17α-ethinyl estradiol; DES, diethylstilbestrol; data from Harmon et al. (1989).

of ICR/JCL and C57BL strain. The treatment induced an increased incidence of polyovular follicles (POF) and follicles containing polynucleated ova (Figure 5.1). The minimum daily dose of DES to cause POF was 0.001 µg in C57BL mice and ten times less in the ICR/JCL strain. The incidence and the number of ova present were dose dependent. Follicles with 2–13 ova were present in groups given the high dose (Iguchi *et al.*, 1986). In the BALB/cCrgl 34 day old mice 0.1 µg increased the frequency from 16% in controls to 100%; the incidence increased to 12.6%. The incidence rose to 34% when the dose was increased to 2 µg. Both the frequency and the incidence are strain dependent (Iguchi *et al.*, 1987).

Forsberg *et al.* (1985) found PF in the ovaries of 21 day old NMRI mice injected with 5 µg of DES.

Uterus

Mice: Mice exposed to DES *in utero* (Chapter 3) or during neonatal life develop permanent proliferation and cornification of the vaginal epithelium, vaginal adenosis and adenocarcinoma (Forsberg, 1976; Newbold and McLachlan, 1982; Plapinger, 1981; Maier *et al.*, 1985). Basal cell carcinoma develops in transplanted vaginal tissue (Takasugi, 1972). The data of Forsberg (1976) are typical. Mice treated daily from the day of birth for 5 days with DES (5 µg) exhibited extensive adenosis and epidermization in the upper vaginal tract (cervical canal) 13 months

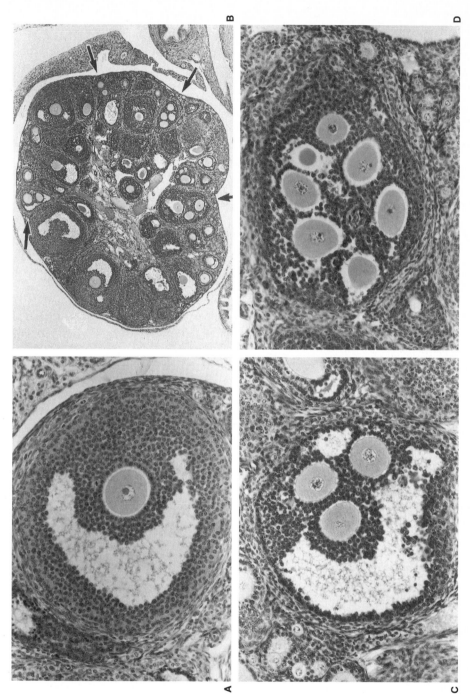

Figure 5.1. Occurrence of Polyovular Follicles in Mice Exposed Neonatally to Diethylstilbestrol. **A** control ovary; **B** ovary of 30 day old BALB/cJCL old mouse treated with 1 µg of diethylstilbestrol for 5 days from birth; note the presence (arrows) of several polyovular follicles, × 52. **C** same treatment as **B** ; 3 oocytes are visible; × 220. **D** same treatment as **B** ; 7 oocytes are present; × 260. Courtesy T Iguchi.

later with suggestive evidence of cancerous development in the glandular epithelium and in the epiderm.

When the dose is low (five injections of 1 μg for the first 5 days of life) squamous metaplasia develops only in the presence of the ovaries. When the daily dose is increased to 3 μg (or more) the metaplasia develops regardless of whether ovaries are present or not (Wong et al., 1982; Walker, 1983; Iguchi et al., 1987).

Epithelial proliferation and cornification (5 μg of DES) was inhibited in the C57BL/Tw strain by simultaneous injections of 200 IU of Vitamin A (Iguchi et al., 1985).

Rats: Slaughter et al. (1977) noted the development of persistent estrus in rats after a single dose of 0.5 μg given on day 5. Branham et al. (1988a) injected 10 μg on the day of birth and continued for 5 days. DES induced first an increase in the weight of the uterus and precocious proliferation of uterine glands. Within a week the uterus atrophied and failed to grow. The treatment decreased the formation of estrogen receptors (Medlock et al., 1988).

Mammary Glands

DES accelerates and increases the incidence of mammary tumor development in female mammary tumor virus (MTV) expressed mice (Mori et al., 1976). Nagasawa et al. (1980) used 5 μg in the SLN strain and found a 90% tumor frequency in the DES group and only a 20% frequency in controls.

Kalland et al. (1980) proposed that elevated PRL concentrations could contribute to the development of malignancies. The idea was based on the observation that pups injected with DES (5 μg) for the first five days after birth responded to estradiol treatment 6 months later with a larger prolactin release than did controls. Lopez et al. (1986) did not confirm the hypothesis; the group was able to increase tumor incidence in the C3H/MTV strain by a dose of 0.001 μg/day without an increase in PRL serum concentration. A daily dose of 2.5 μg led to an increase of PRL concentration in plasma, but tumor frequency was not increased.

5.1.4.2 Effects in Males

Neonatally castrated male rats injected with DES (1 μg for the first 10 days of life, 2 μg for the next 10 days and 4 μg for the last 10 days) developed squamous metaplasia (large papillary epithelial outgrowth) in the epithelium of the coagulating glands and ejaculatory ducts (Arai et al., 1977). Castration increased the degree of changes (Table 5.6). In 20-21 month old males coagulating glands and ejaculatory ducts became malignant in a few animals (Arai et al., 1978).

Table 5.6 Development of Reproductive Tract Abnormalities of Male Rats Exposed Neonatally to Diethylstilbestrol.

Treatment	Squamous metaplasia		Papillary growth in periurethral region of CG and ED	Squamous cell cancer
	Coagulating glands	Ejaculatory ducts		
Neonatal castration[a]	0/4	0/4	0/4	0/4
Control	0/10	0/10	0/10	0/10
Neonatal castration + DES[a]	11/11	11/11	11/11	2/11
Intact + DES	5/8	7/8	0/8	0/8

CG, coagulating glands; ED, ejaculatory ducts; [a] castration during first day of life; data from Arai *et al.* (1978).

5.1.5 Antiestrogens

5.1.5.1 Clomiphene

Fels *et al.* (1971) and Fels (1975) used clomiphene citrate 75–400 µg in baby rats within the first five days of life and found a high percentage of anovulatory females with pyometra and ovarian abscess formation. The exposure had no effect on testicular function in males. Schulz and August (1973), Clark and McCormack (1977; 1980) and Clark and Guthrie (1983) used doses from 10 µg to 500 µg and included the related nafoxidine with doses from 1 µg to 100 µg. They found in 100 day old females a high incidence of estrus smears, cystic atrophic ovaries and hypoplastic uterus in some while in others the gonads and the uteri were enlarged. Some animals exhibited pyometra, hyperplasia of the oviduct, hilus cell tumors, epithelial metaplasia, uterine cystic hyperplasia, and tumors of the uterus. The authors speculate that the agents acted directly on the target tissues.

Branham *et al.* (1988b) reported hypertrophy of uterine epithelium and inhibited uterine gland genesis by injecting daily 10 µg of clomiphene citrate on postnatal days 10–14.

5.1.5.2 Tamoxifen

Tamoxifen, 5 µg on day 1 of life, induced in adult (4 month old) rats atrophic ovaries and uterine and oviductal squamous metaplasia with abscess formation (Chamnes *et al.*, 1979). The abnormalities were the result of inhibited uterine gland genesis (Branham *et al.*, 1985).

Iguchi and Hirokawa (1986) gave 5 daily injections from birth to C57BL/Tw strain mice and evaluated the effect in 150 day old animals. In females a 2 µg dose caused uterine hypoplasia, myometrial involution and suppression of uterine gland genesis. Higher doses (20 µg and 100 µg) caused oocyte degeneration in more than 88% of small follicles, and the ovaries did not respond to stimulation by hCG. Urinary bladder hernia with or without caecum hernia was present in all the females injected with 100 µg (Iguchi *et al.*, 1986b). In the group of males exposed to 100 µg the agent provoked a 60% reduction in the weight of testes. The spermatogenic index (percentage of tubules containing spermatozoa in 100 cross sectioned tubules) was 66.9 ± 6.8 in the 20 μg group and 40.3 ± 7.6 in the 100 μg group. The index was 88.2 ± 1.0 in controls. The highest dose also inhibited body growth.

5.1.5.3 Nafoxidene

Clark and McCormack (1977) found reproductive tract abnormalities in rats exposed neonatally to nafoxidene. The teratologic effect was similar in nature to lesions observed after exposure to clomiphene.

5.1.6 Growth Hormone

A single dose of human growth hormone (200 μg) to 5 day old males increased significantly body growth of adults (body weight increased by 15% and levator ani muscle by 24%); the sex organs were also large (testes 17%, ventral prostate 16%, and seminal vesicles 10%); spermatogenesis was normal. Sexual function and body growth were not affected in females (Kincl, not published).

Strbak *et al.* (1985) reported that in rats stress lowered serum GH concentration during the first 12 days of life. Injecting an antiserum to somatostatin prevented the decrease.

5.1.7 Melatonin

A single injection of melatonin (1 mg) to 5 day old rats inhibited luteinization in 50 day old female rats (Kincl and Dorfman, 1967). The effect may be mediated by inhibiting the GnRH response. Melatonin (Martin and Klein, 1976), serotonin, and 5-methoxytryptamine (Martin *et al.*, 1977) prevent (*in vitro*) GnRH induced release of LH. Melatonin also blocked the release of LH in 8 day old rats (Martin *et al.*, 1980) but not in adult females (Kamberi *et al.*, 1970; Moguilevsky *et al.*, 1976).

5.1.8 Prostaglandins

In humans PGs exerted no observable effects during the first few days of life on a variety of physiological functions (blood chemistry, heart rate, lung function) in infants born to mothers in whom labor was induced by prostaglandins (Blackburn *et al.*, 1973).

5.1.9 Insect Hormones

Ecdysone, the invertebrate molting hormone, induces no specific endocrine changes in adult mammalian species or in neonatal rats. A dose of 100 μg injected within 24 h of birth had no effect on body growth and gonadal function of 3½ month old animals. In males adrenal weights were slightly higher (22.2 ± 4.1 mg (SD) vs 17.9 ± 3.8 mg in controls), and the weights of the thymus were lower (279 ± 78 mg) than in controls (366 ± 89 mg) (Csaba et al., 1978).

5.1.10 Plant Hormones

Csaba et al. (1977) injected rats once with 1 mg of gibberellin within 24 h of birth. In 4 month old males the plant growth hormone caused moderate testes atrophy: 1740 ± 150 mg vs 2070 ± 290 mg (± SD) in controls. In females the ovaries were heavier (55.1 ± 8.9 and 44.5 ± 8.2, respectively). The report is silent on the morphological aspects of the gonads.

5.2 Psychotropic Drugs

The recognition that various psychotropic drugs "protect" neonatal rats against the deleterious effects of androgen (Chapter 4) has prompted evaluation of the agents alone.

Dörner et al. (1977) used monoamine oxidase inhibitor, pargyline, and reported precocious vaginal opening and diminished male mounting behaviors but no clear-cut effects on other aspects of reproduction.

5.2.1 Chlorpromazine

In rats chlorpromazine injections up to the 10th day of life may result in irregular cycles (Kawashima, 1964). Caviezel et al. (1966a) used 400 µg in 4 day old pups and reported delayed sexual maturation and increased pituitary weights (5.7 ± 0.4 mg (SE) vs 4.1 ± 0.2 mg in controls) but no other effects. Arai and Gorski (1968) found no effects on LH levels in females. Ladosky et al. (1970) injected 10 day old males with 20 µg and reported spermatogenesis to be more advanced at the age of 45 days. Injecting the same dose on days 1, 5, 8, 12, or 15 inhibited testicular development.

Hogarth and Chalmers (1973) found accelerated sexual maturation in adolescent female rats (30 days old) after a single injection (20 µg) between days 4 and 10.

5.2.2 Reserpine

Injection of reserpine (causing monoamine depletion) into neonatal rats results in delayed puberty, irregular cycles and decreased pituitary reserves of gonadotropins (Kawashima, 1964; Caviezel et al., 1966b; Dörner et al., 1977).

5.2.3 Phenobarbital

Phenobarbital (40 mg per kg BW) fed daily to lactating rats for seven days caused alterations in the hepatic microsomal mixed-function oxidase (MFO) enzyme system of 37 week old male offspring. As a result nucleotide and protein formation induced by a hepatocarcinogen, aflatoxin B, was increased by about 40%. The authors speculated that exposure could be a contributing factor in cancer induction (Farris and Campbell, 1981).

5.3 Various Agents

5.3.1 Amino Acids

Acidic and sulfur containing amino acids (aspartic, cystic, homocystic) are neurotoxic (Olney et al., 1971; see Kizer et al., 1978). Pharmacological doses of neurotransmitter (L-DOPA, 6-hydroxy DOPA, 5-hydroxytryptophane) accelerate eye opening, cause decreases in serum TSH concentration, and increase PRL in the blood of males and lower PRL in females (Bakke et al., 1978 a).

5.3.1.1 Monosodium Glutamate

A taste improving food additive, monosodium glutamate (MSG), is neurotoxic in neonatal rodents and primates: mice (Olney, 1969; 1971; Lemkey-Johnston and Reynolds, 1974; Pizzi et al., 1977), rats (Trentini et al., 1974; Nemeroff et al., 1977 a,b; Bakke, 1978 a,b; Matswazana et al., 1979; Rodriguez-Sierra et al., 1980) and hamsters (Bunyan et al., 1976; 1980; Lamperti and Blaha, 1976; 1980; Lamperti et al., 1980; Tafelski and Lamperti, 1977). Five to ten subcutaneous injections of 4-8 mg/g body weight given daily, or on alternate days, beginning usually within 1 or 2 days of life are sufficient to produce toxic effects (Dyer et al., 1981).

A direct infusion (5 μl to 10 μl of 1% solution, 59 mM, at the rate of 1 μl/min) into the rostral hypothalamus of baby rats caused persistent hyperphagia and body weight gains of about 3 g daily. The agent destroyed neuron bodies but spared axon development. Infusion of L-tryptophan (30 mM), D-tryptophan (59 mM) and glycine (100 mM) also induced obesity (Simpson et al., 1977).

The agent damages the retina and small optic nerves (Potts et al., 1960). The reduction in the size of the optic nerve and chiasma is so large it is macroscopically visible (Saphier and Dyer, 1981). Microscopic lesions appear in the hypothalamus, primarily in the arcuate nucleus (Olney, 1969; Holzwarth et al., 1974; Lamperti and Blaha, 1976; Nemeroff et al., 1977), which is reduced to about 20% of its normal size (Holzwarth-McBride et al., 1976). The neural connection between the preoptic area and the mediobasal hypothalamus was deficient in female rats fed 4 mg/kg BW on days 2, 4, 6, and 10 of life; the neurons in the hypothalamus received fewer inhibitory inputs from the preoptic area (Saphier and Dyer, 1981).

Destruction of the hypothalamic nucleus leads to retardation of puberty and growth; the adults are stunted and obese. In females estrus cycles are prolonged, fertility is reduced, the gonads are smaller and the pituitaries and adrenals atrophied. The mechanism that triggers the release of GnRH from the hypothalamus becomes defective. Dopamine neurons in the arcuate nucleus are destroyed and/ or the terminals become hypersensitive to stimulation by K^+ ions and prostaglandin E_2 (DePaolo and Negro-Vilar, 1982). This leads to an impaired estrogen-induced release of PRL and LH. Prolactin release triggered by serotonergic stimuli is not blocked (Clemens et al., 1978).

The damage to dopaminergic neurons is reflected by a 50% decrease in dopamine concentration in the mediobasal hypothalamus in both sexes. In males the concentration was 4.31 ± 0.31 ng/mg protein ($\pm$SE) in controls and 2.09 ± 0.16 ng/mg in the MSG group; in females the values were 3.42 ± 0.19 and 1.32 ± 0.17, respectively. There were no differences in norepinephrine concentrations (Conte-Devolx et al., 1981).

Rats

Injections of increasing amounts (2, 2.5, 2.75, 3, 3.5 mg/g BW) into neonatal rats from birth led in the adult to a reduction of pituitary and serum TSH, weight reduction of the pituitary, thyroid and adrenal glands and atrophy of the gonads (Bakke et al., 1978b). The concentration of serum LH and FSH and the stores of GnRH in the hypothalamus are within normal limits while the prolactin concentration is higher (Badger et al., 1982). Nemeroff et al. (1981) confirmed no changes in LH and FSH in blood and found decreased levels of estradiol in 60 day old females (Table 5.7). Hong et al. (1981) reported reduced neuropeptide concentrations in the hypothalamus and Krieger et al. (1979) reduced stores of ACTH and endorphins. Growth hormone secretion is decreased (Dada et al., 1984).

Ovariectomized females respond to estradiol benzoate stimulation by LH release in the same manner as do untreated animals. However, the pulsatile release of LH was less frequent (Dyer et al., 1981).

In male rats the concentration of GnRH in the hypothalamus (Badger et al., 1982) and of LH (Bakke et al., 1978; Clemens et al., 1978; Greeley et al., 1978) and of FSH (Badger et al., 1982) in peripheral blood were reported to be within normal limits. Nemeroff et al. (1981) found increases (Table 5.8). Males respond to GnRH stimulation by releasing LH in the same manner as do control animals. However, in castrates LH concentration and the pulse amplitude are lower (Badger et al., 1982).

Table 5.7 Hormone Concentrations in Adult Female Rats Exposed Neonatally to Monosodium Glutamate (MSG).

Treatment	Hormone concentration, ml serum $\pm$ SE		
	LH, ng	FSH, ng	Estradiol, pg
Controls	62.8 ± 9.9	149 ± 32	481 ± 74
MSG	39.3 ± 8.6	$97.6 \pm 12.7*$	$154 \pm 33*$

* Indicates a significant difference; data from Nemeroff et al. (1981).

Table 5.8 Hormone Concentrations in Adult Male Rats Exposed Neonatally to Monosodium Glutamate (MSG).

Treatment	Hormone concentration, ng/ml serum $\pm$ SE		
	LH	FSH	Testosterone
Controls	32.1 ± 4.2	309 ± 18	2.8 ± 0.17
MSG	$23.1 \pm 3.2^*$	$200 \pm 10^*$	$0.9 \pm 0.17^*$

* Indicates a significant difference; data from Nemeroff *et al.* (1981).

Neonatal MSG has no effect on the maturation of the hypothalamus-pituitary-adrenal axis. The response to stress (ACTH and β-endorphin concentrations in blood) and basal concentrations of ACTH, α-MSH and corticosterone in plasma are within control limits (Conte-Devolx *et al.*, 1981).

Hamsters
Hamsters are sensitive to MSG. In male hamsters (Table 5.9) neonatal MSG given on days 7 and 8 caused decreases in the concentration of FSH and testosterone, but not LH, in blood. The concentration in peripheral plasma was 117 ± 29 ng/ml ($\pm$ SE); in controls the concentration was 539 ± 221 ng/ml. The pituitary stores of FSH were within normal limits. In most males spermatogenesis was inhibited, and the activity of 3β-hydroxy-delta 5-steroid dehydrogenase in Leydig cells was low. Only 14% of the neurons in the arcuate nucleus were judged to be morphologically intact (Lamperti and Blaha, 1977; 1980).

In adult female hamsters (Table 5.10) neonatal exposure provoked irregular estrus cycles and a decrease in FSH pituitary stores and blood concentration. Plasma concentration decreased from 556 ± 99 ng/ml ($\pm$ SE) to 183 ± 60 ng/ml.

Table 5.9 Body and Organ Weights of Male Hamsters Exposed Neonatally to Monosodium Glutamate (MSG).

Treatment	Body wt., g $\pm$ SE	Organ weights, mg $\pm$ SE		
		Testes	Seminal vesicles	Pituitary[a]
Control	102 ± 3	3463 ± 122	390 ± 53	3.5 ± 0.2
MSG[b]	104 ± 7	$1008 \pm 260^*$	$175 \pm 43^*$	$1.7 \pm 0.1^*$

[a] anterior lobe only; [b] 8 mg/g BW on day 7 and 8 of the neonatal period; * indicates statistically significant difference; data from Lamperti and Blaha (1980).

Table 5.10 Effects of Monosodium Glutamate on Endocrine Functions in Female Hamsters.

Treatment	Body wt. g $\pm$ SE	Pituitary (μg/mg)		Plasma (ng/ml)		Organ weights, mg $\pm$ SE		
		LH	FSH	LH	FSH	Ovaries	Uterus	Pituitary[b]
MSG[a]	85 ± 11	2.9 ± 0.4	$0.9 \pm 0.3^*$	17.8 ± 4.5	183 ± 60	$28.4 \pm 5.0^*$	$69 \pm 17^*$	$3.1 \pm 0.6^*$
Control	103 ± 4	2.5 ± 0.4	4.9 ± 1.6	20.2 ± 1.3	556 ± 99	33.0 ± 1.6	337 ± 22	5.5 ± 0.6

[a] MSG, 8 mg injected on days 7 and 8; [b] anterior lobe only; * indicates statistically significant difference; data from Lamperti and Blaha (1980).

Table 5.11 Effect of Monosodium Glutamate on Fertility and Endocrine Functions in Female and Male Mice.

Fertility	Litter size[b]	Body weight g ± SE	Organ weight, mg ± SE				
			Pituitary	Ovaries	Thyroid	Adrenals	Testes
Females							
MSG[a]	3/10	4 ± 0.3	42.4 ± 1.3	1.3 ± 0.1	21.9 ± 3.2	3.4 ± 0.2	8.4 ± 0.6	
Controls	9/10	10.7 ± 0.4	33.2 ± 1.4	2.9 ± 0.4	29.4 ± 1.4	4.2 ± 0.2	10.6 ± 0.5	
Males							
MSG	6/11	12.2 ± 0.8	46.3 ± 2.3	1.6 ± 0.1		3.0 ± 0.3	5.4 ± 0.4	186 ± 11.4
Controls	12/12	10.1 ± 0.5	37.2 ± 1.0	2.4 ± 0.2		4.2 ± 0.4	5.5 ± 0.4	233 ± 11.4

[a] MSG, 2.2 to 4.2 µg per g of body weight daily from day 2 to day 11; autopsy day 295–302 of life; [b] successful insemination only; data from Pizzi *et al.* (1977).

Table 5.12 Corpora Lutea Formation in Adult Female Rats (age 260 days) Treated Neonatally with Three DDT Analogues.

Treatment	Number of rats with CL (total number)
Control	13/13
o,p′-DDD	7/12
p,p′-DDT	2/13
o,p′-DDE	0/13

Data from Gellert and Heinricks (1975).

The pituitary concentration was 4.9 ± 1.6 µg/mg in controls and 0.9 µg/mg in MSG-treated animals. LH activity was not affected. The ovaries contained many small follicles but no corpora lutea. The arcuate nucleus contained only 7% of morphologically intact neurons. The ovaries responded to a stimulatory dose of hCG after pretreatment with PMSG to induce follicle formation (Lamperti and Blaha, 1976).

Mice
MSG during the neonatal period decreased fertility in male and female mice, decreased body weights and caused atrophy of the pituitary, thyroid and gonads (Table 5.11).

5.3.2 Urethane

Taggart Davis (1977) injected into neonatal mice 0.5 mg or 0.75 mg of urethane per g of body weight on days 7,10,13,16,19 and 22 after birth. In the adult (age 180 days) the treatment caused thymic lymphosarcoma, ovarian atrophy and lack of follicular and corpora lutea development. Some 57% of urethane-treated males developed mild azoospermia and testicular atrophy.

5.3.3 Insecticides

The apparent uterotrophic effect (Bitman and Cecil, 1970) of the insecticide DDT and its analogues prompted one group to evaluate their activity in the neonatal rat (Heinrichs *et al.*, 1971; Gellert *et al.*, 1972; 1974). Three injections (1 mg each) on days 2, 3 and 4 of life led to a gradual development of persistent estrus. Four month old females had ovaries which contained large vesicular follicles but no corpora lutea. Prior to the age of four months the treated females cycled normally. Neonatal male rats were insensitive to the effect of DDT. Gellert and Heinrich (1975) described the effect of the three DDT analogues (Table 5.12).

References

Arai Y, Gorski RA (1968) Protection against neural organization effect of exogenous androgens in the neonatal rat. Endocrinology 82: 1005–1009

Arai Y, Gorski RA (1974) Possible participation of pituitary testicular feedback regulation in the sexual differentiation of the brain in the male rat. In: Kawakami (ed) Biological rhythms in neuroendocrine activity. Igaku Shoin, Tokyo, pp 232–240

Arai Y, Suzuki Y, Nishizuka Y (1977) Hyperplastic and metaplastic lesions in the reproductive tract of male rats induced by neonatal treatment with diethylstilbestrol. Virchows Arch [A] 376: 21–28

Arai Y, Chen C-Y, Nishizuka Y (1978) Cancer development in male reproductive tract in rats given diethylstilbestrol at neonatal age. Gann 69: 861–862

Azizi F, Vagenakis AG, Bollinger J, Reichlin S, Braverman LE, Ingbar SH (1974) Persistent abnormalities in pituitary function following neonatal thyrotoxicosis in the rat. Endocrinology 94: 1681–1688

Badger TM, Millard WJ, Martin JB, Posenblum PM, Levenson SE (1982) Hypothalamic-pituitary function in adult rats treated neonatally with monosodium glutamate. Endocrinology lll: 2031–2038

Bakke JL, Lawrence NL, Robinson S (1972) Late effects of thyroxine injected into the hypothalamus of the neonatal rat. Neuroendocrinology 10: 183–195

Bakke JL, Lawrence NL, Wilbur JF (1974) The late effects of neonatal hyperthyroidism upon the hypothalamic-pituitary-thyroid axis in the rat. Endocrinology 96: 406–411

Bakke JL, Lawrence NL, Robinson S, Bennett J (1975) Endocrine studies of the untreated progeny of thyroidectomized rats. Pediatr Res 9: 742–748

Bakke JL, Lawrence NL, Robinson S, Bennett J (1976) Lifelong alterations in endocrine function resulting from brief perinatal hypothyroidism in the rat. J Lab Clin Med 88: 3–13

Bakke JL, Lawrence NL, Robinson SA, Bennett J, Bowers C (1978a) Late endocrine effects of L-dopa, 5-HTP, and 6-OH-dopa administered to neonatal rats. Neuroendocrinology 25: 291–302

Bakke JL, Lawrence N, Bennette J, Robinson S, Bowers CY (1978b) Late endocrine effects of administering monosodium glutamate to neonatal rats. Neuroendocrinology 26: 220–228

Balazs R, Brooksbank BWL, Davison AN, Eayrs JF, Wilson DA (1969) The effects of neonatal thyroidectomy on myelination in the rat brain. Brain Res 15: 219–232

Baum R (1987) Assessing chemical risk to reproduction not easy. Chem Eng News, April 17: 31–36

Bitman J, Cecil HC (1970) Estrogenic activity of DDT analogs and polychlorinated biphenyls. Agic Food Chem 18: 1108–1112

Blackburn MG, Mancusi-Ungaro HR Jr, Orzalesi MM, Hobbins JC, Anderson GG (1973) Effects on the neonate of the induction of labor with prostaglandin $F_{2\alpha}$ and oxytocin. Am J Obstet Gynecol 116: 847–853

Bradbury JT, Gaensbauer F (1939) Masculinization of the female rat by gonadotrophic extracts. Proc Soc Exp Biol Med 41: 128–131

Branham WS, Sheehan DM, Zehr DR, Medlock KL, Nelson CJ, Ridlon E (1985) Inhibition of rat uterine gland genesis by tamoxifen. Endocrinology 117: 2238–2248

Branham WS, Zehr DR, Chen JJ, Sheehan DM (1988a) Postnatal uterine development in the rat: estrogen and antiestrogen effects on luminal epithelium. Teratology 38: 29–36

Branham WS, Zehr DR, Chen JJ, Sheehan DM (1988b) Uterine abnormalities in rats exposed neonatally to diethylstilbestrol, ethynylestradiol, or clomiphene citrate. Toxicology 51: 201–212

Bunyan J, Murrell EA, Shah PP (1976) The induction of obesity in rodents by means of monosodium glutamate. Br J Nutr 35: 25–39

Caviezel F, Carraro A, Fochi M (1966a) Effetti endocrini tardivi di trattamenti neonatali con cloropromazina. Atti Accad Med Lomb XXI:l-4

Caviezel F, Carraro A, Fochi M (1966b) Effetti del trattamento neonatale con reserpina sui meccanismi di feedback. Atti Accad Med Lomb XXI:l-4

Chamness GC, Bannayan GA, Landry LA Jr., Sheridan PJ, McGuire WL (1979) Abnormal reproductive development in rats after neonatally administered antiestrogen (Tamoxifen). Biol Reprod 21: 1087–1090

Clark JH, Guthrie SC (1983) The estrogenic effects of clomiphene during the neonatal period in the rat. J Steroid Biochem 18: 513–517

Clark JH, McCormack S (1977) Clomid or nafoxidine administered to neonatal rats causes reproductive tract abnormalities. Science 197: 164–165

Clark JH, McCormack SA (1980) The effect of clomid and other triphenylethylene derivatives during pregnancy and the neonatal period. J Steroid Biochem 12: 47–52

Clemens JA, Roush ME, Fuller RW, Shaar CJ (1978) Changes in luteinizing hormone and prolactin control mechanisms produced by glutamate lesions of the arcuate nucleus. Endocrinology 103: 1304–1312

Conte-Devolx B, Girand P, Castauas E, Boudouresque F, Orlando M, Gillioz P, Oliver C (1981) Effect of neonatal treatment with monosodium glutamate on the secretion of α-MSH, β-endorphin and ACTH in the rat. Neuroendocrinology 33: 207–211

Crutchfield FL, Dratman MB (1980) Growth and development of the neonatal rat: particular vulnerability of males to disadvantageous conditions during rearing. Biol Neonate 38: 203–209

Csaba G, Darvas S, László V (1977) Effects of treatment with the plant hormone gibberellin on neonatal rats. Acta Biol Med Germ 36: 1487–1488

Csaba G, Darvas S, László V, Juvancz I, Vargha P, Bodrogi L, Feher T (1978) Long-lasting effect of single ecdysone injections in newborn rats. Biol Neonate 33: 170–173

Dada MO, Campbell GT, Blake CA (1984) Effects of neonatal administration of monosodium glutamate on somatotrophs and growth hormone secretion in prepubertal male and female rats. Endocrinology 115: 996–1003

Dainat J, Rebiere A (1976) Variations in the in vivo incorporation of L-[³H]leucine into the proteins of the cerebellum of normal, hypo-, and hyperthyroid rats during the first ten days of postnatal life. J Neurochem 26: 935–940

D'Albora H, Carlevaro E, Riboni L, de los Reyes L, Zipitria D, Dominguez R (1974) Advanced puberty in female guinea pigs treated with human chorionic gonadotropin (HCG) or testosterone enantate (TE) at birth. Horm Res 5: 344–350

DePaolo LV, Negro-Vilar A (1982) Neonatal monosodium glutamate treatment alters the response of median eminence luteinizing hormone-releasing hormone new terminals to potassium and prostaglandin E. Endocrinology 110: 835–841

Diaz J, Samson HH (1980) Impaired brain growth in neonatal rats exposed to ethanol. Science 208: 751–753

Dorner G, Hinz G, Docke F, Tonjes R (1977) Effects of psychotrophic drugs on brain differentiation in female rats. Endokrinologie 70: 113–123

Dyer RG, Weick RF, Mansfield S, Corbet H (1981) Secretion of luteinizing hormone in ovariectomized adult rats treated neonatally with monosodium glutamate. J Endocrinol 91: 341–346

Faris RA, Campbell TC (1981) Exposure of newborn rats to pharmacologically active compounds may permanently alter carcinogen metabolism. Science 211: 719–721

Ficková M, Macho L (1981) Insulin receptors in isolated adipocytes from rats with different neonatal nutrition. Endocrinol Exp 15: 259–268

Fisher DA, Dussault JH, Sack J, Chopra I (1977) Ontogenesis of hypothalamic-pituitary-thyroid function and metabolism in man, sheep, and rat. Recent Prog Horm Res 33: 59–116

Forsberg J-G (1976) Adenosis and clear-cell carcinomas of vagina and cervix. Animal model: estrogen-induced adenosis of vagina and cervix in mice. Am J Pathol 84: 669–672

Gellert RJ, Heinrichs WLR (1975) Effects of DDT homologs administered to female rats during the perinatal period. Biol Neonate 26: 283–290

Gellert RJ, Bakke JL, Lawrance NL (1971) Persistent estrus and altered estrogen sensitivity in rats treated neonatally with clomiphene citrate. Fertil Steril 22: 244–249

Gellert RJ, Heinrichs WL, Swerdloff RJ (1972) DDT homologues: estrogen-like effects on the vagina, uterus and pituitary of the rat. Endocrinology 91: 1095–1100

Gellert RJ, Heinrichs WL, Swerdloff R (1974) Effects of neonatally-administered DDT homologs on reproductive function in male and female rats. Neuroendocrinology 16: 84–94

Goddard RF (1948) Anatomic and physiologic studies in young rats with propylthiouracil induced dwarfism. Anat Rec 101: 539–548

Greeley GH, Nicholson GF, Nemeroff CB, Youngblood WW, Kizer JS (1978) Direct evidence that the arcuate nucleus-median eminence tuberoinfundibular system is not of primary importance in the feedback regulation of luteinizing hormone and follicle stimulating hormone secretion in the castrated rat. Endocrinology 103: 170–175

Hahn P, Kirby L (1973) Immediate and late effects of premature weaning and of feeding a high fat or high carbohydrate diet to weanling rats. J Nutr 103: 690–696

Harmon JR, Branham WS, Sheehan DM (1989) Transplacental estrogen responses in the fetal rat: increased uterine weight and ornithine decarboxylase activity. Teratology 39: 253–260

Heinrichs WL, Gellert RJ, Bakke JL, Lawrence NL (1971) DDT administered to neonatal rats induces persistent estrus syndrome. Science 173: 642–643

Hodgson PA, Ellefson RD, Elveback LR, Hanis LE, Nelson RA, Weidman WH (1976) Comparison of serum cholesterol in children fed high, moderate or low cholesterol milk diets during neonatal period. Metab Clin Exp 25: 739–746

Hogarth PJ, Chalmers P (1973) Effect of the neonatal administration of a single dose of chlorpromazine on the subsequent sexual development of male mice. J Reprod Fertil 34: 539–541

Holzwarth-McBride M, Hurst EM, Knigge KM (1976) Monosodium glutamate induced lesions of the arcuate nucleus: endocrine deficiency and ultrastructure of the median eminence. Anat Rec 186: 185–196

Hong J, Lowe C, Squibb RE, Lamartiniere CA (1981) Monosodium glutamate exposure in the neonate alters hypothalamic and pituitary neuropeptide levels in the adult. Regul Pept 2: 347–352

Hughes AM (1944) Cretinism in rats induced by thiouracil. Endocrinology 34: 69–76

Iguchi T (1985) Occurrence of polyovular follicles in ovaries of mice treated neonatally with diethylstilbestrol. Proc Jpn Acad 61 Ser B:288–291

Iguchi T, Hirokawa M (1986) Changes in male genital organs of mice exposed neonatally to tamoxifen. Proc Jpn Acad 62 (Ser B): 157–160

Iguchi T, Iwase Y, Kato H, Takasugi N (1985) Prevention by vitamin A of the ocurrence of permanent changes in ovariectomized adult mice treated neonatally with diethylstilbestrol and its nullification in the presence of ovaries. Exp Clin Endocrinol 85: 129–137

Iguchi T, Takase M, Takasugi N (1986a) Persistent anovulation in the ovary of mice treated with human chorionic gonadotropin starting at different early postnatal ages. IRCS Med Sci 14: 187–188

Iguchi T, Hirokawa M, Takasugi N (1986b) Occurrence of genital tract abnormalities and bladder hernia in female mice exposed neonatally to tamoxifen. Toxicology 42: l–ll

Iguchi T, Takasugi N, Bern HA, Mill KT (1986c) Frequent occurrence of polyovular follicles in ovaries of mice exposed neonatally to diethylstilbestrol. Teratology 34: 29–35

Iguchi T, Ohta Y, Fukazawa Y, Takasugi N (1987) Strain differences in the induction of polyovular follicles by neonatal treatment with diethylstilbestrol in mice. Med Sci Res 15: 1407–1408

Kalland T, Forsberg J-G, Sinha YN (1980) Long-term effects of neonatal DES treatment on plasma prolactin in female mice. Endocrinol Res Comm 7: 257–166

Kamberi IA, Mical RS, Porter IC (1970) Effect of anterior pituitary perfusion and intraventricular injection of catecholamines and indole amines on LH release. Endocrinology 87: l–12

Kawashima S (1964) Inhibitory action of reserpine on the development of the male pattern of secretion of gonadotropins in the rat. Ann Zool Jpn 37: 79–85

Kikuyama S (1966) Influence of thyroid hormone on the induction of persistent estrus by androgen in the rat. Sci Pap Coll Gen Ed Univ Tokyo 16: 265–270

Kikuyama S (1969) Alteration by neonatal hypothyroidism of the critical period for the induction of persistent estrus in the rat. Endocrinol Jpn 16: 269–273

Kincl FA, Henzl M, Rudel HW (1970) The influence of neonatal injection of cortisol in rats. In: Kazda S, Denenber VH (eds) The postnatal development of phenotype. Academia, Prague, pp 307–318

Kizer JS, Nemeroff CB, Youngblood WW (1978) Neurotoxic amino acids and structurally related analogs. Pharmacol Rev 29: 301–326

Krieger DT, Liotta AS, Nicholsen G, Kizer JS (1979) Brain ACTH and endorphin reduced in rats with monosodium glutamate induced arcuate nucleus lesions. Nature 278: 562–563

Ladosky W, Kesikowski WM, Gaziri IF (1970) Effect of a single injection of chlorpromazine into infant male rats on subsequent gonadotrophin secretion. J Endocrinol 48: 151–156

Lamperti A, Blaha G (1976) The effects of neonatally-administered monosodium glutamate on the reproductive system of adult hamsters. Biol Reprod 14: 362–369

Lamperti A, Blaha G (1980) Further observations on the effects of neonatally administered monosodium glutamate on the reproductive axis of hamsters. Biol Reprod 22: 687–693

Lamperti A, Pupa L, Tafelski T (1980) Time-related effects of monosodium glutamate on the reproductive neuroendocrine axis of the hamster. Endocrinology 106: 553–558

Lemkey-Johnston N, Reynolds WA (1974) Nature and extent of brain lesions in mice related to ingestion of monosodium glutamate. J Neuropathol Exp Neurol 33: 74–97

Lengvari I, Branch BJ, Taylor AN (1977a) The effect of perinatal thyroxine treatment on the development of the plasma corticosterone diurnal rhythm. Neuroendocrinology 24: 65–73

Lengvari I, Branch BJ, Taylor AN (1977b) Effect of perinatal thyroxine treatment on some endocrine functions of male and female rats. Neuroendocrinology 24: 129–139

Li JR, Bale LK, Kottke BA (1980) Effect of neonatal modulation of cholesterol homeostasis on subsequent response to cholesterol challenge in adult guinea pig. J Clin Invest 65: 1060–1068

Lopez J, Ogren L, Talamantes F (1986) Neonatal diethylstilbestrol treatment: response of prolactin to dopamine or estradiol in adult mice. Endocrinology 119: 1020–1027

Maier DB, Newbold RR, McLachlan JA (1985) Prenatal diethylstilbestrol exposure alters murine uterine responses to prepubertal estrogen stimulation. Endocrinology 116: 1878–1886

Martin JE, Klein DC (1976) Melatonin inhibition of the neonatal pituitary response of luteinizing hormone-releasing factor. Science 191: 301–302

Martin JE, Engel JN, Klein DC (1977) Inhibition of the in vitro pituitary response to luteinizing hormone-releasing hormone by melatonin, serotonin, and 5-methoxytryptamine. Endocrinology 100: 675–680

Martin JE, McKellar S, Klein DC (1980) Melatonin inhibition of the in vivo pituitary response to luteinizing hormone-releasing hormone in the neonatal rat. Neuroendocrinology 31: 13–17

Matswzana Y, Yonetani S, Takasaki Y, Iwata S, Sekine S (1979) Studies on reproductive endocrine function in rats treated with monosodium-1-glutamate early in life. Toxicol Lett 4: 359–362

Medlock KL, Sheehan DM, Nelson CJ, Branham WS (1988) Effects of postnatal DES treatment on uterine growth, development, and estrogen receptor levels. J Steroid Biochem 29: 527–532

Meisami E, Valcana T, Timiras PS (1970) Effects of neonatal hypothyroidism on the development of brain excitability in the rat. Neuroendocrinology 6: 160–167

Meserve LA, Leathem JH (1974) Neonatal hyperthyroidism and maturation of the rat hypothalamo-hypophyseal-adrenal axis. Proc Soc Exp Biol Med 147: 510–512

Meserve LA, Leathem JH (1981) Development of hypothalamic-pituitary-adrenal response to stress in rats made hypothyroid by exposure to thiouracil from conception. J Endocrinol 90: 403–409

Moguilevsky JA, Scacchi P, Deis R, Siseles NO (1976) Effect of melatonin on the luteinizing hormone release induced by luteinizing hormone-releasing hormone. Proc Soc Exp Biol Med 151: 663–666

Mori T, Bern HA, Mills KT, Young PN (1976) Long-term effects of neonatal steroid exposure on mammary gland development and tumorigenesis in mice. J Natl Cancer Inst 57: 1057–1061

Nagasawa H, Mori T, Nakajima Y (1980) Long-term effects of progesterone or diethylstilbestrol with or without estrogen after maturity on mammary tumorigenesis in mice. Eur J Cancer 16: 1583–1589

Nemeroff CB, Konkol RJ, Bissette G, Youngblood W, Martin JB, Brazeau P, Rone MS, Prange AJ, Breese GR, Kizer JS (1977a) Analysis of the disruption in hypothalamic-pituitary regulation in rats treated neonatally with monosodium-L-glutamate (MSG): evidence for the involvement of tuberoinfundibular cholinergic and dopaminergic systems in neuroendocrine regulation. Endocrinology 101: 613–622

Nemeroff CB, Grant LD, Bissette G, Ervin GN, Harrell LE, Prange AJ (1977b) Growth, endocrinological and behavioral deficits after monosodium L-glutamate in the neonatal rat: possible involvement of arcuate dopamine neuron damage. Psychoneuroendocrinology 2: 179–184

Nemeroff CB, Lamartiniere CA, Mason GA, Squibb RE, Hong JS, Bondy SC (1981) Marked reduction in gonadal steroid hormone levels in rats treated neonatally with monosodium-L-glutamate: further evidence for disruption of hypothalamic-pituitary-gonadal axis regulation. Neuroendocrinology 33: 265–267

Neumann HO, Buchholz R (1971) Androgen-Sterilisierung der juvenilen weiblichen Ratte nach Behandlung mit humanem Chorion Gonadotropin (HCG). Arch Gynak 209: 416–419

Newbold RR, McLachlan JA (1982) Vaginal adenosis and adenocarcinoma in mice exposed prenatally or neonatally to diethylstilbestrol. Cancer Res 42: 2003–2008

Olney JW (1969) Brain lesions, obesity and other disturbances in mice treated with monosodium glutamate. Science 164: 719–721

Olney JW (1971) Glutamate-induced neuronal necrosis in the infant mouse hypothalamus. J Neuropathol Exp Neurol 30: 75–90

Olney JW, Ho OL, Rhee V (1971) Cytotoxic effects of acidic and sulphur containing amino acids on the infant mouse central nervous system. Exp Brain Res 14: 61–76

Pascual-Leone AM, Garcia MD, Hervas F, Morreale de Escobar G (1976) Decreased pituitary growth hormone content in rats treated neonatally with high doses of L-thyroxine. Horm Metab Res 8: 215–217

Pasqualini JR, Kincl FA (1985) Hormones and the fetus. Pergamon Press, Oxford, p 97

Payne DW, Katzenellenbogen JA (1979) Binding specificity of rat alpha-fetoprotein for a series of estrogen derivatives: studies using equilibrium and nonequilibrium techniques. Endocrinology 105: 743–753

Phelps CP, Leathem JH (1976) Effects of postnatal thyroxine administration on brain development, response to postnatal androgen and thyroid regulation in female rats. J Endocrinol 69: 175–182

Pizzi WJ, Barnhart JE, Fanslow DJ (1977) Monosodium glutamate administration to the newborn reduces reproductive ability in female and male mice. Science 196: 452–454

Plapinger L (1981) Morphological effects of diethylstilbestrol on neonatal mouse uterus and vagina. Cancer Res 41: 4667–4677

Potts AM, Modrell RW, Kingsbury C (1960) Permanent fractionation of the electroretinogram by sodium glutamate. Am J Ophthalmol 50: 900–905

Reiser R, Sidelman Z (1972) Control of serum cholesterol homeostasis by cholesterol in the milk of the suckling rat. J Nutr 102: 1009–1016

Rodriguez-Sierra JF, Sridaran R, Blake CA (1980) Monosodium glutamate disruption of behavioral and endocrine function in the female rat. Neuroendocrinology 31: 228–235

Saphier DJ, Dyer RG (1981) Effects of neonatal exposure to monosodium glutamate on the electrical activity of neurones in the mediobasal hypothalamus and on the plasma concentrations of thyroid-stimulating hormone and prolactin, following stimulation of the rostral hypothalamus in adult female rats. J Endocrinol 89: 379–387

Schulz K-D, Harland A, Haarmann H (1972) Female endocrine control mechanisms during the neonatal period II. The effect of luteinizing hormone. Acta Endocrinol (Copenh) 71: 431–442

Schwark WS, Keesey RR (1975) Effects of thyroid dysfunction on norepinephrine metabolism in the developing rat brain. Pharmacologist 17: 179

Schwark WS, Keesey RR (1976) Cretinism: influence on rate-limiting enzymes of amine synthesis in rat brain. Life Sci 19: 1699–1704

Scow RO, Marx W (1945) Response to pituitary growth hormone of rats thyroidectomized on the day of birth. Anat Rec 91: 227–232

Scow RO, Simpson ME (1945) Thyroidectomy in the newborn rat. Anat Rec 91: 209–216

Shah GV, Shyr SW, Grosvenor CE, Crowley WR (1988) Hyperprolactinemia after neonatal prolactin (PRL) deficiency in rats: evidence for altered anterior pituitary regulation of PRL secretion. Endocrinology 122: 1883-1889

Sheehan DM, Branham WS (1981) The lack of estrogen control of rodent alphafetoprotein levels. Teratogenesis Carcinog Mutagen 1: 383-388

Sheehan DM, Young M (1979) Diethylstilbestrol and estradiol binding to serum albumin and pregnancy plasma of rat and human. Endocrinology 104: 1442-1446

Shyr SW, Crowley WR, Grosvenor CE (1986) Effect of neonatal prolactin deficiency on prepubertal tuberoinfundibular and tuberohypophyseal dopaminergic neuronal activity. Endocrinology 119: 1217-1222

Simson EL, Gold RM, Standish LJ, Pellett PL (1977) Axon-sparing brain lesioning technique: the use of monosodium L-glutamate and other amino acids. Science 198: 515-517

Slaughter M, Wilen R, Ryan KJ, Naftolin F (1977) The effects of low dose diethylstilbestrol administration in neonatal female rats. J Steroid Biochem 8: 621-623

Štrbák V, Greer MA (1979) Acute effects of hypothalamic ablation on plasma thyrotropin and prolactin concentrations in the suckling rat: evidence that early postnatal pituitary-thyroid regulation is independent of hypothalamic control. Endocrinology 105: 488-492

Štrbák V, Michaličková J (1984) Hypothalamic-pituitary-thyroid system during suckling period in rat and man. Endocrinol Exp 18: 183-196

Štrbák V, Jurčovičová J, Vigás M (1981) Thyroliberin (TRH) induced growth hormone (GH) release: test of maturation of hypothalamo-pituitary axis in postnatal rat. Endocrinol Exp 15: 245-249

Štrbák V, Macho L, Alexandrová M, Ponec J (1983) TRH transport to rat milk. Endocrinol Exp 17: 343-350

Štrbák V, Jurčovičova J, Vigaš M (1985) Maturation of the inhibitory response of growth hormone secretion to ether stress in postnatal rat. Neuroendocrinology 40: 377-380

Tafelski TJ, Lamperti AA (1977) The effects of a single injection of monosodium glutamate on the reproductive neuroendocrine axis of the female hamster. Biol Reprod 17: 404-411

Taggart Davis NE (1977) Ovarian atrophy resulting from urethane injection of neonatal mice. J Reprod Fertil 57: 159-161

Takasugi N, Mitsuhashi Y (1972) Effects of gonadotropins on the occurrence of permanent changes in the testis of mice given neonatal estrogen injections. Endocrinol Jpn 19: 423-428

Taylor AN, Lengvari I (1977) Effect of combined thyroxine and corticosterone treatment on the development of the diurnal pituitary-adrenal rhythm. Neuroendocrinology 24: 74-79

Theodoropoulos T, Braverman LE, Vagenakis AG (1979) Thyrotropin releasing hormone is not required for thyrotropin secretion in the perinatal rat. J Clin Invest 63: 588-594

Toh YC (1979) Effect of neonatal administration of thyroxine on the rate of sebum production in rats. J Endocrinol 83: 199-203

Trentini GP, Botticelli A, Botticelli CS (1974) Effect of monosodium glutamate on the endocrine glands and on the reproductive function of the rat. Fertil Steril 25: 478-483

Vannier B, Raynaud J-P (1975) Effect of estrogen plasma binding on sexual differentiation of the rat fetus. Mol Cell Endocrinol 3: 323-337

von Fels E (1975) Keimdrüsenfunktion der Ratte nach post- oder pränataler Injektion von Clomiphen. Endokrinologie 65: 126-132

von Fels E, Moguilevsky J, Rubinstein L, Libertum C (1968) Die Reaktion des androgenisierten Ovars auf Gonadotrophine (Parabioseuntersuchung). Endokrinologie 52: 352-355

von Fels E, Bosch LR, Libertum C (1972) Die Reaktion des androgenisierten Rattenovars auf die Verabfolgung von Gonadotrophinen. Endokrinologie 59: 197-202

Walker BE (1983) Uterine tumors in old female mice exposed prenatally to diethylstilbestrol. J Natl Cancer Inst 70: 477-481

Walker P, Courtin F (1985) Transient neonatal hyperthyroidism results in hypothyroidism in the adult rat. Endocrinology 116: 2246-2250

Whitworth NS, Grosvenor CE (1978) Transfer of milk prolactin to the plasma of neonatal rats by intestinal absorption. J Endocrinol 79: 191-196

Wong LM, Bern HA, Jones LA, Mills KT (1982) Effect of later treatment with estrogen on reproductive tract lesions in neonatally estrogenized female mice. Cancer Lett 17: 115-120

6 Psychosexual Orientation

The experience during fetal, neonatal and adolescent life determines adult behavior. The experience begins early. Before birth babies in the womb hear the heartbeat, noises of the digestive tract and mother's voice; they swallow and taste the amniotic fluid. This prenatal background is vital. Infant rats begin to suckle only after the mother smears her nipples with amniotic fluid. Maternal stress may influence the reproductive ability and behavior of the offspring.

After birth, a bond forms between the mother and the baby essential for the growth and well being of the offspring. Premature babies thrive better if they are touched, rubbed and handled. The skin-to-skin contact enhances neurological development and boosts baby's growth. Mother rat licks her pups to stimulate urination which they are unable to do alone. Stroking with a wet brush substitutes for the tactile stimulation by the mother.

In addition to external stimuli internal signals influence learning and behavior. Pituitary, gonadal and adrenal hormones have been implicated in the process: ACTH, MSH and various polypeptides (vasopressin) act directly on the brain to influence learning and conduct. Gonadal and adrenal hormones and emotional crisis (stress) sway reproductive actions.

Many patterns, even not directly associated with reproduction, are either more prevalent or only present in one sex. Application of sex steroids often induces behavior resembling the opposite sex. Feeding behavior is sex oriented. Female rats show a higher preference to sweet taste (Wade and Zucker, 1969a); ovariectomy or neonatal treatment with testosterone propionate abolishes the preference (Wade and Zucker, 1969b). Yet prior experience may nullify sex dimorphism. Females with much experience with saccharin maintain their preference even after castration. Other consequences of sex hormones are non-specific to a given sex. In adult rats (Bell and Zucker, 1971; Wade, 1972) and hamsters (Zucker *et al.*, 1972) pharmacological doses of testosterone increase food consumption while estradiol has the opposite effect. In guinea pigs the appetite suppressive consequence of estradiol is more pronounced in ovariectomized females than in castrated males (Goy and Goldfoot, 1973). Table 6.1 illustrates examples of dimorphic functioning which can be influenced by androgens or estrogens.

Claims have been made that sex dimorphism may be present in intellectual spheres of human endeavor. Cole and Zuckerman (1987) documented that women scientists in the USA publish less than men; marriage and family obligations do not account for the gender difference. Of course, quantity does not equate quality, and an increased number of reports produced by men may be the exhibition of greater aggressiveness in seeking research grants.

Table 6.1 Examples of Behavioral Traits Which Can Be Influenced in the Adult by Treatment with Gonadal Hormones.

Behavior	Species	Sex dominance	Hormonal dependence	Reference
Psychosexual manifestations				
Lordosis	rat	seen in both sexes		see text
	hamster	seen in both sexes		
Mounting	rats	male	Androgen dependent?; E_2 induced in castrated males but not females	see text
	guinea pig	male	E_2 induced in castrated females but not males	see text
licking the vulva of estrus bitch	dogs	male	TP induces in castrated females	Beach and Kuehn (1970)
Facial expression				
Yawning	rhesus monkey	male	T stimulated	Goy and Resko (1972)
Flehman lip curl	bovine	male	T stimulated	Greene *et al.* (1978)
Territorial defence				
Fighting	hamsters	female	T stimulates in males	Dieterlen (1959)
Territory marking	gerbils	male	T stimulates in females	Thiessen and Lindzey (1970)
Feeding behavior				
Preference for sweet	rat	female	Castration decreases the preference (see text)	Wade and Zucker (1969a)
	guinea pig	female	EB depresses more in castrated females than in castrated males	Goy and Goldfood (1973)
Others				
Playing	rhesus monkey	male	Influenced by T in utero	Goy and Goldfood (1973)
Urinary posture	dog	male	TP induces in females	Martins and Valle (1948)

E_2 = estradiol; EB = estradiol benzoate; T = testosterone; TP = testosterone propionate.

6.1 Behavior Patterns in the Adult

In most adult mammals castration causes the abolition, or reduction, of many aspects of behavior patterns but "... the effects cannot be easily deduced. Their variation from species to species and from individual to individual of the same species ... are great ..." (Young, 1961). To reestablish the conduct gonadal hormones must be given in correct amounts, and in the proper sequence; otherwise the animals display atypical patterns. In addition to the effect of gonadal hormones the adrenal glands, the thyroid, the parathyroid and hyperinsulinism contribute to the expression of a "normal" conduct. Dysfunctions of the parathyroid, the thyroid, the pituitary, the adrenals and hyperinsulinism also generate abnormalities (Young, 1961). The discussion of these aspects is beyond the scope of this section.

Reproductive behavior is expressed in part after hormonal stimuli and in part is governed by psychologic and background factors. The action of one partner may stimulate a hormonal and therefore behavioral response of the other partner. Increased anogenital licking induced by the presence of androgens in neonatal urine of males, self grooming in males, and copulation induced hormonal changes are a few examples which contribute to the modulation of reproductive demeanor (see Komisaruk *et al.*, 1986). The testing is subject to errors. Diamond *et al.* (1973) cautioned that often the direction of the observed change is dependent on the parameter measured. Harlan *et al.* (1980) detected circadian variations in the response of both sexes.

6.1.1 Males

Androgens induce aggression needed to establish territorial claims, to attain dominance within a group, and to induce mating behavior culminating in sexual union. Only androgens which can be aromatized (those possessing 3-keto-4-ene moiety) are active. 5α-Dihydro derivatives (DHT) which are not converted into estrogens do not facilitate mating behavior (McDonald *et al.*, 1970; Feder, 1971; see also Section 6.4.2). DHT will induce masculine behavior, mounting and intromission in castrated males, but only if estradiol (EB) has been added. It has been postulated that EB influences the brain activity while DHT acts peripherally to stimulate androgen dependent tissue (Baum and Vreeburg, 1973; Larsson *et al.*, 1973 a,b; Feder *et al.*, 1974).

Androgens which block aromatization induce abnormal response (Section 6.4.2).

6.1.2 Females

In most females mating behavior takes place during the estrus, the period of elevated estrogen concentrations in blood. In cyclic breeders both ovarian hormones, estradiol and progesterone, are required for full mating conduct. Species which need both hormones include guinea pigs (Dempsey *et al.,* 1936), which exhibit an atypical response if treated with estrogen alone (Boling *et al.,* 1938), rats (Boling and Blandau, 1939; Beach, 1942), mice (Ring, 1944), hamsters (Frank and Fraps, 1945) and cows (Melampy *et al.,* 1957). In induced ovulators (rabbit, cat, ferret) estrogens alone are sufficient to provoke mating behavior, but the conduct is more effectively activated if progesterone is added to the treatment (Sawyer and Everett, 1959). In seasonal breeders (sheep) progesterone must precede estrogens (Dutt, 1953; Robinson, 1954), a reversal of the sequence in rodents.

6.1.3 Copulatory Tests

Male rats or hamsters are tested for mounting and attempted intromission by placing an estrus female in the cage of the male. An elaborate score system has been developed to classify the response.

Female rats and hamsters exhibit readily a period of receptivity, the lordosis (presentation posture), in response to fingering (Lisk *et al.*, 1983). The same response can be obtained in castrate, estradiol primed animals in response to progesterone injection (Table 6.2). The usual regimen is a daily injection of estradiol benzoate in oil (10 μg) for 3 days followed by progesterone injection (1.25 mg in oil). Parsons *et al.* (1982) state that two periods of estradiol treatment with a total exposure of 2 h during a 24 h period are sufficient to activate the lordosis reflex by stimulating the formation of progesterone receptors in the medial hypothalamus-preoptic area.

Guinea pig studies usually involve adult virgin females, ovariectomized 1 week before the test and primed with estrogens. A single injection of 50 μg of progesterone will induce a copulatory response in more than 50% of the animals, and a dose response curve may be obtained between 25 and 100 μg (see Kincl and Dorfman, 1961; Kincl, 1964).

6.1.4 Specificity of the Response

Rodents (hamsters, rats) maintained in the laboratory react bisexually; males behave often as females, and *vice versa*. Thus any studies of a presumed dimorphic response measures subjectively differences of the relationship between a given stimulus and a motor reaction. In one strain of Wistar rats lordosis, a typical female breeding posture, can be induced in intact males by manual stimulation (Södersten, 1978a), or in castrated males by treatment either with testosterone pro-

Table 6.2 Amounts of Estrogens and Progesterone Required for Sexual Receptivity in Rodent Species.

Species	Estrogen dose		Progesterone dose, mg	Reference
Guinea pig	EB	5 RU	0.1	Dempsey *et al.* (1936) Collins *et al.* (1938)
	E$_1$	0.3 μg	0.05-0.1	Byrnes and Shipley (1955) Kincl and Dorfman (1961)
Mouse	EB	10 RU	0.05	Ring (1944)
Rat	EB	10 RU	0.4	Boling and Blandau (1939)
	E$_2$	1 μg		Beyer *et al.* (1976)
	E$_1$	4 μg		
Hamster	EB	33 RU	0.05	Frank and Fraps (1945)
	E$_2$	2 μg	2	Ciaccio (1970)

EB, Estradiol benzoate; E$_2$, estradiol; E$_1$, estrone; given once a day for two days; RU, rat unit, total dose needed to double the uterine weight in castrated rats (or mice), equivalent to 0.05 μg of estradiol? (see Emmens, 1962).

pionate or estradiol benzoate (Södersten and Larsson, 1974). The response does not correlate with the amounts of estradiol and testosterone produced by the testes and found circulating in peripheral plasma (Södersten *et al.*, 1974). The group classified intact male rats according to the degree of displayed lordosis and arranged the animals in "lordosis" and "no lordosis" groups. The "no lordosis" group animals who performed poorly as females had higher estradiol production than "good" performers. The latter group displayed a male response (mounting) with about the same frequency as the "no lordosis" group (Table 6.3).

Gonadal hormones are non-specific in inducing behavioral responses. Intact, adult male rats do behave as females (Beach, 1945). Castration abolishes the response; androgens (Beach 1942b) or estrogens (Ball 1937; 1939; Beach, 1942a) restore the reaction.

Davidson (1969) induced lordosis in castrated adult male rats with 2–10 µg of estradiol benzoate provided the animals were tested within one week after the operation. When the same amount of the estrogen was used 6 weeks after castration the estrogen provoked an "androgenic" response; the frequency of intromission was lower.

Ciaccio induced either masculine or feminine behavior in male hamsters by changing the hormonal milieu. He castrated adults, allowed a 6 week period of rest, and then primed the animals with a sustained delivery estradiol preparation (dimethylpolysiloxane implant) followed by an injection of either progesterone (200 µg) or testosterone (200 µg). The males responded with a masculine behavior if testosterone was used, and as females when progesterone was injected. If the males were injected only with estradiol, they responded as males (Table 6.4).

Table 6.3 Concentrations of Estradiol and Testosterone in Testicular Venous and Peripheral Plasma in Male Rats.

Group	Estradiol, pg/ml		Testosterone, ng/ml	
	testicular	peripheral	testicular	peripheral
Lordosis[a]	2.4 ± 1.3	1.4 ± 0.2	33.6 ± 7.2	1.9 ± 0.1
No lordosis[a]	8.3 ± 3.8	1.7 ± 0.4	26.5 ± 7.7	0.8 ± 0.1
No lordosis[b]	33.2 ± 13.2	3.1 ± 0.6	98.7 ± 8.2	5.9 ± 0.5

[a] Danish Wistar strain; [b] Dutch hooded strain; data from Södersten *et al.* (1974).

Table 6.4 Copulatory Response of Castrated Adult Male Hamsters Primed with Sustained Release Form of Estradiol.

Hormonal treatment	Copulatory response	
	Male-like	Female-like
None*	0/7	0/7
Testosterone, 200 mg*	7/8	0/8
Testosterone, 200 mg	7/7	2/7
Progesterone, 200 mg	3/7	7/7

* Not primed with estradiol.

Often estrogens promote one aspect of the response and androgens another. In cattle estradiol (in a sustained release form) was more effective in promoting in females head-butting, vulvar licking and mounting (male behavior) than testosterone. Testosterone (but not estradiol) was an effective promoter of Flehman lip curl, seen most often in bulls (Greene *et al.*, 1978).

6.1.5 Neural Pathways Regulating Mating Behavior

6.1.5.1 Males

In males the spinal nucleus of the bulbocavernosus (SNB) is important for copulatory behavior. The neurons innervate the perineal musculus bulbocavernosus, the levator ani muscle and the anal sphincter; females have only the last muscle. A sexual dimorphism exists in the number of motoneurons present. The nucleus contains 3–4 times more neurons in adult males than in females (Nordeen *et al.*, 1985). The growth of SNB neurons during prenatal and postnatal life is androgen dependent. In the adult the number decreases after castration. The decrease can be repaired by a four week treatment with testosterone (Kurz *et al.*, 1986).

6.1.5.2 Females

In rats and mice males trigger lordosis, an essential part of the reproductive conduct of female rodents. During mounting the male grasps the female by the forepaws. The pressure and pelvic thrusts stimulate the skin at the tailbase and the perineum and trigger electrical activity of the sensory neurons which are transmitted via an ascending superspinal loop to the medullary reticular formation, the lateral vestibular nucleus and the dorsal mid-brain and mid-brain central gray. A descending pathway controls the motor neurons of axial muscles necessary to maintain postural tone. The neurons in the mid-brain need to be stimulated by connections originating in the ventromedial nucleus of the hypothalamus, which in turn respond to estradiol-progesterone stimulation. Transection of connections descending from the ventromedial nucleus to the midbrain will abolish, and electrical stimulation will facilitate, the lordosis response. The neurons in the hypothalamus, primarily in the medial preoptic area, ventromedial, and anterior hypothalamus are sensitive to estradiol. The cells accumulate radioactivity following administration of labelled estradiol. In castrated females implantation next to the ventromedial aspect will result in increased electrical activity and biochemical potential of the cells. Lesions in the area will decrease lordosis responses (see Pfaff, 1983).

Progesterone influences mainly neurons in the medial preoptic area and ventromedial hypothalamus (Takahashi and Lisk, 1985; Takahashi *et al.*, 1985).

Brain sex dimorphism is covered in Chapter 2 and the influence of hormones on hormone receptors in the CNS in Chapter 4.

6.2 Influence of Sex Hormones During Fetal Life

Sex linked differences have been demonstrated to be present in the morphological structure and biochemical capacity of the hypothalamus in rodents: neural connections (Raisman and Field, 1971; 1973; Raisman, 1974), nuclear volume (Dörner and Staudt, 1968; 1969a,b; Staudt et al., 1973), protein content (Scacchi et al., 1970), serotonin concentration (Ladosky and Gaziri, 1970), RNA metabolism (Gorski and Barraclough, 1963) and oxidative metabolism (Moguilevsky, 1966; Moguilevsky et al., 1966; 1968; 1969; Moguilevsky and Rubinstein, 1967).

6.2.1 Behavior Changes in Animals

Sex hormones are teratogenic and cause fetal malformations (Chapter 3) yet not always behavioral changes. Sexual patterns, spontaneous motor activity, lordosis and copulation response in female rats (Revesz et al., 1963; Clemens et al., 1978) and hamsters (Nucci and Beach, 1971) masculinized in utero (evident by virilized external genitalia) were comparable to those measured in intact females. The experiments of Nucci and Beach are typical: this group injected gravid hamsters with 2 mg of TP (days 11-13, 12-14, or 13-15; the gestation period in hamsters is 16 days), allowed the pups to be born, and tested adult females. Females were born virilized; vaginal opening was absent, the clitoris was enlarged, but the ovaries were apparently normal. Their receptive behavior triggered by estradiol and progesterone was normal. TP did not enhance the masculine response.

Male rats, born to mothers treated during gestation with androgens, do not display abnormal sexual behavior (Beach et al., 1969).

Masculinization of genetic female guinea pigs reduced the lordosis response and increased mounting behavior (Phoenix et al., 1959; Goy et al., 1964; Goldfoot and van der Werff ten Bosch, 1975).

In rabbits masculinization decreased maternal nest building behavior (Fuller et al., 1970; Anderson et al., 1970).

Prenatal androgenization of ewes increased the sensitivity of the females to testosterone. Castrated, testosterone treated ewes, were aggressive in a manner similar to males. Non-androgenized animals were not aggressive. Estrogens induced estrus behavior in androgenized ewes albeit the capacity of the animals to respond was decreased when compared with untreated controls (Clarke, 1977; 1978; Clarke et al., 1976).

Dörner et al. (1977) found decreased sexual behavior in pigs of both sexes born to sows injected 100 mg EB during the 8th and 11th week of gestation.

Rhesus monkeys born to mothers treated with high doses (25 mg daily from postcoital day 40) displayed during the second year of life frequencies of mounting (2.24) closely resembling those of normal males (2.54). The incidence of female presenting (2.36) was higher than that of either control males (2.03) or control females (1.87) (Young et al., 1964; Roy and Goldfoot, 1975).

Gladue and Clemens (1978) used in pregnant rats an inhibitor of androgen biosynthesis (flutamide), 5 mg daily during days 10 to 22 of gestation. Castrated adult males and females responded to estradiol benzoate stimulation with greater frequency. The authors speculated that test animals were more sensitive to estrogens.

6.2.2 Psychological Changes in Humans

Conduct deviations, a result of deviant hormone production, have been credited to patients with chromosomal abnormalities. Klinefelter's syndrome patients (XXY) with prepubertal hypogonadism are often classified as passive, dependent and mentally slow and shallow. Individuals with XYY aberration have been categorized as hostile and aggressive, yet no correlation exists between the amounts of circulating testosterone and various measures of hostility and aggressive behavior (Meyer-Bahlburg, 1982). One possible reason for the absence of any differences could have been methodological. Testosterone had usually been measured in a single sample taken only once a day. Such a method prevents the recognition of possible changes in pulsatile and diurnal release (see Chapter 2). Meyer-Bahlburg (1981; 1984) concluded that there was little consistency between behavior aberrations and androgen concentration in blood and stated "...it seems likely that androgens play only a limited role among many other factors in the development of (pathological) aggressive behavior..."

In some women changes in feeling, social interaction, dysmenorrhea and premenstrual tension occur in relation to the menstrual cycle. Estradiol-progesterone imbalance has been thought to be the causative agent, but the disturbances occur rarely during the follicular phase when circulating estrogen levels are much higher than progesterone (see Rubin *et al.*, 1981).

Money and Erhardt (1972) described changes in virilized girls, tomboyism, energy expanded in recreation and aggression, preference in clothing and adornment, materialism, choosing of career vs marriage, romantism and homosexuality. Virilization has been thought to increase the potential (in either sex) for physical aggression (Reinisch and Karow, 1977), but the conclusion was challenged by Meyer-Mahlburh (1981).

Exposure to barbiturates, preparations often prescribed during pregnancy, may lead in both sexes to learning disabilities, decreased IQ, performance deficits and in boys to demasculinization of gender identity and sex role behavior (Reinisch and Sanders, 1982).

Attempts to correlate changes in sex dimorphic conduct to non drug related complications in pregnancy yielded equivocal results for girls and was positive for boys (Meyer-Mahlburg *et al.*, 1984).

6.2.2.1 Steroid Hormones

Babies were exposed *in utero* to a variety of synthetic steroid hormones, principally constituents of oral contraceptives, and in support of treated abortion. The evaluation of effects is difficult since the treatment schedule varied from patient to patient, between physicians, in the drugs and amounts used; often several separate agents of different composition and amounts were prescribed to the same patient during the course of pregnancy.

Oral Contraceptives

Reinisch (1977) compared the personality scores of girls (average age 12 years) divided into two groups. One was exposed to more progestational agent and low, or no, estrogen. The mean ratio was 548 progestin to 1 mg estrogen; the mean progestin intake of the mother was 3000 mg (range 529-9890). Girls in the other group were exposed to more estrogen than the first group (ratio progestin: estrogen 4:1); the mean estrogenic dose was 6100 mg (range 3500-13900). Reinisch concluded that "...prenatal hormone exposure significantly affected subjects' response to personality measure..." Girls in the first group (21 members) were more independent, sensitive, individualistic, self-assured and self-sufficient. Those exposed to more estrogens were more group oriented and group dependent. Prenatal exosure had no effect on measured intelligence (Fuertes de la Haba *et al.*, 1976; 1977).

Girls exposed to progestational agents having anti-androgenic properties (medroxyprogesterone acetate) showed a degree of stereotype feminity (preference for attractive, stylish dresses). Boys may be less energetic and less interested in marriage and having children (Meyer-Bahlburg *et al.*, 1977; 1984; Erhardt *et al.*, 1977; 1985; Sanders and Reinisch, 1985).

Estrogens

Erhardt *et al.* (1985) interviewed 30 women aged 17-30 with proven prenatal exposure to varying doses of DES (210-10475 mg); ninety percent had vaginal adenosis (Chapter 3). Most were judged to have less well established sex partner relationships and to be lower in sexual desire, enjoyment and excitability.

6.2.2.2 Stress

Stress during pregnancy affects the reproductive performance and behavior of the offspring (Chapter 3); stress during the neonatal period produces less severe changes (Chapter 4).

Several groups described disturbed behavior in offspring born to stressed rats (Thompson, 1957; Hockman, 1961; Keeley, 1962; and Morra, 1965). Ward (1977) noted that males become feminized. Their plasma testosterone concentrations are low. In humans prenatal maternal stress may increase the incidence of death, influence the health of the child and produce behavioral abnormalities (Herrenkohl, 1986).

Dörner (Dörner *et al.*, 1980) proposed that genetic males may develop homosexual tendencies if androgen production is insufficient during fetal life and suggested war stress as an etiological factor. The group established that most homosexual males in Germany (865) were born during the war years, 1941-1947. The highest number within the group (81%) was born in 1945. The authors speculated that "...stressful prenatal (or perinatal) events may represent a...factor for homosexuality in human males..." The authors report that paternal neglect as a possible contributing factor had not been evaluated. In support the authors cited studies in rodents (rats) showing that male rats deprived of androgen (by castration) during the "sexual maturation of the brain" (Dörner and Hinz, 1967; Dörner, 1976) or exposed to prenatal stress (Lieberman, 1963; Ward, 1972) responded to male partners.

The hypothesis has not been generally accepted. The majority of heterosexual and homosexual men do not differ in peripheral androgen concentrations (Meyer-Bahlburg, 1981; 1984). The rat model is at variance with the proposed hypothesis: prenatally androgenized female rats show a clear predominance of female behavior and become "homosexual" only after ovariectomy and administration of testosterone, a condition at variance with a human lesbian. A further problem in relying upon animal data has to do with anatomical changes of the external genitalia; prenatal exposure to sex hormones in rodents results in structural changes of the genitalia (Section 6.4; Chapter 3). Abnormally developed genital organs, especially in males, may make it impossible to achieve a "normal" sexual union.

6.3 Changes Induced by Sex Hormones During the Neonatal Period

In rodents neonatal exposure to androgens influences even dimorphic behavior not directly associated with reproduction. Modifications were recorded in open-field performance in hamsters (Swanson, 1967) and rats (Blizard and Denef, 1973; Scouten et al., 1975), in running (Harris, 1964) and sleep patterns (Branchey et al., 1973), exploratory behavior (Quandagno et al., 1972), play and threat (Young et al., 1964; Goy, 1970; Neumann et al., 1970), emergence (Swanson, 1967; Pfaff and Zigmond, 1971; Quandango et al., 1972), avoidance acquisition (Scouten et al., 1975) and saline intake (Krecek, 1973).

Most studies pertaining to the effect of neonatal steroids have dealt with aberration of sexual behavior in response to hormonal treatment in adulthood; the results were often compared with reactions seen in gonadectomized animals. Usually, hormonal treatment was conventional: a single injection of testosterone propionate (TP) to test male-like behavior, or 2–3 injections of estradiol benzoate (EB), followed by progesterone (P) when testing for a female response. The conventional protocol has severe drawbacks. Neonatal exposure to steroid hormones changes steroid metabolism and metabolic clearance rates in the adult (Chapter 4). The variation influences hormone availability, concentration and binding in target tissues, elimination of hormones from blood and formation of hormone receptors. A dose which will elicit a certain response in controls may or may not elicit the same response in test animals, because the hormone is being metabolized and disposed of differently. Another drawback is the castration procedure used; animals were usually gonadectomized prepubertally and tested after a long lapse of time – a procedure shown to alter liver metabolism. The manner in which steroid hormones are dispensed is important. Either male- or female-like behavior is induced in castrated hamsters if the hormones are furnished as a long acting preparation (Section 6.1.4).

Antiandrogens (Neumann and Elger, 1965; see also Neumann and Steinbeck, 1974), progesterone (Hull *et al.*, 1980; Hull, 1981; Diamond *et al.*, 1973) and anti-estrogens (Södersten, 1978b) have been shown to modify behavior, if given in sufficient doses. Arendash and Gorski (1982) transplanted portions of preoptic tissue obtained from neonatal male rats into female litter mates and claimed in the adult enhanced lordosis and copulatory responses to estradiol.

6.3.1 Androgenized Animals

6.3.1.1 Rats

Females
Both male and inexperienced (nulliparous) females become capable and exhibit nearly all dimensions of maternal behavior if presented with newborn pups for several days (Bridges *et al.*, 1974).

In infant female rats androgens suppress estrogen-induced receptivity in the adult. The reaction has been termed "defeminization" (Wilson, 1943; Whalen and Nadler, 1963; Hendricks and Weltin, 1976). Androgenized females exhibit male-like behavior: mounting and attempted intromission (Barraclough and Gorski, 1962; Whalen *et al.*, 1969; Sheridan *et al.*, 1973; Sachs *et al.*, 1973; Södersten, 1973; Harlan and Gorski, 1978), male-like feeding pattern (Nance, 1976), a "male" pattern of maternal behavior (Quadagno and Rockwell, 1972; Bridges *et al.*, 1973), decreased locomotor activity in an open field test (Magalhaes and Aranjo-Carlini, 1974) and running-wheel activity (Gentry and Wade, 1976).

A high dose of TP given to female pups changes profoundly the response in the adult. A dose of 1.25 mg renders the females sexually unresponsive to males. Progesterone or a combination of estradiol and progesterone does not restore lordosis patterns. A low dose is less damaging. Rats injected with 10 µg of TP on day 5 of life are sexually receptive even though they exhibit long periods of estrous smears (Barraclough and Gorski, 1962; Whalen and Rezek, 1974). Others confirmed these findings (Kennedy, 1964; Feder *et al.*, 1966; Thomas and Gerall, 1969; Whalen *et al.*, 1969; Pfaff and Zigmond, 1971). Doses as low as 1 µg of TP were stated to interfere with the normal sexual response of the adult (Clemens *et al.*, 1969; Sheridan *et al.*,1973).

Injection of pentobarbital (0.5 mg) 4–5 h prior to TP injection blocks the development of the anovulatory syndrome in 45 and 90 day old rats. However, if the barbiturates are injected directly into the hypothalamus, and TP is injected systemically, pentobarbital is ineffective in blocking the action of the androgen (Gorski, 1974). The observation indicates that the CNS active drug may exert its effect peripherally, not by acting at the CNS level.

Males
Castration of male neonates results in the impairment of a "complete" copulatory response (Beach and Holz, 1946; Grady *et al.*, 1965) and produces "feminine" males. Paradoxically, testosterone propionate, given in "high" amounts (38 mg divided between days 1 and 28) diminishes mounting, intromission and ejacula-

tion (Wilson and Wilson, 1943). Whalen (1964), Harris and Levine (1965) and Feder (1967) failed to reduce male behavior by a single neonatal injection of TP. Diamond et al. (1973) injected 5 mg of TP on day 3 of life and noted in the adult decreased frequency of mounting, intromission and ejaculations (from 1.33 in controls to 0.75 in tested group). The authors speculated that a testosterone metabolite (an estrogen?) might have been the causative factor.

Goldfoot et al. (1969) found increased female behavior after neonatal treatment with androstenedione.

Estrogens given in pharmacological amounts to castrated newborn males will arouse the development of ejaculatory capacity in the adult (Booth, 1977; Södersten and Hansen, 1978).

6.3.1.2 Mice

Male mice, raised in isolation, will fight if placed into the same cage; females do not fight.

Females
Neonatal androgen treatment of female mice with testosterone propionate will induce aggressive behavior (Edwards, 1968; Peters and Sorensen, 1971). Lee and Grifo (1973) suggested that a pheromone, present in the urine, triggers the reaction. Edward (1970) has established that an early "critical period" to exposure was not involved in the response. Castrated adolescent (30 days old) female mice, injected daily with TP (100 µg) for 20 days, fought 45 days later 75% of the time while control animals fought only 25% of the time.

Measurement of other conduct suggests that a "critical" period may be present. Edwards and Burge (1971) concur that the estrogen-progesterone procedure only rarely induces the lordosis response in androgenized females.

Nest building seems to be associated with the absence of neonatal androgens (Lisk et al., 1973).

Males
Fighting of males is androgen dependent. Castration prior to puberty abolishes the aggressive behavior; exogenous androgens restore the response (Beeman, 1947). Fighting is increased four-fold in males castrated at birth and injected with TP (0.35 mg/g BW) on day 2 or 4 of life. The incidence during multiple encounter was 27% in androgenized males injected with TP and only 6% in castrated, noninjected males (Bronson and Desjardins, 1969).

6.3.1.3 Hamsters

In hamsters sexual receptivity declines with age in both untreated and androgenized animals of both sexes (Farrell et al., 1977). Both androgens and estrogens will induce a masculine response in the female (Paup et al., 1972).

Females
The hamster female is the aggressive partner and attacks the male except when sexually receptive (Dieterlen, 1959). Androgenization of females increases the hostility (Crossley and Swanson, 1968; Payne, 1974; Whalen and Etgen, 1978). Doty *et al.* (1971) suggested that an olfactory factor may control the belligerence. The identity of a possible component remains to be established.

Androgenization decreases the degree of sexual receptivity (Gerall and Kenney, 1970; Swanson, 1970; Farrell *et al.*, 1977; Gerall, 1979) and shortens the duration of the lordosis response (Schwartz and Gerall, 1979). DeBold and Whalen (1975) report that a dose as low as 1 µg injected into 1 day old pups decreases the lordosis response in the adult. Swanson (1970) proposed that "...the low level of receptivity is not so much a deficiency in ovarian secretion as a decreased responsiveness to female hormones..." A synthetic androgen (fluoxymesterone), 100 µg, on days 2–4 induced mounting behavior (Vomachka and Clemens, 1981).

Males
Neonatal androgenization of males induces in the adult aggression similar to that seen in females (Payne and Swanson, 1972; 1973). Payne (1977) reported that androstenedione (300 µg on day 1) effectively induced belligerence whereas dihydrotestosterone did not.

6.3.1.4 Ferrets

Females
Baum (Baum, 1976; Baum and Erskine, 1984) induced masculine behavior by providing testosterone in a sustained release form for 15 days beginning on the day of birth, leaving the potential to display feminine sexual behavior intact (Erskine and Baum, 1982). Two major metabolites, dihydrotestosterone and estradiol, also induced male-like conduct provided the dose used was higher; the steroids failed to induce defeminization (Baum *et al.*, 1982).

Males
Early castration (day 5 of life) decreases the masculine breeding response to a greater degree than if males are castrated on day 20 or 35 (Baum and Erskine, 1984).

6.3.2 Estrogenized Animals

Neonatal estrogens affect adult behavior in both sexes. During estrus females are more active (Kawashima and Shinoda, 1968). Levine and Mullins (1964) reported that 100 µg of EB after birth abolished the sexual receptivity of adult females and decreased the overall spontaneous locomotor activity (wheel-running test). The normal cyclic pattern was restored by progesterone treatment (Kawashima, 1972).

6.3.2.1 Rats

In estrogenized males ejaculation, intromission, and posterior mounting are suppressed (Levine and Mullins, 1964; Harris and Levine, 1965). The males only rarely achieve intromission (Whalen, 1964) possibly due to a decrease in penis size (9.1 mm in control) to 8 mm caused by the neonatal treatment (Whalen, 1968); Beach (1945; 1975) likewise suggested that the decrease in penis size caused by neonatal hormones was causal in decreasing sexual performance. Influence of other factors could not be excluded (Diamond *et al.*, 1973).

In female pups estrogens (EB) induce a male-like behavior resembling that produced by androgens (Pfeiffer, 1936; Gorski and Wagner, 1965; Dörner, 1968).

6.3.2.2 Mice

Neonatal estrogens stimulate aggressive behavior in adult female mice to a similar degree as do androgens (Edwards and Herndon, 1970).

6.3.2.3 Hamsters

In hamsters EB induces both androgenization and defeminization (Paup *et al.*, 1972; Coniglio *et al.*, 1973). A dose of 50 μg (female pups) enhanced the potential to display both masculine and feminine behavior in adulthood (Whalen and Etgen, 1978).

6.4 Regulation of Mating Behavior

The many facets which control sexual behavior responses in rodents remain to be clarified. Several hypotheses have been advanced to explain the development and regulation of behavior; only the "Bisexual Brain Hypothesis" and a more recently developed refinement of this hypothesis, the "Aromatization Hypothesis", explain a sufficient proportion of the current experimental evidence to be considered in this brief presentation.

6.4.1 Bisexual Brain Hypothesis

Phoenix *et al.* (1959) proposed that the presence of fetal hormones determines future behavior. Accordingly, the presence or absence of androgens during embryonic differentiation determines the pattern of sexual behavior displayed by an individual as an adult. A German endocrinologist Dörner (1976) expanded the hypothesis by suggesting that two centers in the brain regulate dimorphic responses. In the male the main regulatory center, which controls mounting, intromission and ejaculation, is the medial preoptic/anterior hypothalamus region. Lordosis in the female is controlled by the ventromedial nucleus. The differentiation of the centers into one or the other sex pattern is influenced by the presence of sex hormones. In genetic males the production of testosterone by the fetal testes will lead to the predominant "organization" of the "male" center. The development of the femaleness is a passive process. If testosterone is low the "female" center will become dominant.

If androgens are low in a genetic male that individual will be predisposed towards homosexuality ("central nervous homosexuality"). If genetic females are exposed to androgens the "male" center becomes dominant, and she will be predisposed to homosexuality.

The hypothesis is difficult to reconcile with several observations:

(i)   Intact rodents (rats) are capable of both male and female conduct; females will mount available partners and males will accept the mount of a partner and assume lordosis. In females it is possible to augment mounting without suppressing lordosis.

(ii) It is not clear what is meant by "organization" and which is the product of fetal testes that induces the "organization". Testosterone stimulates the differentiation of Wolffian ducts but not of the Müllerian anlage, and the function of the anti-Müllerian substance, if any, on future sexual conduct has not been established.

(iii) In rodents (and other mammals) exposure to sex hormones influences not only behavior but also the structures of the external genitalia (Chapter 3), and anatomical deformities play an important role. Beach (1975) was the main proponent of the hypothesis that anatomically deficient male rats and hamsters (Beach and Pauker, 1949) do not display male behavior (intromission and ejaculation) because of an inadequate penis rather than because of deficient brain organization.

Anatomical structures of male rats are androgen dependent. Penile papillae (spines) located in the adult on the penis epidermis constitute the main stimulatory organelles. In the adult the spines atrophy after castration. The disappearance of the structures parallels a decline in sex behavior (Beach and Levison, 1950). Castration of 1 day old pups prevents the development of the spines and induces other abnormalities in the formation of the penis. Prolonged treatment with testosterone does not restore the normal morphology (Beach *et al.*, 1969).

(iv) Virilized female rats exhibit "homosexuality" (male patterns) only after ovariectomy; if they are left intact (a condition of human female homosexuality) they show only a slight increase in mounting behavior but a very clear predominance of female actions.

(v)   The genitalia of genetic male rat pseudohermaphrodites develop as females. In the absence of fetal androgen stimulation the behavior of the animals would be expected to be feminine. Yet, primed with the appropriate hormonal regimen (estradiol and progesterone) the animals exhibit neither masculine nor feminine behavior (Shapiro *et al.*, 1976). In genetic hermaphroditic female guinea pigs treatment with testosterone propionate will not suppress a female sexual response (Goy and Goldfoot, 1973). Both the observations suggest that female conduct requires an active imprinting.

In cattle the importance of the prenatal influence of androgens on the differentiation of the brain appears minimal. Freemartins, genetic females whose genitalia are masculinized to varying degrees by sharing blood circulation *in utero* with a male sibling, do not differ in behavior from control females. Responses of adult freemartins to testosterone (or estradiol) stimulation provoke a male-like behavior (head-to-head fighting, vulvar interest and mounting) to a same degree as does the treatment in control females (Greene *et al.*, 1978).

6.4.2 Aromatization Hypothesis

A further refinement of the hypothesis states that not androgens but estrogens arising from *in vivo* in the brain exert an "organizing" effect of neural tissue mediating patterns of sexual behavior of adult males. Reddy *et al.* (1974) stated ". . .the androgen effect on brain differentiation is via estrogenic metabolites formed at the site in the brain. . ." McEwen *et al.* (1977) and Parsons *et al.* (1982) proposed that the effect of testosterone is mediated in the ventromedial nucleus.

Several observations favor the hypothesis. Only androgens which can be aromatized (those possessing 3-keto-4-ene moiety) are active. Dihydrotestosterone (and derivatives) which are not converted into estrogens do not facilitate mating behavior in adduct castrated rats (Luttge and Whalen, 1970; Arai, 1972; Larsson *et al.*, 1973a; McDonald and Doughty, 1974) except if combined with estradiol (Baum and Vreeburg, 1973; Larsson *et al.*, 1973b). Those androgens which block aromatization (1,4,6-androstatriene-3,17-dione (ATD), 4-hydroxy-4-androstene-3,17-dione) induce a feminine response in adult males. McEwen *et al.* (1977) provided ATD (in a PDS implant) during the first few days of life. In males ATD increased the lordosis response, and in females the inhibitor blocked the effect of TP injected the following day (see also Lieberburg *et al.*, 1977; Gladue *et al.*, 1978). Vreeburg *et al.* (1977) did not observe male feminization in pups injected with 1 mg ATD when born. Concurrent administration of estradiol (as benzoate) prevented the inhibitory effect of aromatase blockers in males (Davies *et al.*, 1979). The use of an antiestrogen (MER-25) prevented androgenization of neonatal females by TP (McDonald and Doughty, 1973; see also McEwen *et al.*, 1977).

Estrogens may be needed to stimulate CNS activity and possibly protein (enzyme) synthesis. The view is supported by the observation that in females estradiol induced mating behavior is blocked by agents which inhibit RNA and protein synthesis (see McEwen *et al.*, 1982).

The hypothesis does not explain several experimental findings:

(i)   Not all androgens which aromatize provoke the same response. Testosterone induces male behavior while androstenedione permits somatic and genital development of males but does not block the evolution of feminine behavior traits.

(ii) The nature of the estrogen(s) arising in the CNS of male fetuses has not been determined, and information pertaining to the concentration of androgen/estrogens and their receptor molecules in the fetal brain is lacking.

(iii)   The metabolic yield of estrogens from androgens in the CNS is low, less than 1% (Naftolin *et al.*, 1972; Whalen and Etgen, 1978). In fetal male rats testosterone production peaks on days 18-19 and then rapidly declines. Testosterone is however also present in female embryos. The maximal average concentration in males reaches 2200 pg/ml and in females 1300 pg/ml (Weisz and Ward, 1980). Conversion of the difference (900 pg/ml) would therefore contribute an additional 9-10 pg of estrogens in the CNS of males. It has not been established whether this difference would influence brain "differentiation" towards "maleness".

(iv)   Estrogens of maternal origin persist in newborn rat pups for several days. Friend (1977) detected in the peripheral plasma of 1 day old pups 170 pg/ml of estradiol, and 50 pg/ml in the plasma of 5 day old pups of either sex. The biological half-life of estradiol is long: in the neonate the $t_{1/2}\alpha$ is about 1 day and $t_{1/2\beta}$ about 3 days (Friend, 1977).

(v)   The difference in metabolic activity (measured as uptake of labelled testosterone) in the hypothalamus of female pups is three times higher than that in the hypothalamus of male pups. The uptake in the hypothalamus is low. In female animals the adrenals concentrate 30 times more, and the ovaries 20 times more, testosterone than the hypothalamus or the brain (Kincl, 1970).

(vi)   Androgenization, or estrogenization, has a negligible effect on the uptake of testosterone in adult male rats (Kincl and Chang, 1970).

(vii)   If estrogens are the active hormone then exogenous estradiol should be significantly more active than testosterone in inducing mounting behavior. Whalen and Etgen (1978) report that estradiol (EB) is only 3 times more active than testosterone (TP).

(viii) In female ferrets it is the neonatal exposure to testosterone itself, not estradiol, which brings on behavioral masculinization (Baum *et al.*, 1982). Five day old female pups were not masculinized by testosterone released from a DPS implant which liberated more androgen than was found in intact male rats between postnatal day 5 to 40 (Baum and Erskine, 1984).

(ix)   Treatment of castrated experienced adult male rats with a combination of dihydrotestosterone and estradiol elicits both masculine and feminine mating conduct (Baum and Vreeburg, 1973; Larsson *et al.*, 1973 a,b; Feder *et al.*, 1974).

References

Anderson CO, Zarrow MX, Denenberg VH (1970) Maternal behavior in the rabbit: effects of androgen treatment during gestation upon the nest-building behavior of the mother and her offspring. Horm Behav 1: 337-345

Arai V (1972) Effect of 5α-dihydrotestosterone on differentiation of masculine pattern of the brain in the rat. Endocrinol Jpn 19: 389-393

Arendash GW, Gorski RA (1982) Enhancement of sexual behavior in female rats by neonatal transplantation of brain tissue from males. Science 217: 1276-1278

Ball J (1937) Sex activity of castrated male rats increased by estrin administration. J Comp Physiol 24: 135-144

Ball J (1939) Male and female mating behavior in prepubertally castrated male rats receiving estrogens. J Comp Physiol 28: 273-283

Barraclough CA, Gorski RA (1962) Studies on mating behaviour in the androgen-sterilized female rat in relation to the hypothalamic regulation of sexual behaviour. J Endocrinol 25: 175-182

Baum MJ (1976) Effects of testosterone propionate administered perinatally on sexual behavior of female ferrets. J Comp Physiol Psychol 90: 399-404

Baum MJ, Erskine MS (1984) Effect of neonatal gonadectomy and administration of testosterone on coital masculinization in the ferret. Endocrinology 115: 2440-2444

Baum MJ, Vreeburg JTM (1973) Copulation in castrated male rats following combined treatment with oestradiol and dihydrotestosterone. Science 182: 283-285

Baum MJ, Gallagher CA, Martin JT, Damassa DA (1982) Effects of testosterone, dihydrotestosterone, or estradiol administered neonatally on sexual behavior of female ferrets. Endocrinology lll: 773-780

Beach FA (1942a) Copulatory behavior in prepuberally castrated male rats and its modification by estrogen administration. Endocrinology 31: 679-683

Beach FA (1942b) Male and female mating behavior in prepuberally castrated female rats treated with androgens. Endocrinology 31: 673-678

Beach FA (1945) Bisexual mating behavior in the male rat: effects of castration and hormone administration. Psychol Zool 18: 390-402

Beach FA (1975) Hormonal modification of sexually dimorphic behaviour. Psychoneuroendocrinology l: 3-23

Beach FA, Holz AM (1946) Mating behavior in male rats castrated at various ages and injected with androgen. J Exp Zool 101: 91-142

Beach FA, Kuehn RE (1970) Coital behaviour in dogs. X. Effects of androgenic stimulation during development on feminine mating responses in females and males. Horm Behav l: 347-367

Beach FA, Levinson G (1950) Effect of androgen on the glans penis and mating behavior of castrated male rats. J Exp Zool 114: 159-171

Beach FA, Merari A (1970) Coital behavior in dogs. V. Effects of estrogen and progesterone on mating and other forms of social behavior in the bitch. J Comp Physiol Psychol 70: 1-22

Beach FA, Pauker RS (1949) Effects of castration and subsequent androgen administration upon mating behavior in the male hamster (*Mesocricetus auratus*). Endocrinology 45: 211-221

Beach FA, Noble RG, Orndoff RK (1969) Effects of perinatal androgen treatment on responses of male rats to gonadal hormones in adulthood. J Comp Physiol Psychol 68: 490-497

Beckhardt S, Ward IL (1983) Reproductive functioning in the prenatally stressed female rat. Dev Psychobiol 16: 111-117

Beeman EA (1947) The effect of male hormone on aggressive behavior in mice. Physiol Zool 20: 373-405

Bell DD, Zucker I (1971) Sex differences in body weight and eating: organization and activation by gonadal hormones in the rat. Physiol Behav 7: 27-34

Beyer C, Moralli G, Larsson K, Södersten P (1976) Steroid reguulation of sexual behavior. J Steroid Biochem 7: 1171-1176

Blizard D, Denef C (1973) Neonatal androgen effects on open field activity and sexual behavior in the female rat: the modifying influence of ovarian secretions during development. Physiol Behav ll: 65-69

Boling JC, Blandau RJ (1939) The estrogen-progesterone induction of mating responses in the spayed female rat. Endocrinology 25: 359-364

Boling JL, Young WC, Dempsey EW (1938) Miscellaneous experiments on the estrogen-progesterone induction of heat in the spayed guinea pig. Endocrinology 23: 182-187

Booth JE (1977) Sexual behaviour of neonatally castrated rats injected during infancy with oestrogen and dihydrotestosterone. J Endocrinol 72: 135-141

Bottiglioni F, Collins WP, Flamigni C, Neumann F, Sommerville IF (1971) Studies on androgen metabolism in experimentally feminized rats. Endocrinology 89: 553-559

Branchey L, Branchey M, Nadler RD (1973) Effects of sex hormones on sleep patterns of male rats gonadectomized in adulthood and in the neonatal period. Physiol Behav ll: 609–611

Bridges RS, Zarrow MX, Denenberg VH (1973) The role of neonatal androgen in the expression of hormonally induced maternal responsiveness in the adult rat. Horm Behav 4: 315–322

Bridges RS, Zarrow MX, Goldman BD, Denenberg VH (1974) A developmental study of maternal responsiveness in the rat. Physiol Behav 12: 149–151

Bronson FH, Desjardins C (1969) Aggressive behaviour and seminal vesicle function in mice: differential sensitivity to androgen given neonatally. Endocrinology 85: 971–974

Byrnes WW, Shipley EG (1955) Guinea pig copulatory reflex in response to adrenal steroids and similar compounds. Endocrinology 57: 5–9

Ciaccio LA (1971) Estrogen and progesterone: factors involved in the regulation of estrous behavior in the femal hamster (Mesocricetus auratus). Thesis, Princeton University, NJ

Clarke IJ (1977) The sexual behaviour of prenatally androgenized ewes observed in the field. J Reprod Fertil 49: 311–315

Clarke IJ (1978) Induction of male behaviour in ovariectomized ewes and ovariectomized-androgenized ewes chronically implanted with oestradiol-17 or testosterone. Anim Reprod Sci l: 305–312

Clarke IJ, Scaramuzzi RJ, Short RV (1976) Sexual differentiation of the brain: endocrine and behavioural responses of androgenized ewes to oestrogen. J Endocrinol 71: 175–176

Clemens LG, Hiroi M, Gorski RA (1969) Induction and facilitation of female mating behavior in rats treated neonatally with low doses of testosterone propionate. Endocrinology 84: 1430–1438

Clemens LG, Gladue BA, Coniglio LP (1978) Prenatal endogenous androgenic influences on masculine sexual behavior and genital morphology in male and female rats. Horm Behav 10: 40–53

Cole JR, Zuckerman H (1987) Marriage, motherhood and research performance in science. Sci Am 256(2): 119–125

Collins VJ, Boling JL, Dempsey EW, Young WC (1938) Quantitative studies of experimentally induced sexual receptivity in the spayed guinea-pig. Endocrinology 23: 188–196

Coniglio LP, Paup DC, Clemens LG (1973) Hormonal factors controlling the development of sexual behavior in the male golden hamster. Physiol Behav 10: 1087–1094

Crossley DA, Swanson HH (1968) Modification of sexual behavior of hamsters by neonatal administration of testosterone propionate. J Endocrinol 41: 8–9

Davidson JM (1969) Effects of estrogen on the sexual behaviour of male rats. Endocrinology 84: 1365–1372

Davis PG, Chaptal CV, McEwen BS (1979) Independence of the differentiation of masculine and feminine sexual behavior in rats. Horm Behav 12: 12–19

DeBold JF, Whalen RE (1975) Differential sensitivity of mounting and lordosis control systems to early androgen treatment in male and female hamsters. Horm Behav 6: 197–209

Dempsey EW, Hertz R, Young WC (1936) The experimental induction of oestrus (sexual receptivity) in the normal and ovariectomized guinea-pig. Am J Physiol 116: 201–209

Diamond M, Llacuna A, Wong CL (1973) Sex behavior after neonatal progesterone, testosterone, estrogen or antiandrogens. Horm Behav 4: 73–88

Dieterlen F (1959) Das Verhalten des syrischen Goldhamster. Z Tierpsychol 16: 47–103

Dörner G (1968) Hormonal induction and prevention of female homosexuality. J Endocrinol 42: 163–164

Dörner G (1976) Hormones and brain differentiation. Elsevier, Holland

Dörner GV, Fatschel J (1970) Wirkungen neonatal verabreichter Androgene und Antiandrogene auf Sexualabverhalten und Fertilität von Rattenweibchen. Endokrinologie 56: 29–34

Dörner G, Hinz G (1967) Homosexuality of neonatally castrated male rats following androgen substitution in adulthood. Germ Med Mont 12: 281–283

Dörner G, Staudt J (1968) Structural changes in the preoptic anterior hypothalamic area of the male rat following neonatal castration and androgen substitution. Neuroendocrinology 3: 136–140

Dörner G, Staudt J (1969a) Perinatal structural sex differentiation of the hypothalamus in rats. Neuroendocrinology 5: 103–106

Dörner G, Staudt J (1969b) Structural changes in the hypothalamic ventromedial nucleus of the male rat following neonatal castration and androgen treatment. Neuroendocrinology 4: 278–281

Dörner G, Hinz G, Schlenker G (1977) Demasculinizing effect of prenatal oestrogen on sexual behaviour in domestic pigs. Endokrinologie 69: 347–350

Dörner G, Geier Th, Ahrens L, Krell L, Munx G, Sieler H, Kittner E, Muller H (1980) Prenatal stress as possible aetiogenetic factor of homosexuality in human males. Endokrinologie 75: 365–368

Doty RL, Carter CS, Clemens LG (1971) Olfactory control of sexual behavior in the male and early-androgenized female hamster. Horm Behav 2: 325–335

Dutt RH (1953) Induction of estrus and ovulation in anestrual ewes by use of progesterone and pregnant mare serum. J Anim Sci 12: 515–523

Edwards DA (1968) Mice: fighting by neonatally androgenized females. Science 161: 1027–1028

Edwards DA (1970) Post-neonatal androgenization and adult aggressive behavior in female mice. Physiol Behav 5: 465–467

Edwards DA, Burge KG (1971) Early androgen treatment and male and female sexual behavior in mice. Horm Behav 2: 49–58

Edwards DA, Herndon J (1970) Neonatal estrogen stimulation and aggressive behavior in female mice. Physiol Behav 5: 993–995

Erhardt AA, Grisanti GC, Meyer-Bahlburg HFL (1977) Prenatal exposure to medroxyprogesterone acetate (MAP) in girls. Psychoneuroendocrinology 2: 391–398

Erhardt AA, Meyer-Bahlburg HFL, Feldman JF, Ince SE (1984) Sex dimorphic behavior in childhood subsequent to prenatal exposure to exogenous progestogens and estrogens. Arch Sex Behav 13: 457–477

Erhardt AA, Meyer-Bahlburg HFL, Rosen LA, Feldman JF, Veridian NP, Zimmerman I, McEwen BS (1985) Sexual orientation after prenatal exposure to exogenous estrogen. Arch Sex Behav 14: 57–77

Erskine MS, Baum MJ (1982) Plasma concentrations of testosterone and dihydrotestosterone during perinatal development in male and female ferrets. Endocrinology lll: 767–772

Farrell A, Gerall AA, Alexander MJ (1977) Age related decline in receptivity in normal neonatally androgenized female and male hamsters. Exp Aging Res 3: 117–128

Feder HH (1971) The comparative actions of testosterone propionate and 5α-androstan-17β-ol-3-one propionate on the reproductive behavior, physiology and morphology of male rats. J Endocrinol 51: 241–252

Feder HH, Phoenix CH, Young WC (1966) Suppression of feminine behavior by administration of testosterone propionate to neonatal rats. J Endocrinol 34: 131–132

Feder HH, Naftolin F, Ryan KJ (1974) Male and female sexual responses in male rats given estradiol benzoate and 5α-androstan-17β-ol-3-one propionate. Endocrinology 94: 136–141

Frank AH, Fraps RM (1945) Induction of estrus in the ovariectomized golden hamster. Endocrinology 37: 357–361

Friend JP (1977) Persistence of maternally derived ^{3}H-estradiol in fetal and neonatal rats. Experientia 33: 1235–1236

Fuertes de la Haba A, Santiago G, Bangdiwala IS (1976) Measured intelligence in offspring of oral and nonoral contraceptive users. Am J Obstet Gynecol 125: 980–983

Fuertes de la Haba A, Santiago G, Bangdiwala IS, Roure CA (1977) Intelligence quotient in offspring of oral and non-oral contraceptive users: second part. Bol Asoc Med P R 69: 10–14

Fuller GB, Zarrow MX, Anderson CO, Denenberg VH (1970) Testosterone propionate during gestation in the rabbit - effect on subsequent maternal behaviour. J Reprod Fertil 23: 285–290

Gentry RT, Wade GN (1976) Sex differences in sensitivity of food intake, body weight, and running-wheel activity to ovarian steroids in rats. J Comp Physiol Psychol 90: 747–754

Gerall AA, Kenney AMcM (1970) Neonatally androgenized females' responsiveness to estrogen and progesterone. Endocrinology 87: 560–566

Gladue BA, Clemens LG (1978) Androgenic influences on feminine sexual behavior in male and female rats: defeminization blocked by prenatal antiandrogen treatment. Endocrinology 103: 1702–1709

Gladue BA, Gary DP, Lynwood CG (1978) Hormonally mediated lordosis in female rats: actions of flutamide and an aromatization inhibitor. Pharmacol Biochem Behav 9: 827–832

Goldfoot DA, Feder HH, Goy RW (1969) Development of bisexuality in the male rat treated neo-
    natally with androstenedione. J Comp Physiol Psychol 67: 41–45
Goldfoot DA, van der Werff ten Bosch JJ (1975) Mounting behavior of female guinea pigs after
    prenatal and adult administration of the propionates of testosterone, dihydrotestosterone, and
    androstanediol. Horm Behav 6: 139–148
Gorski RA (1974) Barbiturates and sexual differentiation of the brain. In: Zimmermann E,
    George R (eds) Narcotics and the hypothalamus. Raven Press, New York, pp 197–211
Gorski RA, Barraclough CA (1963) Effects of low dosages of androgen on the differentiation of
    hypothalamic regulatory control of ovulation in the rat. Endocrinology 73: 210–216
Gorski RA, Wagner JW (1965) Gonadal activity and sexual differentiation of the hypothalamus.
    Endocrinology 76: 226–239
Goy RW (1970) Experimental control of psychosexuality. Philos Trans R Soc Lond [Biol] 259:
    149–162
Goy RW, Goldfoot DA (1973) Hormonal influences on sexually dimorphic behavior. In:
    Greep RO (ed) Handbook of physiology, endocrinology II, part I. The Physiological Society,
    Washington DC, pp 169–186
Goy RW, Goldfoot DA (1975) Neuroendocrinology: animal models and problems of human sex-
    uality. Arch Sex Behav 4: 405–420
Goy RW, Phoenix CH (1971) The effects of testosterone propionate administered before birth on
    the development of behavior in genetic female rhesus monkeys. In: Sawyer C, Gorski R (eds)
    Steroid hormones and brain functions. Univ Cal Press, Berkeley, pp 193–202
Goy RW, Resko JA (1972) Gonadal hormones and behavior of normal and pseudohermaphrodit-
    ic nonhuman female primates. Rec Prog Horm Res 28: 707–733
Goy RW, Bridson WE, Young WC (1964) Period of maximal susceptibility of the prenatal female
    guinea pig to masculinizing actions of testosterone propionate. J Comp Physiol Psychol 57:
    166–174
Grady KL, Phoenix CH, Young WC (1965) Role of the developing testis in differentiation of the
    neural tissues mediating mating behavior. J Comp Physiol Psychol 59: 176–182
Greene WA, Mogil L, Foote RH (1978) Behavioral characteristics of freemartins administered
    estradiol, estrone, testosterone, and dihydrotestosterone. Horm Behav 10: 71–84
Harlan RE, Gorski RA (1978) Dissociation between release of luteinizing hormone and prolactin
    and induction of female sex behavior in normal and androgenized rats. Biol Reprod 19:
    439–446
Harlan RE, Shivers BD, Moss RL, Shryne JE, Gorski RA (1980) Sexual performance as a func-
    tion of time of day in male and female rats. Biol Reprod 23: 64–71
Harris GW (1964) Sex hormones, brain development and brain function. Endocrinology 75:
    627–648
Harris GW, Levine S (1965) Sexual differentiation of the brain and its experimental control. J
    Physiol 181: 379–400
Herrenkohl LR (1986) Prenatal stress disrupts reproductive behavior and physiology in offspring.
    In: Komisaruk BR, Siegel HI, Cheng M-F, Feder HH (eds) Reproduction: a behavioral and
    neuroendocrine perspective. Ann NY Acad Sci 474: 120–128
Hockman CH (1961) Prenatal maternal stress in the rat: its effects on emotional behavior in the
    offspring. J Comp Physiol Psychol 54: 679–684
Hull EM (1981) Effects of neonatal exposure to progesterone on sexual behaviour of male and
    female rats. Physiol Behav 26: 401–405
Hull EM, Franz JR, Snyder AM, Nishita JK (1980) Perinatal progesterone and learning, social
    and reproductive behaviour in rats. Physiol Behav 24: 251–256
Kawashima S (1972) Restoration of cyclic locomotor activity by progesterone in neonatally
    estrogenized persistent-estrous rats. Proc Jpn Acad 48: 191–196
Kawashima S, Shinoda A (1968) Spontaneous activity of neonatally estrogenized female rats.
    Endocrinol Jpn 15: 305–312
Keeley K (1962) Prenatal influence on behavior of offspring of crowded mice. Science 135: 44–45
Kennedy GC (1964) Mating behaviour and spontaneous activity in androgen-sterilized female
    rats. J Physiol 172: 393–399
Kincl FA (1964) Copulatory reflex response to steroids. In: Dorfman RI (ed) Methods in hor-
    mone research, vol III. Academic Press, New York, pp 477–484

Kincl FA (1970) Neonatal sterilization with steroid hormones. 1. The uptake of radioactivity from injected estradiol or testosterone in various organs of 5-day old male and female rats. Endocrinol Exp 4: 139–141

Kincl FA, Chang CC (1970) Neonatal sterilization of rodents with steroid hormones. 2. Distribution of intravenously injected testosterone in adult male rats treated at the age of 5 days with testosterone propionate or estradiol benzoate. Endocrinol Exp 4: 207–213

Kincl FA, Dorfman RI (1961) Copulatory reflex in guinea pigs induced by progesterone and related steroids. Acta Endocrinol (Copenh) 38: 257–261

Komisaruk BR, Siegel HI, Cheng M-F, Feder HH (eds) (1986) Reproduction: a behavioral and neuroendocrine perspective. Ann NY Acad Sci 474: 1–460

Křeček J (1973) Sex differences in salt taste: the effect of testosterone. Physiol Behav 10: 683–688

Ladosky W, Gaziri LCJ (1970) Brain serotonin and sexual differentiation of the nervous system. Neuroendocrinology 6: 168–174

Larsson K, Södersten P, Beyer C (1973a) Induction of male sexual behaviour by oestradiol benzoate in combination with dihydrotestosterone. J Endocrinol 57: 563–564

Larsson K, Södersten P, Beyer C (1973b) Sexual behavior in male rats treated with estrogen in combination with dihydrotestosterone. Horm Behav 4: 289–299

Lee CT, Griffo W (1973) Early androgenization and aggression pheromone in inbred mice. Horm Behav 4: 181–189

Levine S, Mullins RF Jr (1966) Hormonal influences on brain organization in infant rats. Science 152: 1585–1592

Lieberburg I, Wallach G, McEwen BS (1977) The effects of an inhibitor of aromatization (1,4,6-androstatriene-3,17-dione) and an anti-estrogen (Cl-628) on in vivo formed testosterone metabolites recovered from neonatal rat brain tissues and purified cell nuclei. Implications for sexual differentiation of the rat brain. Brain Res 128: 176–181

Lieberman MW (1963) Early developmental stress and later behavior. Science 141: 824–825

Lisk RD (1969) Cyclic fluctuations in sexual responsiveness in the male rat. J Exp Zool 171: 313–320

Lisk RD, Russell JA, Kahler SG, Hanks JB (1973) Regulation of hormonally mediated maternal nest structure in the mouse (Mus musculus) as a function of neonatal hormone manipulation. Anim Behav 21: 296–301

Lisk RD, Ciaccio LA, Cantazaro C (1983) Mating behaviour of the golden hamster under deminatural conditions. Anim Behav 1: 659–666

Luttge WG, Whalen RE (1970) Dihydrotestosterone, androstenedione, testosterone: comparative effectiveness in masculinizing and defeminizing reproductive systems in male and female rats. Horm Behav 1: 265–281

Magalhães HM, de Araujo Carlini EL (1974) Effects of perinatal testosterone treatment on body weight, open field behavior and Lashley III maze performance of rats. Acta Physiol Lat Am 24: 317–327

Martins T, Rocha A (1931) The regulation of hypophysis by the testicles and some problems of sexual dynamics. Endocrinology 15: 421–429

Martins T, Valle JR (1948) Hormonal regulation of the micturition behavior of the dog. J Comp Physiol Psychol 41: 301–311

McDonald PG, Doughty C (1973) Androgen sterilization in the neonatal female rat and its inhibition by an estrogen antagonist. Neuroendocrinology 13: 182–188

McDonald PG, Doughty C (1974) Effect of neonatal administration of different androgens in the female rat: correlation between aromatization and the induction of sterilization. J Endocrinol 61: 95–103

McDonald P, Beyer C, Newton F, Brien B, Baker R, Tan HS, Sampson C, Kitching P, Greenhill R, Pritchard D (1970) Failure of 5α-dihydrotestosterone to initiate sexual behaviour in the castrated male rat. Nature 227: 964–965

McEwen BS, Lieberburg I, Chaptal C, Krey LC (1977) Aromatization: important for sexual differentiation of the neonatal rat brain. Horm Behav 9: 249–263

McEwen BS, Biegon A, Davis PG, Krey LC, Luine VN, McGinnis MY, Paden CM, Parsons B, Rainbow TC (1982) Steroid hormones: humoral signals which alter brain cell properties and functions. Recent Prog Horm Res 38: 41–92

Meyer-Bahlburg HFL (1977) Sex hormones and male homosexuality in comparative perspective. Arch Sex Behav 6: 297–325

Meyer-Bahlburg HFL (1979) Sex hormones and female homosexuality: a critical examination. Arch Sex Behav 8: 101-119

Meyer-Bahlburg HFL (1981) Androgens and human aggression. In: Brain PF, Benton D (eds) The biology of aggression. Sythoff and Noordhoff International Publ, pp 263-290

Meyer-Bahlburg HFL (1984) Psychoendocrine research on sexual orientation. Current status and future options. Prog Brain Res 61: 375-397

Meyer-Bahlburg HFL, Ehrhardt AA (1982) Prenatal sex hormones and human aggression: a review, and new data on progestogen effects. Aggres Behav 8: 39-62

Meyer-Bahlburg HFL, Grisanti GC, Erhardt AA (1977) Prenatal effects of sex hormones on human male behavior: medroxyprogesterone acetate (MAP). Psychoneuroendocrinology 2: 383-390

Meyer-Bahlburg HFL, Feldman JF, Ehrhardt AA, Cohen P (1984) Effects of prenatal hormone exposure versus pregnancy complications on sex-dimorphic behavior. Arch Sex Behav 13: 479-495

Moguilevsky JA (1966) Effect of testosterone in vitro on the oxygen uptake of different hypothalamic areas. Acta Physiol Lat Am 16: 353-356

Moguilevsky JA, Rubinstein L (1967) Glycolytic and oxidative metabolism of hypothalamic areas in prepubertal and androgenized rats. Neuroendocrinology 2: 213-221

Moguilevsky JA, Schiaffini O, Foglia V (1966) Effect of castration on the oxygen uptake of different parts of hypothalamus. Life Sci 5: 447-452

Moguilevsky JA, Libertun C, Schiaffini O, Szwarcfarb B (1968) Sexual differences in hypothalamic metabolism. Neuroendocrinology 3: 193-199

Moguilevsky JA, Libertun C, Schiaffini O, Scacchi P (1969) Metabolic evidence of the sexual differentiation of hypothalamus. Neuroendocrinology 4: 264-269

Money J, Erhardt AA (1972) Gender dimorphic behaviour and fetal sex hormones. Recent Prog Horm Res 28: 735-763

Morra M (1965) Level of maternal stress during two pregnancy periods of rat offspring behaviors. Psych Sci 3: 7-9

Naftolin F, Ryan KJ, Petro Z (1972) Aromatization of androstenedione by the anterior hypothalamus of adult male and female rats. Endocrinology 90: 295-298

Nance DM (1976) Sex differences in the hypothalamic regulation of feeding behavior in the rat. In: Friesen AH, Thompson RF (eds) Advances in psychobiology. J. Wiley & Sons, New York, 3: 75-123

Neumann F, Elger W (1965) Physiological and psychical intersexuality of male rats by early treatment with an antiandrogenic agent (1,2-methylene-6-chloro- $\Delta^6-17\alpha$-progesterone acetate). Acta Endocrinol (Copenh) (Suppl) 100: 1-174

Neumann F, Steinbeck H (1974) Antiandrogens. In: Eichler O, Farah A, Herken H, Welch AD (eds) Handbook of experimental pharmacology, vol 35/2. Springer, Berlin Heildelberg New York, pp 235-484

Nordeen EJ, Nordeen KW, Senglaub DR, Arnold AP (1985) Androgens prevent normally occurring cell death in a sexually dimorphic spinal nucleus. Science 229: 671-673

Nucci LP, Beach FA (1971) Effects of prenatal androgen treatment on mating behavior in female hamsters. Endocrinology 88: 1514-1515

Parrot RF (1974) Effects of 17β-hydroxy-4-androsten-19-ol (19-hydroxytestosterone) and 5α-androstan-17β-ol (dihydrotestosterone) on aspects of sexual behaviour in castrated male rats. J Endocrinol 61: 105-115

Parsons B, McEwen BS, Pfaff DW (1982) A discontinuous schedule of estradiol treatment is sufficient to activate progesterone facilitated feminine sexual behavior and to increase cytosol receptors for progestins in the hypothalamus of the rat. Endocrinology 110: 613-619

Parsons B, Rainbow TC, McEwen BS (1984) Organizational effects of testosterone via aromatization on feminine reproductive behavior and neural progestin receptors in rat brain. Endocrinology 115: 1412-1417

Paup DC, Coniglio LP, Clemens LG (1972) Masculinisation of the female golden hamster by neonatal treatment with androgen or estrogen. Horm Behav 3: 123-131

Payne AP (1974) Neonatal androgen administration and aggression in the female golden hamster during interactions with males. J Endocrinol 63: 497-506

Payne AP (1977) Changes in aggressive and sexual responsiveness of male golden hamsters after neonatal androgen administration. J Endocrinol 73: 331-337

Payne AP, Swanson HH (1972) Neonatal androgenization and aggression in the male golden hamster. Nature 239: 282–283

Payne AP, Swanson HH (1973) The effects of neonatal androgen administration on the aggression and related behavior of male golden hamster during interactions with females. J Endocrinol 58: 627–636

Peters H, Sörensen IN (1971) Fertility and mating behaviour of androgenized mice. J Endocrinol 51: 589–594

Pfaff DW, Zigmond RE (1971) Neonatal androgen effects on sexual and non-sexual behavior of adult rats tested under various hormone regimes. Neuroendocrinology 7: 129–145

Pfeiffer CA (1936) Sexual differences of the hypophysis and their determination by the gonads. Am J Anat 58: 195–225

Phoenix CH, Goy RW, Gerall AA, Young WC (1959) Organizing action of prenatally administered testosterone propionate on the tissues mediating mating behavior in the female guinea pig. Endocrinology 65: 369–382

Quadagno DM, Rockwell J (1972) The effect of gonadal hormones in infancy on maternal behavior in the adult rat. Horm Behav 3: 55–62

Quadagno DM, Shryne J, Anderson C, Gorski RA (1972) Influence of gonadal hormones on social, sexual, emergence, and open field behaviour in the rat *(Rattus norvegicus)*. Anim Behav 20: 732–740

Raisman G (1974) Evidence for a sex difference in the neuropil of the rat preoptic area and its importance for the study of sexually dimorphic function. Aggression 52: 42–51

Raisman G, Field PM (1971) Sexual dimorphism in the preoptic area of the rat. Science 173: 731–733

Raisman G, Field PM (1973) Sexual dimorphism in the neuropil of the preoptic area of the rat and its dependence on neonatal androgen. Brain Res 54: 1-29

Reddy VVR, Naftolin F, Ryan KJ (1974) Conversion of androstenedione to estrone by neural tissues from fetal and neonatal rats. Endocrinology 94: 117–121

Reinisch JM (1977) Prenatal exposure of human foetuses to synthetic progestin and oestrogen affects personality. Nature 266: 561–562

Reinisch JM (1981) Prenatal exposure to synthetic progestins increases potential for aggression in humans. Science 211: 1171–1173

Reinisch JM, Karow WG (1977) Prenatal exposure to synthetic progestins and estrogens: effects on human development. Arch Sex Behav 6: 257–288

Reinisch JM, Sanders SA (1982) Early barbiturate exposure: the brain, sexually dimorphic behavior and learning. Neurosci Biobehav Rev 6: 311–319

Revesz C, Kernaghan D, Bindra D (1963) Sexual drive of female rats "masculinized" by testosterone during gestation. J Endocrinol 25: 549–550

Ring JR (1944) The estrogen-progesterone induction of sexual receptivity in the spayed female mouse. Endocrinology 34: 269–275

Robinson TJ (1954) The necessity for progesterone with estrogen for the induction of recurrent estrus in the ovariectomized ewe. Endocrinology 55: 403–408

Rodriguez-Sierra JF, Sridaran R, Blake CA (1980) Monosodium glutamate disruption of behavioral and endocrine function in the female rat. Neuroendocrinology 31: 228–235

Rubin RT, Reinisch JM, Haskett RF (1981) Postnatal gonadal steroid effects on human behavior. Science 211: 1318–1324

Sachs BD, Pollack EI, Krieger MS, Barfield RJ (1973) Sexual behavior: normal male patering in androgenized female rats. Science 181: 770–772

Sanders SA, Reinisch JM (1985) Behavioral effects on humans of progesterone-related compounds during development and in the adult. Curr Top Neuroendocrinol 5: 175–205

Sawyer CH, Everett JW (1959) Stimulatory and inhibitory effects of progesterone on the release of pituitary ovulating hormone in the rabbit. Endocrinology 65: 644–651

Scacchi P, Moguilevsky JA, Libertun C, Christot J (1970) Sexual differences in protein content of the hypothalamus in rats. Proc Soc Exp Biol Med 133: 845–848

Schwartz BD, Gerall AA (1979) Influence of photoperiod and neonatally administered androgen on estrous cycles and behavior in hamsters. Biol Reprod 21: 1115–1124

Scouten CW, Grotelueschen LK, Beatty WW (1975) Androgens and the organization of sex differences in active avoidance behavior in the rat. J Comp Physiol Psychol 88: 264–270

Shapiro BH, Goldman AS, Steinbeck HF, Neumann F (1976) Is feminine differentiation of the brain hormonally determined? Experientia 32: 650–651

Sheridan PJ, Zarrow MX, Dennenberg VH (1973) Androgenization of the neonatal female rat with very low doses of androgen. J Endocrinol 57: 33–45

Södersten PJ (1973) Increased mounting behavior in the female rat following a single neonatal injection of testosterone propionate. Horm Behav 4: 1–17

Södersten P (1978a) Lordosis behaviour in immature male rats. J Endocrinol 76: 233–240

Södersten P (1978b) Effects of anti-oestrogen treatment of neonatal male rats on lordosis behaviour and mounting behaviour in the adult. J Endocrinol 76: 241–249

Södersten P, de Jong FH, Vreeburg JTM, Baum MJ (1974) Lordosis behavior in intact male rats: absence of correlation with mounting behavior or testicular secretion of estradiol-17β and testosterone. Physiol Behav 13: 803–808

Staudt VJ, Dorner G, Doll R, Blose J (1973) Geschlechtsspezifische morphologische Unterschiede im Nucleus arcuatus und Pramamillaris ventralis der Ratte. Endokrinologie 62: 234–236

Stern JJ (1969) Neonatal castration, androstenedione, and the mating behavior of the male rat. J Comp Physiol Psychol 69: 608–612

Swanson HH (1967) Alteration of sex-typical behavior of hamsters in open field and emergence tests by neo-natal administration of androgen or oestrogen. Anim Behav 15: 209–216

Swanson HH (1970) Effects of castration at birth in hamsters of both sexes on luteinization of ovarian implants, oestrous cycles and sexual behaviour. J Reprod Fertil 21: 183–186

Takahashi LK, Lisk RD (1985) Diencephalic sites of progesterone action for inhibiting aggression and facilitating sexual receptivity in estrogen-primed golden hamsters. Endocrinology 116: 2393–2399

Takahashi LK, Lisk RD, Burnett II AL (1985) Dual estradiol action in diencephalon and the regulation of sociosexual behavior in female golden hamsters. Brain Res 359: 194–207

Thiessen DD, Lindzey G (1970) Territorial marking in the female mongolian gerbil: short-term reactions to hormones. Horm Behav 1: 157–160

Thomas TR, Gerall AA (1969) Dissociation of reproductive physiology and behavior induced by neonatal treatment with steroids. Endocrinology 85: 781–784

Thompson WR (1957) Influence of prenatal maternal anxiety on emotionality in young rats. Science 125: 698–699

Vomachka AJ, Clemens LG (1981) Fluoxymesterone and the development of sexual behavior in the golden hamster *Mesocricetus auratus*. Dev Psychobiol 14: 333–342

Vreeburg JTM, van der Vaart PDM, van der Schoot P (1977) Prevention of central defeminization but not masculinization in male rats by inhibition neonatally of oestrogen biosynthesis. J Endocrinol 74: 375–382

Wade GN (1972) Gonadal hormones and behavior regulation of body weight. Physiol Behav 8: 523–534

Wade GN, Zucker I (1969a) Hormonal and developmental influences on rat saccharin preferences. J Comp Physiol Psychol 69: 291–300

Wade GN, Zucker I (1969b) Taste preferences of female rats: modification by neonatal hormones, food deprivation and prior experience. Physiol Behav 4: 935–943

Ward IL (1972) Prenatal stress feminizes and demasculinizes the behavior of males. Science 175: 82–84

Ward IL (1977) Exogenous androgen activates female behavior in noncopulating, prenatally stressed male rats. J Comp Physiol Psychol 91: 464–471

Weisz J, Ward IL (1980) Plasma testosterone and progesterone titers of pregnant rats, their male and female fetuses and neonatal offspring. Endocrinology 106: 306–316

Whalen RE (1964) Hormone induced changes in the organization of sexual behavior in the male rat. J Comp Physiol Psychol 57: 175–182

Whalen RE (1968) Differentiation of the mechanisms which control gonadotropin secretion and sexual behavior. In: Diamond M (ed) Perspective in reproduction and sexual behavior. Indiana Univ Press, Bloomington, pp 303–340

Whalen RE, Etgen AM (1978) Masculinization and defeminization induced in female hamsters by neonatal treatment with estradiol benzoate and RU-2858. Horm Behav 10: 170–177

Whalen RE, Nadler RD (1963) Suppression of the development of female mating behavior by estrogen administered in infancy. Science 141: 273–274

Whalen RE, Rezek DL (1974) Inhibition of lordosis in female rats by subcutaneous implants of testosterone, androstenedione or dihydrotestosterone in infancy. Horm Behav 5: 125-128

Whalen RE, Peck CK, LoPiccolo J (1966) Virilization of female rats by prenatally administered progestin. Endocrinology 78: 965-970

Whalen RE, Edwards DA, Luttge WG, Robertson TR (1969) Early androgen treatment and male sexual behavior in female rats. Physiol Behav 4: 33-39

Wilson JG (1943) Reproductive capacity of adult female rats treated prepuberally with estrogenic hormone. Anat Rec 86: 341-359

Wilson JG, Wilson HC (1943) Reproductive capacity in adult male rats treated prepuberally with androgenic hormone. Endocrinology 33: 353-360

Young WC (1961) The hormones and mating behaviour. In: Young WC (ed) Sex and internal secretions, vol 2. Williams and Wilkins, Baltimore, pp 1173-1239

Young WC, Goy RW, Phoenix CH (1964) Hormones and sexual behavior. Science 143: 212-218

Zucker I, Wade GN, Ziegler R (1972) Sexual and hormonal influences on eating, taste preferences, and body weight of hamsters. Physiol Behav 8: 101-111

7 Author's Overview

Pharmacological doses of many hormones during uterine life or shortly after birth cause severe disturbances in reproductive functions, affect the development of sex organs and sex behavior, and predispose some tissue to the development of neoplastic lesions. The neonate is very sensitive to steroid hormone insult. A single injection of as little as 1 µg of estradiol benzoate and 5–10 µg of testosterone propionate is sufficient to produce sterility in the adult. Females exposed to such small amounts may sexually mature and ovulate for several weeks, or even a few months, before becoming sterile. The rodent neonate was held to be sensitive to the insult only during a "critical" period defined to last about 10 days after birth.

A hypothesis evolved stating that prepubertal exposure of female rodents to androgens permanently "masculinized" those hypothalamic centers which in the adult control ovulating mechanism(s); in this view a female "cyclic" hypothalamus is the result of steroid hormone absence during a "critical" period while males acquire a "tonic" gonadotropin release due to an androgen imprint.

Both androgens and estrogens are teratogenic, but the action of each is distinct. In males estrogens produce sterility while androgens do not. In the female both hormones produce superficially the same anovulatory syndrome, albeit differently: luteinization can be induced more readily in androgen sterilized females while the estrogen sterilized ovary is refractive.

7.1 Behavior

Adults, exposed to steroid hormones as neonates, behave abnormally. The observation has not yielded clear-cut evidence to implicate the CNS alone. Tested under artificial laboratory conditions rats and hamsters behave bisexually; males exhibit readily female behavior, and vice versa. Female (lordosis) response to manual stimulation is strain dependent. A strain may show lordosis while another strain does not although endogenous levels of testosterone and estradiol are the same. Male pseudohermaphrodites exhibit neither male nor female sexual behavior when primed with the proper sex hormones. The lack of response is due to the deficiency in the target organ, not the brain.

Male mice are aggressive; androgens induce aggressiveness in females. Edwards obtained aggressive mice by chronic administrations of testosterone to adolescents (30 days old) which led him to conclude "...the results do not indicate the potential for androgen induced differentiation of hormone sensitive mechanism for aggression is totally absent after the first few weeks of birth; rather indicate that the capacity for a single injection of a given amount of androgen to facilitate aggressive behavior in the female is (only) diminished by the tenth day of life..."

Another difficulty in interpreting behavior data resides in the need of steroid hormones during the formative stages for proper development of external genitalia. Beach and others feel that in males proper penile development is imperative to male copulatory behavior; they disregard central nervous imprinting. The effects of steroid hormones on the development of female genitalia, except for detailed studies relating to teratological effects on vaginal and uterine epithelium, are unknown.

7.2 The Brain

In the neonate stereotaxically implanted testosterone at the level of the anterior commissure in the dorsal preoptic-anterior hypothalamus produces persistent estrus syndrome. Yet, implants in the cortex are also effective provided the concentration of the hormone is higher. The question whether the hormone, present in higher amounts, has diffused into a greater volume of the brain, including the hypothalamic centers, has not been answered.

Unimpaired hypothalamic function early in life is essential for the future development of gonadal functions. Blocking of GnRH release and/or function in neonatal males (1 or 5 day old) by a specific antibody results in decreased testes weight, oligospermia and reduced fertility in the adult, but plasma levels of FSH, LH, PRL, and testosterone are not affected. The chance that such treatment causes a decrease in gonadotropin binding sites in the gonads remains a distinct possibility, especially in view of the observation that the ontogenesis of the binding sites may be altered by neonatal manipulation (Kolena).

Two groups proposed that estrogens (estradiol) rather than androgens provide the "masculinizing" effect in males. Support for this hypothesis came from observations that inhibitors of aromatizing enzymes prevent androgenization effects; unfortunately, the critical experiments were terminated at an early age (75 days), and the possibility that such a protection was only temporary, resulting in a delayed anovulatory syndrome, cannot be excluded. The hypothesis also does not explain the activity of synthetic estrogens, of natural hormones not found in rodents (estriol), and the inactivity of compounds such as estrone which are more readily convertible to estradiol than testosterone.

The observation in females that several sedative drugs (reserpine, chlorpromazine, pento- and phenobarbital, progesterone) have a protective action has been used to support the hypothesis claiming sole CNS involvement. However, a general nervous depressant, cold anesthesia, does not protect the neonate. A direct infusion of pentobarbital into the brain does not reverse the deleterious effects of peripherally injected androgens.

Studies measuring the uptake of gonadal hormones in the CNS do not contribute to our understanding of the mechanism. The hypothalamus of neonatal rodents does not accumulate preferentially either testosterone or estradiol; indeed, there is no sex difference or age difference; the uptake of testosterone or estradiol is about the same in 5, 10 or 21 day old animals. The pituitary gland and the sex organs accumulate more radioactivity than the brain (Kincl).

7.3 The Pituitary Gland

The stimulated pituitary of female PE rats releases enough gonadotropins to cause luteinization. The gland is deficient in that it fails to release LH in response to estrogen stimulation (Barraclough, Flerkó, and Gorski). The observation that the pituitary responds to GnRH, yet not to exogenous estrogens, suggests that either the hypothalamic neurons have been altered or that the fate of circulating hormones has been modified.

Studies on prolactin concentration and release indicate general "aging" of the gland. The base levels of this hormone are higher in PE females, resembling old constant-estrus rats. Altered ovarian steroid biosynthesis may be the cause: in senile constant-estrous rats the increased prolactin concentration in the peripheral plasma can be abolished by ovariectomy (Johnson).

7.4 The Gonads

Female gonads respond to treatment with PMSG, hCG, or LH, with the formation of corpora lutea. Some animals ovulate and mate. The corpora lutea in mated animals are maintained for several cycles, and the concentrations of progesterone, the 20α-alcohol and prolactin in peripheral plasma are comparable to pregnant rats (Brown-Grant, Johnson).

7.4.1 Transplantation Experiments

It has been long held that ovarian tissue transplanted into male recipients will become luteinized only if the males are castrated during the "neonatal" period; if males are castrated as adults, or androgenized females are the recipients, luteinization will not take place. Some results indicate otherwise. Several groups noted corpora lutea formation in grafts transplanted into intact males, but it is believed that since these were few in numbers and small, such tissue "...almost certainly represents granulosa or thecal luteinization and does not represent a cyclic release of luteinizing hormone, with consequent ovulation and luteinization..." We have seen luteinization in ovaries transplanted under the kidney capsule of male rats or hamsters, castrated neonatally or as adults. When the males were castrated as adults we had to allow a 30 day period between castration and transplantation. The concentration of progesterone in peripheral plasma correlated with the number of corpora lutea formed, indicating functionality of the tissue.

Transplants placed under the spleen capsule became luteinized in androgenized adult females. Spleen transplants are exposed to a different hormonal milieu; the main venous drainage from the spleen is through the portal vein into the liver. Thus, the hormones produced in the transplant are drained into the liver where the bulk is metabolized during the first pass.

The results show that the hormone milieu in androgenized females, or recently castrated males, prevents formation of corpora lutea.

7.5 Other Endocrine Systems

Other endocrine glands may influence the outcome of transplantation. Neonatal androgenization (or estrogenization) alters both thyroid and adrenal functions. Both may influence graft acceptance. Fujii androgenized neonatal pups with a high dose (1 mg) of testosterone propionate and removed the thyroid in some. Corpora lutea formed in thyroidectomized animals. Irregular cycles continued for about a month.

The effects of unimpaired thymus function remain to be clarified. Neonatally thymectomized animals of both sexes are more sensitive to the toxic effects of gonadal hormones while injection of thymocytes provides a "protective" function. Several investigators have described a "wasting disease" in young rats of both sexes injected with gonadal hormones; estrogens were more reactive in this respect. Sterility and ovarian dysgenesis (characterized by hyperplasia and hypertrophy of interstitial cells and follicular degeneration), accompanied by autoimmune thyroiditis, were reported in adult, neonatally thymectomized mice and rats. In males, neonatal thymectomy will lead to atrophic testes characterized by edematous interstitium, a reduction in the number of seminiferous tubules, and azoospermia.

Neonatal steroids may damage other endocrine and paracrine modalities: the influence on brain opioids, polypeptide hormones of the gonads, binding and transport of hormones in blood and blood-brain permeability has not been clarified.

7.6 Direct Effects

Many observations reveal that the function of some organs is affected directly, independently of the hypothalamus. The sensitivity of ovaries in androgenized females to gonadotropin stimulation is decreased, possibly to one-tenth (Mennin). Others tested GnRH response in ovariectomized, estrogen treated TP females and noted a subnormal response and postulated "... a direct effect of neonatal androgen administration on the pituitary..."

A number of groups (Harris, Takasugi, Wilson, Zeilmaker) have commented on the decrease of androgenized uterine tissue to estrogenic stimulation. Decreased sensitivity most likely results from decreased estradiol binding, possibly due to an increased concentration of a cytosol binding protein which inhibits the translocation of estradiol to the cell nucleus. Decreases of 8S estradiol receptors in prepubertal rats following neonatal estrogen treatment has been reported by Gellert and coworkers.

The Japanese workers (Iguchi, Kawashima, Mori, Takasugi) and Bern have proved convincingly that the vaginal epithelium is lesioned directly. Neonatal estrogenization leads to a damaged epithelium, even if the ovaries are extirpated within a few days after the neonatal insult. The autonomous proliferation can achieve alarming proportions and lead to the formation of cancer.

In males neonatal androgens increase the sensitivity of peripheral tissue to androgens. The action is independent of the brain function (Johnson). Harris commented that the results (of estrogen exposure) are "...target organ lesion rather than solely due to interferences with the CNS function..." In adult androgenized castrated males the seminal vesicles are 30 times more sensitive to testosterone, the levator ani muscle 9 times, and the penis 3 times. The rapid synthesis of RNA in some organs led Bronson *et al.* to propose that neonatal androgens exert an organizing influence on the genome in the adult. Increased sensitivity can also be explained as the result of altered testosterone metabolism and the biological half-life. In estrogenized males lower testosterone production inhibits full development of testes and of accessory sex tissues (Kincl).

In addition neonatal steroid hormones have been shown to alter steroid hormones metabolism in the liver, adrenals and kidneys (Denef, Ghraf, Gustafsson, Stenberg).

7.7 Conclusions

The androgen sterilized rat has been a useful tool helping to map the various hypothalamic nuclei involved in the regulation of gonadal functions. The regulatory role of the central nervous system on hormonal functions was firmly established. The preoccupation with the brain functions obscured in the minds of many the possibility that steroid hormones may modify directly other organs and gave rise to precipitous claims reaching beyond the justification of experimental data.

The consequences associated with the indiscriminate use of diethylstilbestrol during pregnancy are well known. Possibly, this tragedy could have been avoided if the danger signs, demonstrated in rodents, had been taken into account.

In my view the presence of the hormones, in toxic amounts, during the formative stages gives rise to a complex syndrome. In the adult, the syndrome is manifested by disruption of the delicate homeostatic balance between the central nervous system, the pituitary and the gonads, abnormal hormone production, transport in plasma and metabolism, and aberrant cellular development in peripheral organs. The brain is not the sole respondent; unable to cope with altered hormonal feedback, the hypothalamic function is altered; the organism has prematurely aged.

Appendix

The appendix lists trivial names and abbreviations used in text, describes briefly the biological tests used to assess the activity of steroid hormones, and lists the main biological function and strutures of the hormones discussed in the monograph.

1. Abbreviations Used

5-Androstenedione	5-androstene-3,17-dione
Androstenedione	:az4-androstene-3,17-dione
Androsterone	3α-hydroxy-5α-androstan-17-one
BW	body weight
Chlormadione acetate	6-chloro-17-acetoxy-4,6-pregnadiene-3,20-dione
CL	corpus luteum
Clomegestone acetate	6-chloro-16α-methyl-17-acetoxy-4,6-pregnadiene-3,20-dione
Clomid (clomiphene)	2[4-(2-chloro-1,2-diphenylethenyl)-phenoxy]-N,N-diethyl-ethanamine
Corticosterone	11β,21-dihydroxy-4-pregnene-3,20-dione
Cortisone	17,21-dihydroxy-4-pregnene-3,11,20-trione
Cyproterone acetate	1,2α-methylene-6-chloro-17-acetoxy-4,6-pregnadiene-3,20-dione
o,p'-DDD (Mitotane)	1-chloro-2-[2,2-dichloro-1-(4-chlorophenyl)ethyl]benzene
p,p'-DDD	1,1-dichloro-2,2-bis(p-chlorophenyl)ethane
DDT	1,1,1,-trichloro-2-(o-chlorophenyl)-2-(p-chlorophenol)-ethane
Dehydroepiandrosterone (DHA)	3β-hydroxy-5-androsten-17-one
Deoxycorticosterone	21-hydroxy-4-pregnene-3,20-dione
DHT (5α-DHT)	5α-dihydrotestosterone
Dienestrol	4,4'-(1,2-diethylidene-1,2-ethanediyl)bisphenol
Diethylstilbestrol (DES)	(E)-4,4'-(1,2-diethyl-1,2-diethenediyl)bisphenol
5α-Dihydrotestosterone	17β-hydroxy-5α-androstan-3-one
5β-Dihydrotestosterone	17β-hydroxy-5β-androstan-3-one
DOPA (3-hydroxytyrosine)	3-(3,4-dihyroxyphenyl)alanine
EB	estradiol benzoate
ED	estradiol dipropionate
Ethinyl estradiol	17α-ethinyl-1,3,5(10)-estratrine-3,17β-diol
Estradiol	1,3,5(10)-estratriene-3,17β-diol
Estrone	3-hydroxy-1,3,5(10)-estratriene-17-one
Ethamoxytriphetol (MER-25)	α-[4[2-diethylamino)ethoxy]-phenyl]4-methoxy-α-phenylbenzeneethanol
Etiocholanolone	3α-hydroxy-5β-androstan-17-one
Fluoxymesterone	9-fluoro-11β,17β-dihyroxy-17α-methyl-4-androsten-3-one
Flutamide	2-methyl-N-[4-nitro-3(trifluoromethyl]propanamide
FSH	follicle stimulating hormone
GH	growth hormone
GnRH	gonadotropin releasing hormone
hCG	human chorionic gonadotropin
Hexestrol	4,4'-(1,2-diethyl-1,2-ethanediyl)b-isphenol
IU	international unit
ip	intraperitoneal
kD	kilodalton
LH	luteinizing hormone
Medroxyprogesterone acetate	6α-methyl-17-acetoxy-4-pregnene-3,20-dione
Megestrol acetate	6-methyl-17-acetoxy-4,6-pregnadiene-3,20-dione
Mestranol	3-methoxy-17α-ethinyl-1,3,5(10)-estratriene-17β-ol

MSG	monosodium glutamate
Nafoxidine	1-2-[p-3,4-dihydroxy-6-methoxy-2-phenyl-1-naphtyl)phenoxy]-ethyl
Norethandrolone	17α-ethyl-17β-hydroxy-4-estren-3-one
Norethindrone (NET)	17α-ethinyl-17β-hydroxy-4-estren-3-one
Norethyndrol	17α-ethinyl-17β-hydroxy-5(10)-estren-3-one
Norgestrel	d,l-18-methyl-17α-ethinyl-17β-hydroxy-4-estren-3-one
19 Nortestosterone	17β-hydroxy-4-estren-3-one
PDS	polydimethylsiloxane
PMSG	pregnant mares' serum gonadotropin
Pregananolone	3α-hydroxy-5β-pregnan-20-one
Pregnanediol	5β-pregnane-3α,20α-diol
Pregnenolone	3β-hydroxy-5-pregnen-20-one
PRL	prolactin
Progesterone	4-pregnene-3,20-dione
R-2858	11β-methoxy-17α-ethinyl-1,3,5(10)-estratriene-3,17β-diol
Tamoxifen	(Z)-2-[p-(1,2-diphenyl-1-butenyl)-phenoxy]-N,N-dimethyl-ethylamine
Testosterone	17β-hydroxy-4-androsten-3-one
TP	testosterone propionate
TRH	thyrotropin stimulating hormone
TSH	thyroid stimulating hormone

2. Structures and Functions of Principal Hormones

The structure of polypeptides is written beginning with the amino acid having the free amino group and ending with the amino acid having the free carboxyl group. Amino acids are represented either by a three-letter or one-letter symbol (Table A 1).

Table A 1 Amino Acid Symbols.

One-letter	Three-letter	Amino acid
A	Ala	Alanine
C	Cys	Cysteine
D	Asp	Aspartic acid
E	Glu	Glutamic acid
F	Phe	Phenylalanine
G	Gly	Glycine
H	His	Histidine
I	Ile	Isoleucine
K	Lys	Lysine
L	Leu	Leucine
M	Met	Methionine
N	Asn	Asparagine
P	Pro	Proline
Q	Glu	Glutamine
R	Arg	Arginine
S	Ser	Serine
T	Thr	Threonine
V	Val	Valine
W	Trp	Tryptophane
Y	Tyr	Tyrosine

2.1. Hypothalamic Hormones

The polypeptide hormones of the hypothalamus (releasing hormones) are either carried via the hypothalamo-pituitary portal vessels to the anterior pituitary lobe where they stimulate the release of tropic hormones or transported through axonal connections and stored in the posterior lobe (oxytocin, vasopressin). The hormones are synthesized as large precursors (prohormones, preprohormones) on ribosomes associated with the endoplasmic reticulum and are passed into Golgi apparatus for incorporation into secretory granules. The granules are transported by axoplasmic flow to axon terminals from whence they are discharged. The precursors are cleaved into the active fragment during the axonal transport.

Gonadotropin Releasing Hormone (GnRH): also luteinizing hormone-releasing hormone (LH-RH), or releasing factor (RF). The releasing hormone isolated from hypothalami of porcine origin is a decapeptide with the following amino acid sequence: (pyro)Glu-His-Trp-Ser-Tyr-Gly-Leu-Arg-Pro-Gly-NH₂. The structure of the hormone has been confirmed by synthesis. A polypeptide isolated from the human placenta has been shown to be similar, or identical, to GnRH present in the central nervous system (Lee *et al.*, 1981). GnRH precursor (human placental hormone) comprises the GnRH sequence followed by 59 amino acids.

Gonadotropin releasing hormone regulates both the polypeptide synthesis and glycosylation of LH in the pituitary gland (Ramey *et al.*, 1987). $C_{55}H_{75}N_{17}O_{13}$; MW: 1182.33; $[\alpha]_D - 50°$ (1% acetic acid).

Thyrotropin Releasing Hormone (TRH): The tripeptide (pyro-Glu-His-Pro-NH₂) is present in the hypothalamus, median eminence, brain stem and spinal cord. In addition to TSH stimulation, TRH has been shown to stimulate prolactin release and cerebral noradrenaline turnover, enhance the effects of centrally active drugs and depress the activity of central neurones. The TRH prohormone is a protein (MW 29 kD) which contains five copies of the TRH sequence (Lecham *et al.*, 1986).

Corticotropin Releasing Hormone (CRH): A polypeptide which stimulates both ACTH and β-endorphin-like activity has been synthesized and shown to stimulate ACTH secretion in rats. The structure of ovine CRF is: S-Q-E-P-P-I-S-L-D-L-T-F-H-L-L-R-E-V-L-E-M-T-L-A-D-Q-L-A-Q-Q-A-H-S-N-R-K-L-L-A-I-A. The characterization of the hormone has been reviewed by Yasuda *et al.* (1982) and Guillemin *et al.* (1984).

Atrial Natriuretic Factors (ANF): ANF are a family of 17-membered ring peptides (formed by an internal disulfide bond) varying in the length of their N- and C-terminal extensions. The peptides are released from the heart in response to distension of the atrium caused by increased venous return (Katsube *et al.*, 1985; Lang *et al.*, 1985). ANF antagonize angiotensin and vasopressin and thus influence the maintenance of fluid and electrolyte homeostasis. Brain neurons are capable of ANF synthesis. The hormones can alter hypothalamic control by interacting with dopaminergic and peptidergic systems (Samson *et al.*, 1988).

Neuropeptide Y: Neuropeptide Y is a 36 a. a. peptide of known sequence (Tatemoto, 1982). The neuropeptide augmented the GnRH stimulated release of LH from pituitary cells *in vitro* (Crowley *et al.*, 1987).

Enkephalins and Endorphins (Opioids): Peptides which bind to membrane receptors of nerve cells are classified as "opioids" to distinguish this class of compounds from "opiate" alkaloids and related synthetic drugs. Opioids depress the autonomic effector system in the periphery and inhibit central nervous systems pathways from spinal cord to cerebral cortex and hypothalamic and neurohypophyseal hormone release. The hormones arise from the cleavage of larger prohormones (6–8 kD). The origin and structures were reviewed by Akil *et al.* (1984). The smaller (Goldstein *et al.*, 1981; Seizinger *et al.*, 1984) peptides (less than 20 a. a.) present in the CNS and pituitary gland are biologically active (Panula and Lindberg, 1987). Possible regulatory participation of opioids in the function of the anterior and posterior pituitary hormones was reviewed by Clement–Jones and Rees (1982).

2.2 Pituitary Hormones

The hormones of the anterior pituitary, high molecular weight polypeptides (ACTH, PRL, GH) or glycoproteins (FSH, LFH, TSH) are specific tropic hormones which stimulate the target organs: for example the stimulatory influence of gonadotropins (FSH, LH) is needed for the production of sex hormones by the gonads. In the absence of the stimulatory influence the target tissue will atrophy. The hormones are produced when the gland is stimulated by hypothalamic releasing hormones.

The characteristic features of gonadotropins are the presence of sugars and aminosugars (glycoproteins) and their high molecular weight. Each of these hormones contains two peptide subunits, designated α and β. The amino acid sequence of the chain is identical or very similar in all the hormones; isolated, the α-chain lacks biological activity. The amino acid sequence of the β-chain varies from species to species and confers hormonal, and immunospecificity to each hormone. Isolated β-chains may have slight intrinsic biological activity while a full spectrum is only restored upon recombination. Hybrids formed from two dissimilar units (i. e. αLH and βTSH) will exert the biological activity specified by the β-subunit (i. e. thyrotropic activity).

The glycoprotein hormones have as yet to be isolated in a pure state. The activity of macromolecules of uncertain purity must be expressed in terms of units of biological activity related to an accepted standard preparation or must be measured by methods (radioimmunoassays) specific for proteins.

Adrenocorticotropic Hormone (ACTH): ACTH stimulates steroidogenesis in adrenal cortical cells. Increased formation of adrenocorticotropic hormones (cortisol) results from the increased conversion of cholesterol to pregnenolone. Formation of the 3-keto-4-ene moiety in ring A and hydroxylation of the 11, 17 and 21 positions are not rate-limiting steps.

The hormone is a single chain polypeptide containing 39 amino acids: S-T-S-M-E-H-F-A-T-G-L-P-V-G-L-L-A-A-P-V-L-V-T-P-D-A-G-E-D-Q-S-A-E-A-F-P-L-E-F. The first 24 amino acid residues are identical in all species; this portion is required for biological activity. The remaining residues (25 to 39) vary from species to species and impart species immunospecificity.

In addition to tropic action on the adrenals, ACTH promotes lipolysis in fat cells and stimulates amino acid and glucose uptake in muscle.

MW: ca 4500; pI: 4.65–4.80.

Prolactin (PRL): Prolactin (present in many vertebrate phyla) acts directly on tissues and does not regulate the function of a secondary endocrine gland. In mammals prolactin is needed to initiate and maintain lactation; in some species (rats) PRL is also luteotropic. In pigeons the hormone stimulates a "milk" production from the crop sac and in some fishes, production of mucus to feed the young and osmoregulation.

The primary structure of ovine and bovine PRL has been established. Ovine PRL consists of a single chain of 198 amino acid residues and three intrachain disulfide linkages located between resides 4–11, 58–173 and 190–198. The differences in the structure of bovine PRL are minor (see Pasqualini and Kincl, 1985).

MW: 22, 500; pI: (bovine) pH 5.7; $[\alpha]_D$ – 40° (1% at pH 7).

Follicle-Stimulating Hormone (FSH): The primary biological activity includes stimulation of the development of the primordial follicle(s) in females and of spermatogenesis in males. FSH of mammalian species is characterized by a low content of methionine and a high content of acidic residues. The carbohydrate content is about 25%. Ovine α-FSH subunit consists of 96 amino acids and the β-subunit of 119 amino acids (Papkoff et al., 1973). The primary structure of human FSH has been reported by Shome and Parlow (1974). The amino-acid sequence of the α-chain (89 a.a.) does not differ from hLH and differs by only three additional a.a. at the N-terminus. In contrast the hormone specific β-subunit (115 a.a.) contains only seven residues which are identical with the same position of hLHβ. Two heteropolysaccharide units are attached at position 7 and 24.

MW: ca 32,000; pI at pH 3.7–6.1.

The primary structure of human FSH and LH β-chain is shown below:

hLHβ S-R-E-P-L-R-P-W-C-H-P-I-N-A-U-L-A-V-E-K-E-G-C-P-V-C-I-T-F-N-T-T-I-C-A-

hFSHβ N-S-C-E--T-N-I-T-I-A--E-K-E-E-C-R-F-C-I-S-I-N-T-T-T-N-W--E-C-A-G-Y-

hLHβ G-Y-C-P-T-M-R-V-L-Q-A-V-L-P-P-L-P-Q-V-C-T-Y-R-D-V-R-F-E-S-I-R-L-P-G-C-

hFSHβ C-T-R-D-L-V-Y-K-D-P-K-P-R-I-Q-K-T-C-T-F-K-E-L-V-Y-E-T-V-R-V-P-G-C-A-

hLHβ P-R-G-V-D-P-V-V-S--F-P-V-A-L-S-C-R-C-G-P-C-R-R-S-T-S-D-C-G-G-P-K-N-

HFSHβ H-H-A-D-S-L-Y-T-Y-P-V-A-T-Q-C-H--G-K-C-D-S-D-S-T-D-C-T-V-R--L-G-P-

hLHβ H-P-L-T-C-N-Q-P-H-S-K-G

hFSHβ S-Y-C-S-F-G-E-M-Q-K

Luteinizing Hormone (LH): LH promotes primarily luteinization of Graffian follicles in females and Leydig cell function of the testes in males.

Mammalian LHs are characterized by a high content of cystine, proline and other hydrophobic a. a. residues (Papkoff *et al.*, 1973). The carbohydrate content varies from 12 to 16%. Sialic acid is absent in ovine, bovine and porcine LH, while 2.3 residues per mole are found in hLH. The short biological half-life of LH is attributed to the sialic acid content. Two carbohydrate chains are linked to the α-subunit and one to the β-chain in position 13 in ovine, bovine and porcine LH and position 30 in hLH (Ward *et al.*, 1973). The isolation, purification procedures and measurement of activity by various methods have been described by Loeber (1977). The amino acid sequence of both subunits is known for four species. The α-chain contains 89-95 residues and the β-chain, about 115-119. The position of the disulfide bonds remains uncertain.

MW: 30,000-34,000; pI at pH 5.7-8.4.

The primary structure of the α-subunits (bovine LH) is shown below (the structure of bovine TSH is the same):

F-P-D-G-E-F-T-M-Q-G-C-P-E-C-K-L-K-E-N-K-Y-F-S-K-P-D-A-P-I-Y-Q-C-M-C-C-C-F-S-R-A-Y-P-T-P-A-R-S-K-K-T-M-L-V-P-K-N-I-T-S-E-A-T-C-C-V-A-K-A-T-T-K-A-T-V-M-G-N-V-R-V-E-N-H-T-E-C-H-C-S-T-C-Y-Y-H-K-S

Pregnant Mares' Serum Gonadotropin (PMSG): The hormone (produced in endometrial cups of gravid mares) possess both FSH- and LH-like activities. The sialic acid content is very high, 9.4 residues per mole (about the same as in hCG), resulting in a long (270 min) biological half-life. Total carbohydrate content is 44.4%. The structure is uncertain.

MW: about 53,000; pI at pH 1.8.

Thyroid-Stimulating Hormones (TSH): Thyrotropin exerts stimulating effects on thyroid functions: increase in size and vascularity of the gland, increased iodine uptake, synthesis of thyroglobulin and iodothyronines and release of thyroxine and triiodothyronine all result from increased thyrotropin release. The structure of the hormone (β-chain) is similar to LH; the structure of the α-chain and the sialic acid content (16.2%) are also very similar to LH. The β-chain contains 119 amino acid residues.

MW: ca 27,000.

2.3. Hormones of the Gonads

The gonads produce polypeptide hormones (inhibins) and steroid hormones.

Polypeptide Hormones

A variety of polypeptides produced in the gonads modulate the function of specific cells within the goands and affect the release of gonadotropins (FSH) by the anterior lobe of the pituitary gland.

Inhibins: FSH inhibiting polypeptides have been isolated from follicular fluid and Sertoli cell secretion of several species. Inhibins suppress the release of FSH from the pituitary, reduce testosterone production by Leydig cells in males and increase the transformation of pregnenolone to progesterone in the growing follicle.

Inhibins are glycoprotein heterodimers of two subunits bonded by a disulfide bridge. Each subunit is derived from larger precursors (more than 400 a. a.). The molecular size differs from species to species; porcine inhibin α-subunit has a MW of 18 K to 20 K and the β-subunit, 14 K. The bovine α-chain is 44 K and the β-chain about 14 K. The primary structure of bovine (Fukuda *et al.*, 1986), porcine (Mayo 1986), human (Mason *et al.*, 1985) and murine (Esch *et al.*, 1987) inhibins were deduced from sequenced DNA.

Activins: Activins stimulate FSH secretion from the pituitary. The proteins are produced by the growing follicle and also by cells outside the reproductive organs. Activin A is a dimer of two inhibin β-subunits. Activins and TGFβ (see below) also act to regulate granulosa cell function and differentiation (Ying *et al.*, 1986). In females activins augment aromatase activity and inhibit progesterone synthesis. In males activins suppress steroid biosynthesis stimulated by LH.

MW: about 30 kD.

The FSH releasing protein reported by Vale *et al.* (1986) is probably similar to one of the activins.

Transforming Growth Factors (TGFs): Polypeptide factors which stimulate cellular proliferation. The agents elicit mitogenic signals after binding to specific cell surface receptors in target cells. TGFs are homologues (interact with a common receptor type) with epidermal growth factors.

MW: about 23 kD (Massagué, 1985).

TGFs are paracrine regulators. In females TGFβ augments FSH induced aromatase activity and progesterone synthesis.

Follistatin (FS): FS is a single chain glycoprotein present in follicular fluids (more than 300 a. a.) having a molecular weight of about 34 kD. Two forms are known differing probably in sugar content (Ying, 1988). FS inhibits FSH release (*in vitro* data) and FSH-stimulated synthesis by granulosa cells.

Steroid Hormones

All steroid hormones possess the same cyclopentanoperhydrophenantrene nucleus. The hormones possessing a 4-ene-3-keto moiety are characterized by high absorbancy (E mol 16,500–17,500) at 240 nm. Ring A phenolic steroids exhibit a lower intensity (E mol 1800–2200) absorbancy at 280–282 nm. The numbering of carbon atoms in the steroid nucleus is shown in Figure 1. The conformation of a reduced ring A (5α- and 5β-dihydroisomers) are shown in Figure 2.

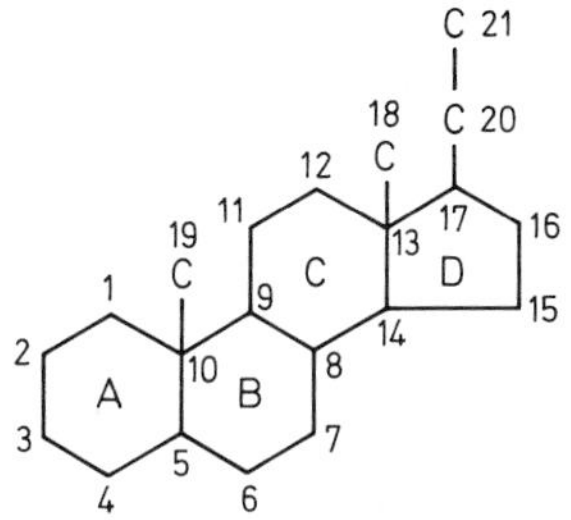

Figure 1. Steroid Hormone Structure. The structure of a pregnane showing the numbering of carbon atoms in the cyclopentanophenanthrene nucleus.

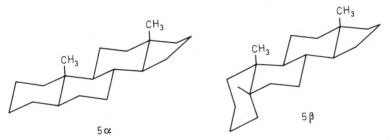

Figure 2. 5α and 5β Configuration. The 5α- and 5β reduced steroid nucleus is shown. In both cases the ring A assumes the thermodynamically more stable chair conformation.

Corticosterone (11β, 21-dihydroxy-4-pregnene-3,20 dione)

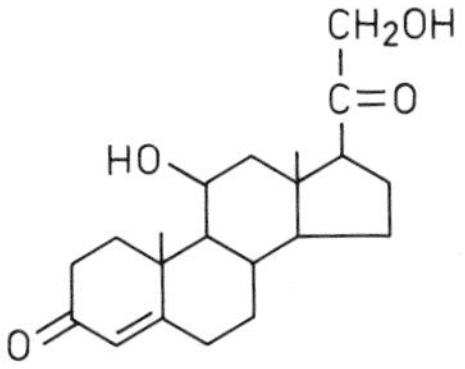

$C_{21}H_{30}O_4$; MW: 346.45
m.p. 180–182 °C; $[\alpha]_D$ + 223° (EtOH)

The principal adrenocortical hormone in some species (rodents).

Cortisol (Hydrocortisone, 11β,17,21-trihydroxy-4-pregnene-3,20-dione)

CH_2OH
$C=O$
HO   --OH

$C_{21}H_{30}O_5$; MW: 362.47
m.p. 217-220°; $[\alpha]_D$ +167° (EtOH)

In many species the principal adrenocortical hormone which influences carbohydrate, nitrogen, mineral and water metabolism.

5α-Dihydrotestosterone (DHT,17β-hydroxy-5α-androstan-3-one)

OH

$C_{19}H_{30}O_2$; MW: 290.41
m.p. 180-181 °C; $[\alpha]_D$ + 33° (EtOH)

A product of testosterone reduction and the active hormone which stimulates the growth and secretion of several organs.

Estradiol (1,3,5(10)-estratrien-3,17β-diol)

OH

HO

$C_{18}H_{24}O_2$; MW: 272.37
m.p. 173-179 °C; $[\alpha]_D$ +79 ±3 (dioxane)

The principal estrogenic hormone in many species, possibly the active principle in all; formed by the ovary, placenta, testis and adrenal cortex.

Estrone (3-hydroxy-1,3,5(10)-estratrien-17-one)

O

HO

$C_{18}H_{22}O_2$; MW: 270.36
m.p. 254-256 °C; $[\alpha]_D$ + 152° (Chlf.)

The most abundant estrogenic hormone in some species; not active *per se*, only when converted into estradiol.

Progesterone (4-pregnane-3,20-dione)

$C_{21}H_{30}O_2$; MW: 314.45
m.p. 127–131 °C; $[\alpha]_D$ +177±5° (dioxane)

The "gestational" hormone produced by the corpus luteum, the placenta, the testis and the adrenals.

Testosterone (17β-hydroxy-4-androsten-3-one)

$C_{19}H_{28}O_2$; MW: 288.41
m.p. 155 °C; $[\alpha]_D$ +109° (EtOH)

The "male" hormone produced by the testis, the ovary, the placenta and the adrenals; the active species which binds to receptor proteins in androgen sensitive tissues is 5α-dihydrotestosterone.

2.4. Prostaglandins

Prostaglandins (PG), PG endoperoxides and thromboxanes are structurally related lipids found in nearly all animal cells. They affect the male and female reproductive systems and the gastrointestinal, the cardiovascular, the renal and the nervous systems. They are released when blood clots and when tissues become inflamed; aspirin and indomethacin inhibit prostaglandin synthesis.

Prostaglandin F₂ (9,11,15-trihydroxy-5,13-prostadien-1-oic acid)

$C_{20}H_{34}O_5$; MW: 345.49
m.p. 25–35 °C; $[\alpha]_D$ +23.5° (THF)

Stimulates smooth muscle contractions and is abortifacient.

2.5. Thyroid Hormones

The hormones of the thyroid gland exert a stimulating effect on metabolism.

Thyroxine (T$_4$)

$C_{15}H_{11}I_4NO_4$; MW: 776.93
m.p. 235–236 °C (dec.); $[\alpha]_D$ +4° NaOH in 70% (EtOH)

Only the L-form is active.

Triiodothyronine (T$_3$)

$C_{15}H_{12}I_3NO_4$; MW: 651.01
m.p. 236–237 °C (dec.); $[\alpha]_D$ +21° (mixture of HCl and EtOH)

T$_3$ is about five times more active that T$_4$.

3. Biological Assays Protocols

Data by Cancer Chemotherapy National Service Center, Endocrine Evaluation Branch

Androgenic and Myogenic Activity

Purpose:
: To compare the androgenic effect of subcutaneously administered test material with that of a reference compound on secondary sex structures and on muscle.

Test Animal:
: Rat, single strain, male, 21-day-old, 40–55 grams.

Standard:
: Testosterone: 0.3, 0.6, 1.2, 2.4 mg (total dose)

Vehicle:
: SV # 17874, 0.5 ml per injection

Procedure:
: Castrate animals on the 21st–24th day of life. Randomize animals. Beginning on the same day, administer treatment once daily for *10* consecutive days and autopsy on the 11th day.

End Points:
: (1) Initial and (2) final body weights to the nearest gram; (3) ventral prostate, (4) seminal vesicle (without coagulating gland and devoid of fluid) and (5) levator ani muscle weights to the nearest 0.5 mg.

Quantitative:

Material administered	Total dose (mg) SV # 17 874 (SC)		No. of animals
	Standard	Unknown	
Vehicle		*	12
Standard	0.3		12
Standard	0.6		12
Standard	1.2		12
Standard	2.4		12
Unknown		X	12
Unknown		2X	12
Unknown		4X	12
Unknown		8X	12
Total needed:	54.0 +	180X +	108

* Vehicle only.

Inhibition of Gonadotropin Production and/or Release (Parabiosis)

Purpose: To compare the suppressing effect of subcutaneously administered test material (except estrogenic compounds) with that of the reference androgen on pituitary gonadotropins measured by ovarian weight changes in intact female parabiont with castrate male rat.

Test Animal: Rat, single strain, male and female, 30–31 day old, 70–80 grams.

Standard: *Testosterone:* 0.3, 0.6, 0.9 mg (total dose)

Vehicle: SV # 17 874, 0.2 ml per injection

Procedure: Male (right castrate parabiont) and female (left intact parabiont) are joined in parabiosis under sterile conditions, and the skin wound closed with metal clips. Randomize pairs. Beginning on the same day, administer treatment once daily for *10* consecutive days and autopsy on the 11th day.

End Points: *Left parabiont* - (1) final body weight to the nearest gram, (2) ovarian weight to the nearest 0.5 mg, and (3) the ovarian response (presence of corpora lutea and follicles).
                Right parabiont - (1) final body weight to the nearest gram, and weights of the (2) ventral prostate, (3) seminal vesicle, and (4) levator ani to the nearest 0.5 mg.

Experimental design:

Preliminary material administered	Total dose (mg) SV # 17 874 (SC)		No. of pairs
	Standard	Unknown	
Vehicle			5
Standard	0.4		5
Standard	0.8		5
Unknown*		0.6	5
Unknown		30.0	5
Total needed:	6.0 +	153.0 +	25

* For compounds with known androgenic activity, use two doses adjusted to be equivalent to the two doses of testosterone.

Quantitative:

Material administered	Total dose (mg) SV # 17 874 (SC)		No. of pairs
	Standard	Unknown	
Vehicle			8
Standard	0.3		8
Standard	0.6		8
Standard	0.9		8
Unknown		x	8
Unknown		2x	8
Unknown		3x	8
Total needed:	14.4 +	48x	56

Uterotrophic Activity – Estrogenic

Purpose:	To compare the estrogenic effect of subcutaneously administered test material with that of the reference compound on secondary sex structures.
Test Animal:	Mouse, single strain, female, 20–23-day-old.
Standard:	Estrone; 0.04, 0.08, 0.16, 0.32 µg
Vehicle:	Sesame oil, 0.1 ml per injection
Procedure:	Randomize animals. Administer treatment once daily for 3 consecutive days and autopsy on the 4th day.
End Points:	(1) Initial and (2) final body weights to the nearest gram; (3) uterus weight to the nearest 0.5 mg.

Experimental design:

Preliminary material administered	Total dose (µg) Sesame oil (SC)		No. of animals
	Standard	Unknown	
Vehicle	*		10
Standard	0.08		5
Standard	0.32		5
Unknown		0.08	5
Unknown		1.60	5
Total needed:	2.00 +	8.40 +	30

Antiuterotrophic Activity

Purpose:	To measure the inhibiting effect of subcutaneously administered test material on the estrogen stimulation of secondary sex structures.
Test Animal:	Mouse, single strain, female, 20–23-day-old.
Standard:	Estrone; 0.32 µg
Vehicle:	For standard compound, sesame oil, 0.1 ml per injection. For test material, SV # 17 874, 0.1 ml per injection.
Procedure:	Randomize animals. Administer treatments at separate sites once daily for 3 consecutive days and autopsy on the 4th day.
End Points:	(1) Initial and (2) final body weights to the nearest gram; (3) uterus weight to the nearest 0.5 mg.

Experimental
design:

Material administered	Total dose (µg)		No. of animals
	Sesame oil (SC) Standard	SV # 17 874 (SC) Unknown	
Vehicle	*	*	10
Standard	0.32	*	8
Standard + unknown	0.32	10	8
Standard + unknown	0.32	300	8
Standard + unknown	0.32	9,000	8
Unknown	*	9,000	8
Total needed:	10.24 +	146,480 +	50

* Vehicle only.

Quantitative:

Material administered	Total dose (µg) Sesame oil (SC)		No. of animals
	Standard	Unknown	
Vehicle		*	12
Standard	0.04		12
Standard	0.08		12
Standard	0.16		12
Standard	0.32		12
Unknown		X	12
Unknown		2X	12
Unknown		4X	12
Unknown		8X	12
Total needed:	7.20 +	180 X	

* Vehicle only.
Graph scale: Dose – logarithmic, response (uterus weight) – logarithmic.

References

Akiel H, Watson SJ, Young E, Lewis ME, Kchachaturian H, Walker JM (1984) Endogenous opioids: biology and function. Ann Rev Neurosci 7: 223–257

Clement-Jones V, Rees LH (1982) Neuroendocrine correlates of the endorphins and enkephalins. In: Besser GM, Martini L (eds) Clinical neuroendocrinology, vol 2. Academic Press, New York, 139–146

Crowley WR, Hassid A, Kalra SP (1987) Neuropeptide Y enhances the release of luteinizing hormone (LH) induced by LH-releasing hormone. Endocrinology 120: 941–945

Esch FS, Shimasaki S, Cooksey K, Mercado M, Mason AJ, Ying S-Y, Ueno N, Ling N (1987) Complementary deoxyribonucleic acid (cDNA) cloning and DNA sequence analysis of rat ovarian inhibins. Mol Endocrinol 1: 388–392

Goldstein A, Fischli W, Lowney LI, Hunkapiller M, Hood L (1981) Porcine pituitary dynorphin: complete amino acids sequence of the biologically active heptadecapeptide. Proc Natl Acad Sci USA 78: 7219–7222

Guillemin R, Brazeau R, Böhlen P, Esch F, Ling N, Wehrenberg WB, Bloch P, Mougin C, Zeytin F, Baird A (1984) Somatocrinin, the growth hormone releasing factor. Recent Prog Horm Res 40: 233-286

Katsube N, Schwartz D, Needleman P (1985) Release of atriopeptin in the rat by vasoconstrictors or water immersion correlates with changes in right atrial pressure. Biochem Biophys Res Commun 133: 937-941

Lang RE, Tholken H, Ganten D, Luft FC, Ruskoaho H, Unger T (1985) Atrial natriuretic factor – a circulating hormone stimulated by volume loading. Nature 314: 264-265

Lechan RM, Wu P, Jackson IMD, Wolf H, Cooperman S, Mandei G, Goodman RH (1986) Thyrotropin-releasing hormone precursor: characterization in rat brain. Science 231: 159-161

Lee JN, Seppala M, Chard T (1981) Characterization of placental luteinizing hormone-releasing factor-like material. Acta Endocrinol (Copenh) 96: 394-397

Loeber JG (1977) Human luteinizing hormone. Structure and function of some preparations. Acta Endocrinol (Copenh) 85 (Suppl 210): 1-130

Mason AJ, Hayflick JS, Ling N, Esch F, Ueno N, Ying SY, Guillemin R, Niall HD, Seeburg PH (1985) Complementary DNA sequences of ovarian follicular fluid inhibin show precursor structure and homology with transforming growth factor-beta. Nature 318: 659-660

Massaque J (1985) Transforming growth factors. Isolation, characterization, and interaction with cellular receptors. Prog Med Virol 32: 142-158

Mayo KE, Cerelli GM, Spiess J, Rivier J, Rosenfeld MG, Evans RM, Vale W (1986) Inhibin A-subunit cDNAs from porcine ovary and human placenta. Proc Natl Acad Sci USA 83: 5849-5851

Panula P, Lindberg I (1987) Enkephalins in the rat pituitary gland: immunohistochemical and biochemical observations. Endocrinology 121: 48-58

Papkoff H, Sairam MR, Farmer SW, Li CH (1973) Studies on the structure and function of interstitial cell-stimulating hormone. Recent Prog Horm Res 29: 563-588

Pasqualini JR, Kincl FA (1985) Hormones and the fetus, vol 1. Pergamon Press, Oxford

Ramey JW, Highsmith RF, Wilfinger WW, Baldwin DM (1987) The effects of gonadotropin-releasing hormone and estradiol on luteinizing hormone biosynthesis in cultured rat anterior pituitary cells. Endocrinology 120: 1503-1513

Samson WK, Aguila MC, Bianchi R (1988) Atrial natriuretic factor inhibits luteinizing hormone secretion in the rat: evidence for a hypothalamic site of action. Endocrinology 122: 1573-1582

Seizinger B, Hollt V, Herz A (1984) Proenkephalin B (prodynorphin)-derived opioid peptides: evidence for a differential processing in lobes of the pituitary. Endocrinology 45: 662-671

Shome B, Parlow AF (1974) Human follicle stimulating hormone: first proposal for the amino acid sequence of the hormone-specific β-subunit (hFSHβ). J Clin Endocrinol Metab 39: 203-205

Tatemoto K (1982) Neuropeptide Y: complete amino acid sequence of the brain peptide. Proc Natl Acad Sci USA 79: 5485-5488

Vale W, Rivier J, Vaughan J, McClintock R, Corrigan A, Woo W, Karr D, Spiess J (1986) Purification and characterization of a FSH releasing protein from porcine ovarian follicular fluid. Nature 321: 776-777

Ward DN, Reichert LE Jr, Liv W-K, Nahm HS, Hsia J, Lamkin WM, Jones NS (1973) Chemical studies of luteinizing hormone from human and ovine pituitaries. Recent Prog Horm Res 29: 533-557

Ying H-Y (1988) Inhibins, activins, and follistatins: gonadal proteins modulating the secretion of follicle-stimualting hormone. Endocrinol Rev 9: 267-293

Ying S-Y, Becker A, Ling N, Ueno N, Guillemin R (1986) Inhibin and beta type transforming growth factor (TGFβ) have opposite modulating effects on the follicle stimulating hormone (FSH)-induced aromatase activity of cultured rat granulosa cells. Biochem Biophys Res Commun 136: 969-973

Ying S-Y, Czvik J, Becker A, Ling N, Ueno N, Guillemin R (1987) Secretion of follicle stimulating hormone and production of inhibin are reciprocally related. Proc Natl Acad Sci USA 84: 4631-4632